EXPLORING DATA ANALYSIS

Exploring Data Analysis

The Computer Revolution In Statistics

Edited by

W. J. DIXON
Department of Biomathematics
University of California, Los Angeles

and

W. L. NICHOLSON
Battelle Pacific Northwest Laboratories
and
National Bureau of Standards

UNIVERSITY OF CALIFORNIA PRESS
Berkeley Los Angeles London

University of California Press
Berkeley and Los Angeles, California

University of California Press, Ltd.
London, England

ISBN: 0-520-02470-2
Library of Congress Catalog Card Number: 73-78549

Printed in the United States of America

CONTENTS

LIST OF TABLES

Chapter 7

Chapter 8

LIST OF FIGURES

LIST OF EXHIBITS

PREFACE

The genesis of this book was a conference on statistical computing, organized as a workshop, to examine the frontiers of data analysis based on computer use. It was held in the Health Sciences Computing Facility (HSCF) at the University of California at Los Angeles in September 1971.

The original impetus for such a workshop came from discussions with Wesley Nicholson during an international meeting in London some years earlier. We were dismayed at the current ivory tower trends in statistics. Mimicking the mathematicians, statisticians were increasingly avoiding the real world of application, and were purifying and separating the field from other sciences. The conference was planned as a counter-revolution to that trend.

The Health Sciences Computing Facility provided an excellent place for the workshop. The facility is dedicated to serving biomedical research through research in mathematics, statistics and computer science. It has an IBM 360/91 and numerous typewriter, character scope, and graphics consoles served by a time-sharing operating system. The system specializes in interactive statistical techniques and the programs to serve them. Of special importance to conference participants was the use of graphical statistical techniques.

Participants were limited to a select group of practicing data analysts. The papers presented real problems and included a discussion of the physical mechanisms involved in generating data for the analyses. With a real problem as a focal point, the analyses pursued the needs of the problem rather than stressing particular techniques of statistics. But any new techniques useful for the analyses were emphasized, and the degree to which the derivation and use of the techniques was dependent on the computer was stressed.

Each paper was available to several critics in advance of the meeting. Their comments are included in this volume as well as additional comments by the authors and other critics that developed during the sessions.

The conference revealed many characteristics of a data analyst at work.

In contrast to the biologist who examines his data with the constructs of his own field in mind, the data analyst examines the data for its apparent similarity to a variety of statistical models he has in mind, letting the results of successive analytical attempts guide the direction he pursues (and refines) as he proceeds. The statistician's approach might be described as one in which he states: "if we assume normality, independence, and perhaps other fundamentals, then the results indicate the validity of certain stated hypotheses with associated probabilities." In contrast, the data analyst may use many of the same techniques, but he will explore (also with statistical techniques) the degree to which these assumptions might be affecting his conclusions, and the consequences to the applicational field of deviations from reality in the analytical assumptions.

The data analyst seems to be more involved in exploration than in refinement. He is slow to make assumptions before he examines the data. He is quite satisfied if any advance is made in the problem area independent of the sophistication of the analysis, the goodness of agreement of his model, or the presentability of the statistical analysis itself.

He is quite prepared to find that one might arrive at the same conclusion using quite different routes and quite different techniques. The data analyst is almost sure to have a deep involvement in computers since he requires computing power for his freedom to use a wide variety of techniques.

Techniques and analysts are not independent. They interact. One obtains a maximum result from interactions rather than from main effects. A particular person who uses certain techniques more powerfully than someone else may obtain better answers using those techniques than others can. On the other hand, another person may use his own techniques and do equally well, that is, there is an interaction in the process.

Even when techniques are mathematically equivalent, different analysts use them in different ways. One may think and do analysis of variance, and another may think regression. They may be doing the same thing but their thought processes and the way they proceed through the analysis of the problem differ because of the way they conceptualize analysis of variance and regression; although the language may differ and even communication may be difficult, the overall analyses may really be very similar.

By the end of the conference it was clear that there is a heavy interaction between analysts and scientists in other fields. In most cases, the analyst has become very involved with the subject matter of the field's basic theories and problems. The statistical research for his data analysis is truly collaborative—in many cases he enlists the cooperation of other statisticians as well. The statistical analysis is not separated and pursued for its mathematical elegance, rather it is oriented toward the needs of the problem.

Perhaps this "team" work and cooperation is the most important and far-reaching revelation of the conference.

A short definition of data analysis was proposed at the conference: Data analysis is the application of one or more techniques to a set of data steered by the problem.

Computer facilities at HSCF were available to participants before and during the conference, and a UCLA "buddy" was assigned to each participant to help in any way necessary. Data presented at the conference is available from HSCF in machine readable form. A data set description containing at least a partial listing of the data from each paper is given in this book.

The computational support was made possible by grant RR-3 of the Biotechnology Resources Branch, Division of Research and

Resources of the National Institutes of Health. The conference itself was supported by grant GJ-29844 from the National Science Foundation.

Acknowledgements are due several members of my staff for their help with the conference and in preparing material for this book. Ed Chen, Dolores Adams and Ellen Sommers assisted in preparations for and during the conference. Ellen Sommers prepared and edited the associated data sets. Lyda Boyer edited, and Betsy Potter typed the manuscripts.

Much of the work or organization of the conference itself and working with the authors on the preparation of their manuscripts was done by Wesley Nicholson.

W. J. Dixon

CHAPTER 1

ADVANCED BREAST CANCER DATA

JAMES DICKEY
Statistics Department, State University of New York at Buffalo

and

JUDY WALRATH
Department of Epidemiology and Public Health, Yale University

The majority of medical data-analysis problems arise from a physician's hope that his records of past cases will yield useful information. The real problems are mathematically vague, but tangible: What lessons are to be learned from past experience for future clinical practice? What patient subpopulations have distinctive behavior patterns? What treatments should be used in what kinds of cases?

In the language of John Tukey (1962, 1970), these are problems of "exploratory data analysis" — problems of how to Find Interesting Reportable Effects (FIRE).

FIRE problems, however, are not the subject of the bulk of statistical theory, which is devised for After The Revelation Orderly Pickling of HYpotheses (ATROPHY), and to Guard Against Silly Selection Effects by Definition (GASSED).

Research for this study was supported by NIGMS-NIH Grant GM 16557.

Linear discrimination procedures have not been very productive in real medical problems (Radhakrishna, 1964). Even the FIRE-problem-motivated stepwise linear procedures (regression and discrimination) deliver linear functions that tend to be almost meaningless as final answers to physicians and statisticians alike, especially linear functions of three or more variables. They may, however, be useful in pointing out the few important variables.

In this paper we strive to concentrate on FIRE problems of clinical-experience data, with the aim of contributing to a general systematic approach involving the use of computer programs as steps in an analytic sequence. We discuss exploratory data analysis for an important class of problems — the prediction of a dichotomized treatment-response variable.

Prof. Wilfrid J. Dixon's (1969, 1970) BMD biomedical computer programs are widely used for practical data analysis. Contributions to a systematized approach, inspired by the BMD programs, are put forth here, together with a few rough predecessor FORTRAN language programs, and programs not yet available.

In the following section we introduce, as concrete motivation, the well-studied (Armitage et al, 1969) advanced breast cancer data analysis, and the clinical-decision problem of Bulbrook et al (1960), and Atkins et al (1968). Each of the remaining sections describes a type of computer program:

- First Look At Graphs (FLAG);
- Subsample Histograms Or Plots (SHOP);
- Shop In Full Totality (SIFT); and
- a discussion of discriminant analysis per se, with an emphasis on recent nonparametric procedures.

A DATA ANALYSIS PROBLEM

The typical medical data set features a few (1 - 10) response variables and many (10 - 100) mixed-type (dichotomous to practically continuous) predictor variables, for a precious few (10 - 1000) observed cases. Missing values abound. The definitions of individual variables are ambiguous and ill-conceived. The data embody histories of undisciplined clerks' misunderstandings. In short, the statistics teacher's nightmare: imperfect data and vague problems.

We consider here a decision problem in the management of advanced breast cancer, and a related data set from Guy's Hospital, London (Atkins et al, 1968), unusual for the painstaking care with which it was collected. This concrete data-analysis problem is put forth as representative of many in being suited to a general systematic approach.

Two hundred and ten advanced breast cancer patients were included in the study. Approximately two-thirds (139/210) of them had undergone attempted cure by radical (116/210) or simple (23/210) mastectomy, and then a year or so later had a recurrence of tumor growth locally or at a distant site. The other one-third (71/210) had been first diagnosed as already advanced. Three-fifths (132/210) began the palliative stage of their treatment with the administration of hormones, which were useful in some cases (17/132) for up to one year in controlling tumor growth.

Then it was a question of whether or not surgery should be used to alter the hormonal environment of the tumors. If so, which of two operations should be performed: bilateral adrenalectomy with oophorectomy (removal of all adrenals and ovaries), or hypophysectomy (removal of pituitary). Each patient underwent an operation, about half each kind (115/210, 95/210).

For one-quarter of the patients (54/210), the surgery was successful (complete remission of symptoms for over six months); for another one-quarter (53/210), intermediate results (partial remission); and for the other half (103/210), failure (no improvement).

Both surgical procedures are radical attempts to prolong life. Hypophysectomy is a more involved and dangerous operation, but its whole-sample remission percentages (28/95 and 24/95) were essentially the same as those for adrenalectomy (26/115 and 29/115).

Natural suggestions for variables related to surgical success include:

1. measures of tumor growth rate

   a) age of patient
   b) extent of disease at mastectomy
   c) time from mastectomy to recurrence;

2. tumor histology;

3. menopausal status;

4. history of mastectomy; and

5. systemic (hence urinary) hormone levels.

In 1960, Dr. R.D. Bulbrook and his coinvestigators at Guy's Hospital developed a linear discriminant function of two 24-hour-urinary-steroid levels, aetiocholanolone (E) and 17-hydroxy-corticosteroid (17 OHCS),

$$80 - 80(17\text{ OHCS}) + E, \qquad (1)$$

positive values of which tend to predict favorable response to surgery. After further prospective studies, Atkins et al (1968) reported "the discriminant function by itself provides an efficient guide to response to hypophysectomy but does not do so for adrenalectomy in this series." They also found small effects for the factors 1. c), 3., and 4. above.

Armitage et al (1969) carried out extensive FIRE-like analyses of these same data. First, each of three response variables was dichotomized and fit by Hills' (1967) stepwise sample-splitting discrimination procedure for dichotomized predictor variables. Then they performed special analyses, each suited to each original response variable.

The response, a clinical assessment of success (as success, intermediate, and failure, defined above), was dichotomized into nonfailure and failure, and then related to various sets of predictor variables. Our discussion is restricted to this choice of a dichotomous response variable and to dichotomized responses in general, thus neglecting other important developments of methodology, for example, survival-time data.

At the suggestion, and through the kindness, of Prof. Marvin Zelen, a card copy of the Armitage et al (1969) data was obtained from John Copas, and a slightly updated version of the original patient records (including 16 new cases) from Dr. R. D. Bulbrook. The updated records of all 210 cases are on file at HSCF under the title "Advanced Breast Cancer Data (J. Dickey)." A complete listing of the cancer data in card image form is given in the Data Set Description at the end of this chapter. This includes a description of the 50 variables associated with each patient, and, parenthetically, single word acronyms which identify variables.

FIRST LOOK AT GRAPHS (FLAG)

Newly punched data will, with high probability, contain mistaken values appearing as

1. overpunches and illegal characters;
2. data-to-format mismatches;
3. nonsense values of a variable
   a) off-range numeric values
   b) meaningless multiple-choice values;
4. nonsense combinations of variable values, e.g., autopsy date preceding date of death;

5. multivariate outliers; and

6. undetectable-per-se mistaken values.

Computer program-processing systems tend to abort program runs when data input contains mistakes of types 1 and 2. Many data-analysis programs abort or deliver unacceptable output from input mistakes of type 3, and less commonly, of type 4.

One of the functions of our computer program, FLAG (Goldman et al, 1971) is to detect, and identify by flagged output, mistaken data values of types 1 - 4 without aborting or otherwise disrupting program functions. The programming of mistakes 1 and 2 is conceived as isolated in system-specific subprograms.

The program has an option for a one-line-per-card printer listing of the data, assuming input is in standard casewise form with a constant number of cards per case. Optionally, special lines on each page should be used to indicate the actual card columns of each data variable.

The principal function of FLAG is to deliver coarse parallel plots of the variables against sequential case numbers with flags for missing values and mistakes 1 - 4. Figure 1 depicts such output for our first nine predictor variables. Each print column corresponds to a value or grouping interval for a variable. The print columns should be headed with the actual data values too, rather than as presently keyed. And special print columns should be used for user-programmed nonsense checks (mistake of type 4).

Prof. Chester I. Bliss stated in conversation with one of the authors that data analysis by desk calculator has the advantage of forcing the analyst into intimate familiarity with the actual data values. We hope FLAG offers some of the same advantages. Runs, trends, and many other univariate patterns are easy to recognize. To a lesser extent, multivariate patterns can be recognized, for example, simultaneously missing values.

Figure 1. First Look At Graphs (FLAG) output.

```
PAGE-        1            THIS IS THE BREAST CANCER DATA


AGEM         ---          AGE AT MASTECTOMY OR WHEN FIRST SEEN

MONTHM       ---          DATE OF MASTECTOMY OR WHEN FIRST SEEN - MONTH

YEARM        ---          DATE OF MASTECTOMY OR WHEN FIRST SEEN - YEAR

TYPE         ---          TYPE OF MASTECTOMY
                               0 MEANS NONE
                               1 MEANS SIMPLE
                               2 MEANS RADICAL

MPSTATEM     ---          MENOPAUSAL STATE AT MASTECTOMY OR WHEN FIRST SEEN
                               1 MEANS PRE-MENOPAUSAL
                               2 MEANS MENOPAUSAL
                               3 MEANS POST MENOPAUSAL
                               4 MEANS HYSTERECTOMY

POSTM        ---          IF POST , THE NUMBER OF YEARS

TUMOR        ---          MAXIMUM DIMENSION OF TUMOR SIZE (TO NEAREST TENTH INCH )

PATHNODES    ---          PATHOLOGICAL NODES
                               1 MEANS NEGATIVE
                               2 MEANS POSITIVE

GRADE        ---               GRADE - 1 , 2 , 3
```

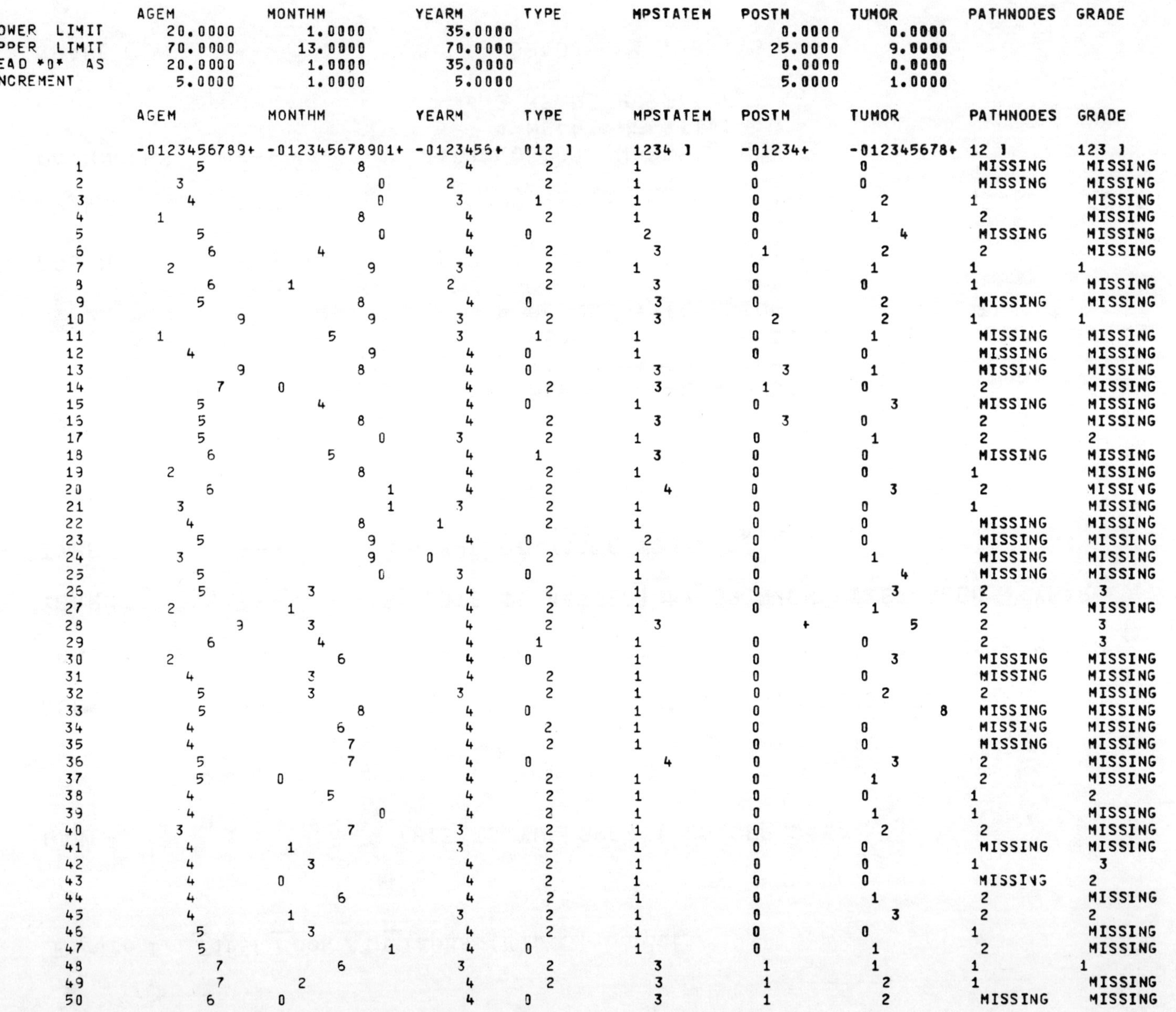

	AGEM	MONTHM	YEARM	TYPE	MPSTATEM	POSTM	TUMOR	PATHNODES	GRADE
LOWER LIMIT	20.0000	1.0000	35.0000			0.0000	0.0000		
UPPER LIMIT	70.0000	13.0000	70.0000			25.0000	9.0000		
READ *0* AS	20.0000	1.0000	35.0000			0.0000	0.0000		
INCREMENT	5.0000	1.0000	5.0000			5.0000	1.0000		

	AGEM	MONTHM	YEARM	TYPE	MPSTATEM	POSTM	TUMOR	PATHNODES	GRADE	
	-0123456789+	-012345678901+	-0123456+	012]	1234]	-01234+	-012345678+	12]	123]	
1	5	8	4	2	1	0	0	MISSING	MISSING	1
2	3	0	2	2	1	0	0	MISSING	MISSING	2
3	4	0	3	1	1	0	2	1	MISSING	3
4	1	8	4	2	1	0	1	2	MISSING	4
5	5	0	4	0	2	0	4	MISSING	MISSING	5
6	6	4	4	2	3	1	2	2	MISSING	6
7	2	9	3	2	1	0	1	1	1	7
8	6	1	2	2	3	0	0	1	MISSING	8
9	5	8	4	0	3	0	2	MISSING	MISSING	9
10	9	9	3	2	3	2	2	1	1	10
11	1	5	3	1	1	0	1	MISSING	MISSING	11
12	4	9	4	0	1	0	0	MISSING	MISSING	12
13	9	8	4	0	3	3	1	MISSING	MISSING	13
14	7	0	4	2	3	1	0	2	MISSING	14
15	5	4	4	0	1	0	3	MISSING	MISSING	15
16	5	8	4	2	3	3	0	2	MISSING	16
17	5	0	3	2	1	0	1	2	2	17
18	6	5	4	1	3	0	0	MISSING	MISSING	18
19	2	8	4	2	1	0	0	1	MISSING	19
20	6	1	4	2	4	0	3	2	MISSING	20
21	3	1	3	2	1	0	0	1	MISSING	21
22	4	8	1	2	1	0	0	MISSING	MISSING	22
23	5	9	4	0	2	0	0	MISSING	MISSING	23
24	3	9	0	2	1	0	1	MISSING	MISSING	24
25	5	0	3	0	1	0	4	MISSING	MISSING	25
26	5	3	4	2	1	0	3	2	3	26
27	2	1	4	2	1	0	1	2	MISSING	27
28	9	3	4	2	3	+	5	2	3	28
29	6	4	4	1	1	0	0	2	3	29
30	2	6	4	0	1	0	3	MISSING	MISSING	30
31	4	3	4	2	1	0	0	MISSING	MISSING	31
32	5	3	3	2	1	0	2	2	MISSING	32
33	5	8	4	0	1	0	8	MISSING	MISSING	33
34	4	6	4	2	1	0	0	MISSING	MISSING	34
35	4	7	4	2	1	0	0	MISSING	MISSING	35
36	5	7	4	0	4	0	3	2	MISSING	36
37	5	0	4	2	1	0	1	2	MISSING	37
38	4	5	4	2	1	0	0	1	2	38
39	4	0	4	2	1	0	1	1	MISSING	39
40	3	7	3	2	1	0	2	2	MISSING	40
41	4	1	3	2	1	0	0	MISSING	MISSING	41
42	4	3	4	2	1	0	0	1	3	42
43	4	0	4	2	1	0	0	MISSING	2	43
44	4	6	4	2	1	0	1	2	MISSING	44
45	4	1	3	2	1	0	3	2	2	45
46	5	3	4	2	1	0	0	1	MISSING	46
47	5	1	4	0	1	0	1	2	MISSING	47
48	7	6	3	2	3	1	1	1	1	48
49	7	2	4	2	3	1	2	1	MISSING	49
50	6	0	4	0	3	1	2	MISSING	MISSING	50

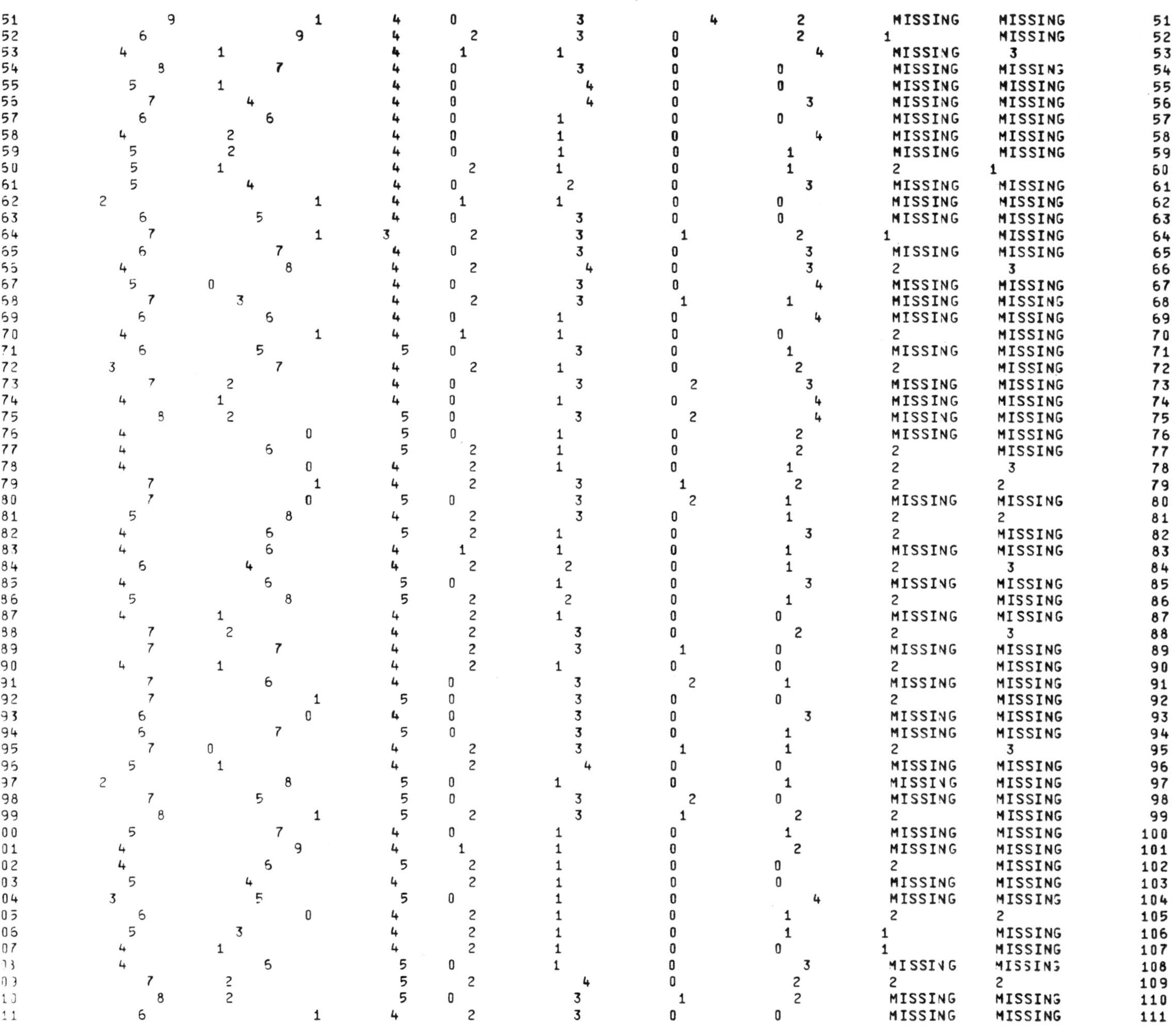

51	9	1	4	0	3	4	2	MISSING	MISSING	51
52	6	9	4	2	3	0	2	1	MISSING	52
53	4	1	4	1	1	0	4	MISSING	3	53
54	8	7	4	0	3	0	0	MISSING	MISSING	54
55	5	1	4	0	4	0	0	MISSING	MISSING	55
56	7	4	4	0	4	0	3	MISSING	MISSING	56
57	6	6	4	0	1	0	0	MISSING	MISSING	57
58	4	2	4	0	1	0	4	MISSING	MISSING	58
59	5	2	4	0	1	0	1	MISSING	MISSING	59
60	5	1	4	2	1	0	1	2	1	60
61	5	4	4	0	2	0	3	MISSING	MISSING	61
62	2	1	4	1	1	0	0	MISSING	MISSING	62
63	6	5	4	0	3	0	0	MISSING	MISSING	63
64	7	1	3	2	3	1	2	1	MISSING	64
65	6	7	4	0	3	0	3	MISSING	MISSING	65
66	4	8	4	2	4	0	3	2	3	66
67	5	0	4	0	3	0	4	MISSING	MISSING	67
68	7	3	4	2	3	1	1	MISSING	MISSING	68
69	6	6	4	0	1	0	4	MISSING	MISSING	69
70	4	1	4	1	1	0	0	2	MISSING	70
71	6	5	5	0	3	0	1	MISSING	MISSING	71
72	3	7	4	2	1	0	2	2	MISSING	72
73	7	2	4	0	3	2	3	MISSING	MISSING	73
74	4	1	4	0	1	0	4	MISSING	MISSING	74
75	8	2	5	0	3	2	4	MISSING	MISSING	75
76	4	0	5	0	1	0	2	MISSING	MISSING	76
77	4	6	5	2	1	0	2	2	MISSING	77
78	4	0	4	2	1	0	1	2	3	78
79	7	1	4	2	3	1	2	2	2	79
80	7	0	5	0	3	2	1	MISSING	MISSING	80
81	5	8	4	2	3	0	1	2	2	81
82	4	6	5	2	1	0	3	2	MISSING	82
83	4	6	4	1	1	0	1	MISSING	MISSING	83
84	6	4	4	2	2	0	1	2	3	84
85	4	6	5	0	1	0	3	MISSING	MISSING	85
86	5	8	5	2	2	0	1	2	MISSING	86
87	4	1	4	2	1	0	0	MISSING	MISSING	87
88	7	2	4	2	3	0	2	2	3	88
89	7	7	4	2	3	1	0	MISSING	MISSING	89
90	4	1	4	2	1	0	0	2	MISSING	90
91	7	6	4	0	3	2	1	MISSING	MISSING	91
92	7	1	5	0	3	0	0	2	MISSING	92
93	6	0	4	0	3	0	3	MISSING	MISSING	93
94	6	7	5	0	3	0	1	MISSING	MISSING	94
95	7	0	4	2	3	1	1	2	3	95
96	5	1	4	2	4	0	0	MISSING	MISSING	96
97	2	8	5	0	1	0	1	MISSING	MISSING	97
98	7	5	5	0	3	2	0	MISSING	MISSING	98
99	8	1	5	2	3	1	2	2	MISSING	99
100	5	7	4	0	1	0	1	MISSING	MISSING	100
101	4	9	4	1	1	0	2	MISSING	MISSING	101
102	4	6	5	2	1	0	0	2	MISSING	102
103	5	4	4	2	1	0	0	MISSING	MISSING	103
104	3	5	5	0	1	0	4	MISSING	MISSING	104
105	6	0	4	2	1	0	1	2	2	105
106	5	3	4	2	1	0	1	1	MISSING	106
107	4	1	4	2	1	0	0	1	MISSING	107
108	4	5	5	0	1	0	3	MISSING	MISSING	108
109	7	2	5	2	4	0	2	2	2	109
110	8	2	5	0	3	1	2	MISSING	MISSING	110
111	6	1	4	2	3	0	0	MISSING	MISSING	111

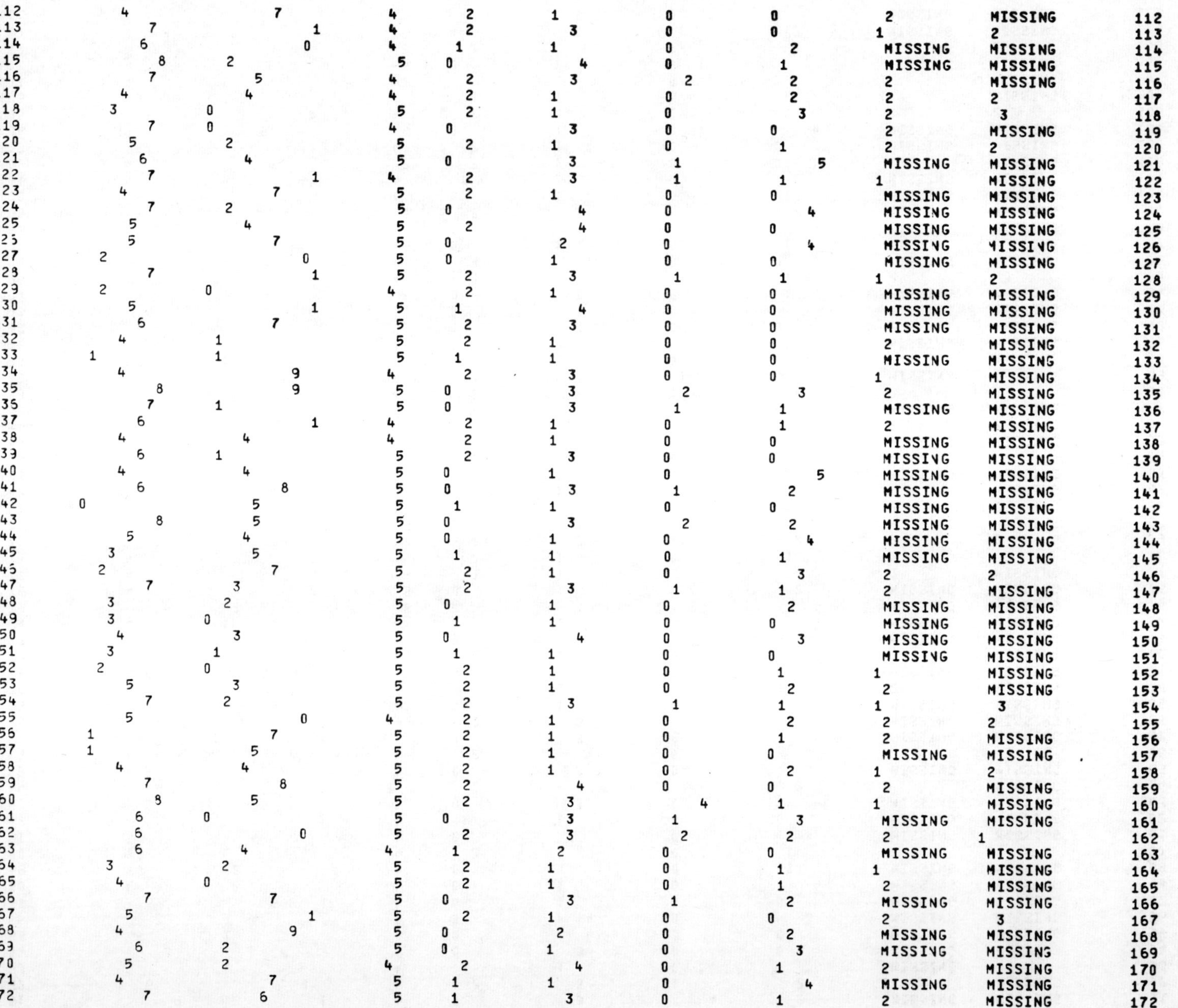

112	4	7	4	2	1	0	0	2	MISSING	112
113	7	1	4	2	3	0	0	1	2	113
114	6	0	4	1	1	0	2	MISSING	MISSING	114
115	8	2	5	0	4	0	1	MISSING	MISSING	115
116	7	5	4	2	3	2	2	2	MISSING	116
117	4	4	4	2	1	0	2	2	2	117
118	3	0	5	2	1	0	3	2	3	118
119	7	0	4	0	3	0	0	2	MISSING	119
120	5	2	5	2	1	0	1	2	2	120
121	6	4	5	0	3	1	5	MISSING	MISSING	121
122	7	1	4	2	3	1	1	1	MISSING	122
123	4	7	5	2	1	0	0	MISSING	MISSING	123
124	7	2	5	0	4	0	4	MISSING	MISSING	124
125	5	4	5	2	4	0	0	MISSING	MISSING	125
126	5	7	5	0	2	0	4	MISSING	MISSING	126
127	2	0	5	0	1	0	0	MISSING	MISSING	127
128	7	1	5	2	3	1	1	1	2	128
129	2	0	4	2	1	0	0	MISSING	MISSING	129
130	5	1	5	1	4	0	0	MISSING	MISSING	130
131	6	7	5	2	3	0	0	MISSING	MISSING	131
132	4	1	5	2	1	0	0	2	MISSING	132
133	1	1	5	1	1	0	0	MISSING	MISSING	133
134	4	9	4	2	3	0	0	1	MISSING	134
135	8	9	5	0	3	2	3	2	MISSING	135
136	7	1	5	0	3	1	1	MISSING	MISSING	136
137	6	1	4	2	1	0	1	2	MISSING	137
138	4	4	4	2	1	0	0	MISSING	MISSING	138
139	6	1	5	2	3	0	0	MISSING	MISSING	139
140	4	4	5	0	1	0	5	MISSING	MISSING	140
141	6	8	5	0	3	1	2	MISSING	MISSING	141
142	0	5	5	1	1	0	0	MISSING	MISSING	142
143	8	5	5	0	3	2	2	MISSING	MISSING	143
144	5	4	5	0	1	0	4	MISSING	MISSING	144
145	3	5	5	1	1	0	1	MISSING	MISSING	145
146	2	7	5	2	1	0	3	2	2	146
147	7	3	5	2	3	1	1	2	MISSING	147
148	3	2	5	0	1	0	2	MISSING	MISSING	148
149	3	0	5	1	1	0	0	MISSING	MISSING	149
150	4	3	5	0	4	0	3	MISSING	MISSING	150
151	3	1	5	1	1	0	0	MISSING	MISSING	151
152	2	0	5	2	1	0	1	1	MISSING	152
153	5	3	5	2	1	0	2	2	MISSING	153
154	7	2	5	2	3	1	1	1	3	154
155	5	0	4	2	1	0	2	2	2	155
156	1	7	5	2	1	0	1	2	MISSING	156
157	1	5	5	2	1	0	0	MISSING	MISSING	157
158	4	4	5	2	1	0	2	1	2	158
159	7	8	5	2	4	0	0	2	MISSING	159
160	8	5	5	2	3	4	1	1	MISSING	160
161	6	0	5	0	3	1	3	MISSING	MISSING	161
162	6	0	5	2	3	2	2	2	1	162
163	6	4	4	1	2	0	0	MISSING	MISSING	163
164	3	2	5	2	1	0	1	1	MISSING	164
165	4	0	5	2	1	0	1	2	MISSING	165
166	7	7	5	0	3	1	2	MISSING	MISSING	166
167	5	1	5	2	1	0	0	2	3	167
168	4	9	5	0	2	0	2	MISSING	MISSING	168
169	6	2	5	0	1	0	3	MISSING	MISSING	169
170	5	2	4	2	4	0	1	2	MISSING	170
171	4	7	5	1	1	0	4	MISSING	MISSING	171
172	7	6	5	1	3	0	1	2	MISSING	172

	AGEM	MONTHM	YEARM	TYPE	MPSTATEM	POSTM	TUMOR	PATHNODES	GRADE	
173	4	1	5	0	1	0	3	MISSING	1	173
174	5	4	5	2	1	0	0	1	MISSING	174
175	5	8	5	2	1	0	0	1	MISSING	175
176	4	9	4	1	1	0	2	MISSING	MISSING	176
177	4	5	5	2	1	0	0	2	MISSING	177
178	7	9	5	0	3	1	4	MISSING	MISSING	178
179	5	5	3	2	4	0	0	MISSING	MISSING	179
180	5	9	5	2	1	0	0	MISSING	MISSING	180
181	4	0	4	1	1	0	0	MISSING	MISSING	181
182	7	1	5	0	3	0	2	2	MISSING	182
183	4	9	5	0	1	0	2	MISSING	MISSING	183
184	3	3	5	0	1	0	2	2	3	184
185	4	1	4	2	2	0	1	2	2	185
186	5	9	5	2	1	0	0	MISSING	MISSING	186
187	4	1	5	0	1	0	2	MISSING	MISSING	187
188	4	1	5	2	1	0	0	2	MISSING	188
189	7	7	5	0	3	1	4	MISSING	MISSING	189
190	6	0	5	0	2	0	4	MISSING	MISSING	190
191	5	7	5	2	4	0	2	2	MISSING	191
192	4	9	4	2	1	0	0	1	MISSING	192
193	4	2	1	1	1	0	0	MISSING	MISSING	193
194	4	2	5	2	1	0	2	2	2	194
195	4	0	6	2	1	0	1	2	2	195
196	7	6	6	0	3	0	5	MISSING	MISSING	196
197	6	0	4	2	3	0	1	1	3	197
198	6	4	5	2	2	0	0	2	1	198
199	6	2	6	2	3	1	0	2	MISSING	199
200	6	5	5	2	MISSING	MISSING	2	2	MISSING	200
201	1	3	5	1	1	0	0	MISSING	MISSING	201
202	7	0	6	0	3	1	0	MISSING	MISSING	202
203	7	2	5	2	3	2	0	MISSING	MISSING	203
204	1	1	6	0	1	0	0	MISSING	MISSING	204
205	4	4	6	2	1	0	0	MISSING	MISSING	205
206	4	8	5	2	1	0	0	MISSING	MISSING	206
207	8	8	6	0	3	1	0	MISSING	MISSING	207
208	7	1	5	2	3	2	0	2	MISSING	208
209	5	4	5	2	1	0	1	2	3	209
210	3	5	3	2	1	0	0	MISSING	MISSING	210
	-0123456789+	-012345678901+	-0123456+	012]	1234]	-01234+	-012345678+	12]	123]	

```
AGEM          MONTHM          YEARM      TYPE        MPSTATEM   POSTM      TUMOR        PATHNODES  GRADE

-0123456789+  -012345678901+  -0123456+  012 ]       1234 ]     -01234+    -012345678+  12 ]       123 ]
```

EACH X REPRESENTS APPROX.   3.20 CASES

```
ACTUAL COUNT

            AGEM         MONTHM          YEARM      TYPE       MPSTATEM   POSTM      TUMOR         PATHNODES  GRADE

            -0123456789+ -012345678901+ -0123456+   012 ]      1234 ]     -01234+    -012345678+  12 ]       123 ]
            0 0 0 0 0 0   0 0 0 0 0 0 0 0 0 0 0 0   0 0 0 0 0 0 0 0 0 0   0 0 0 0 0 0 0 0 0 0 0 0 0 0 0 0   0 0 0 0 0
            0 0 0 0 0 0   0 0 0 0 0 0 0 0 0 0 0 0   0 0 0 0 0 1 0 0 0 0   1 0 0 0 0 0 0 0 0 0 0 0 0 0 0 0   0 0 0 0 0
            0 0 1 4 3 0   1 2 1 1 1 1 0 0 0 1 8 0   2 0 0 0 0 1 6 0 0 0   6 1 0 0 0 0 4 2 0 0 0 2 0 0 0 0   1 0 0 0 0
            0 7 4 2 5 4   7 0 9 6 7 6 0 0 2 5 8 0   3 0 0 0 0 5 7 0 0 0   6 2 2 0 0 0 9 2 4 0 0 9 0 0 0 0   8 0 0 0 0
             0 0 0 0 0 0 0 0 0 0 0 0 0   0 0 0 0 0 0 0 0 0 0   0 0 0 0 0 0 0 0 0 0   0 0 0 0 0   0 0 0 0 0 0 0 0 0 0
             0 0 0 0 0 0 0 0 0 0 0 0 0   0 0 0 0 0 0 1 0 0 0   0 0 0 0 0 0 0 0 0 0   0 0 0 0 0   0 0 0 0 0 0 0 0 0 0
             0 1 5 3 0 0 0 2 1 1 1 1 2   0 0 9 0 0 7 1 0 0 0   1 1 0 0 0 0 2 0 0 0   7 4 1 0 0   6 0 0 0 0 0 1 0 0 0
             1 0 6 2 9 0 0 1 3 6 9 6 0   1 2 5 7 0 1 6 0 0 0   1 6 0 0 0 0 6 2 1 0   8 0 6 0 1   9 0 0 0 0 7 6 0 0 0

PER CENT

            AGEM         MONTHM          YEARM      TYPE       MPSTATEM   POSTM      TUMOR         PATHNODES  GRADE

            -0123456789+ -012345678901+ -0123456+   012 ]      1234 ]     -01234+    -012345678+  12 ]       123 ]
            0 0 0 1 1 0   0 0 0 0 0 0 0 0 0 0 4 0   1 0 0 0 0 5 3 0 0 0   7 0 0 0 0 0 2 1 0 0 0 2 0 0 0 0   4 0 0 0 0
            0 3 6 9 6 1   8 9 9 7 8 7 0 0 0 7 1 0   0 0 0 0 0 5 2 0 0 0   9 5 0 0 0 0 3 0 1 0 0 9 0 0 0 0   3 0 0 0 0
            0 3 6 9 6 3   0 5 0 6 0 6 0 0 9 1 9 0   9 0 0 0 0 0 0 0 0 0   4 7 9 0 0 0 3 4 9 0 0 5 0 0 0 0   9 0 0 0 0
            0 3 6 9 6 0   9 2 4 1 9 1 0 0 5 4 0 0   5 0 0 0 0 2 5 0 0 0   2 4 5 0 0 0 3 7 0 0 0 9 0 0 0 0   0 0 0 0 0
             0 0 2 1 0 0 0 0 0 0 0 0 0   0 0 4 0 0 3 5 0 0 0   0 0 0 0 0 0 1 0 0 0   3 1 0 0 0   7 0 0 0 0 1 3 0 0 0
             0 4 6 5 4 0 0 9 6 7 9 7 9   0 0 5 3 0 3 5 0 0 0   5 7 0 0 0 0 2 0 0 0   7 9 7 0 0   0 0 0 0 0 7 9 0 0 0
             4 7 6 2 2 0 0 9 1 6 0 6 5   4 9 2 3 0 8 2 0 0 0   2 6 0 0 0 0 4 9 4 0   1 0 6 0 4   4 0 0 0 0 0 0 0 0 0
             7 6 6 3 8 0 0 9 9 1 4 1 2   7 5 3 3 0 0 3 0 0 0   6 5 0 0 0 0 4 5 7 0   4 4 1 0 7   0 0 0 0 0 7 2 0 0 0

NUMBER ANSWERING

            AGEM         MONTHM          YEARM      TYPE       MPSTATEM   POSTM      TUMOR         PATHNODES  GRADE
                210          210            210        210        209        209        210            98         41
```

```
STATISTICS FOR VARIABLE - AGEM

   AVERAGE        47.0476
   STDV            8.8349

   EXTREME ORDER STATISTICS

         142       23.0000
         157       25.0000
          11       26.0000
         156       26.0000
         133       26.0000

         207       63.0000
          28       65.0000
          10       66.0000
          51       68.0000
          13       69.0000
```

```
STATISTICS FOR VARIABLE - MONTHM

   AVERAGE         6.4524
   STDV            3.5119

   EXTREME ORDER STATISTICS

          14        1.0000
          37        1.0000
          43        1.0000
          50        1.0000
          67        1.0000

          62       12.0000
          51       12.0000
          47       12.0000
          21       12.0000
          20       12.0000
```

```
STATISTICS FOR VARIABLE - YEARM

   AVERAGE        58.6810
   STDV            4.0033

   EXTREME ORDER STATISTICS

          24       39.0000
         193       42.0000
          22       44.0000
           8       45.0000
           2       47.0000

         196       65.0000
         195       65.0000
         207       66.0000
         204       66.0000
         202       66.0000
```

```
STATISTICS FOR VARIABLE - TYPE

   AVERAGE         1.2143
   STDV             .9190

   EXTREME ORDER STATISTICS

           5        0.0000
           9        0.0000
          12        0.0000
          13        0.0000
          15        0.0000

          10        2.0000
           1        2.0000
           8        2.0000
           7        2.0000
           6        2.0000
```

```
STATISTICS FOR VARIABLE - MPSTATEM

   AVERAGE         1.9234
   STDV            1.0822

   EXTREME ORDER STATISTICS

           1        1.0000
           2        1.0000
           3        1.0000
           4        1.0000
           7        1.0000

          66        4.0000
          56        4.0000
          55        4.0000
          36        4.0000
          20        4.0000
```

```
STATISTICS FOR VARIABLE - POSTM

   AVERAGE         2.3158
   STDV            4.4802

   EXTREME ORDER STATISTICS

           1        0.0000
           2        0.0000
           3        0.0000
           4        0.0000
           5        0.0000

          16       17.0000
          13       18.0000
         160       20.0000
          51       24.0000
          28       27.0000
```

```
STATISTICS FOR VARIABLE - TUMOR

   AVERAGE         1.5248
   STDV            1.4696

   EXTREME ORDER STATISTICS

           1        0.0000
           2        0.0000
           8        0.0000
          14        0.0000
          16        0.0000

         121        5.0000
          28        5.0000
         196        5.5000
         140        5.5000
          33        8.0000
```

```
STATISTICS FOR VARIABLE - PATHNCOES

   AVERAGE         1.7041
   STDV             .4565

   EXTREME ORDER STATISTICS

           3        1.0000
           7        1.0000
           8        1.0000
          10        1.0000
          19        1.0000

           6        2.0000
          17        2.0000
          16        2.0000
          14        2.0000
           4        2.0000
```

```
STATISTICS FOR VARIABLE - GRADE

   AVERAGE         2.2195
   STDV             .7161

   EXTREME ORDER STATISTICS

           7        1.0000
          10        1.0000
          48        1.0000
          60        1.0000
         162        1.0000

          53        3.0000
          28        3.0000
          42        3.0000
          26        3.0000
          29        3.0000
```

As well as the horizontal alignment of the plots by cases, there is vertical alignment of the print columns with the cells of histograms following the plots, followed in turn by the actual cell counts. Counts should also be given of the various flags.

The usual univariate statistics are given: means, standard deviations, and extreme order statistics with their case numbers (5 lowest and 5 highest values).

Conceptually, FLAG does not require internal computer storage for the full data set. Hence medians and other fractiles, obtainable by sorting, were judged to require too much internal storage or too many passes of the data.

Reruns of FLAG can be useful after correcting data as suggested by earlier runs. The same is true after other transformations, or "transgenerations" of data. We use a general input-output and data-passing routine as program or subprogram, inserting ad hoc FORTRAN code as needed for data transformations. This is done in the spirit of Dixon's (1969) programs BMDX77 (Transgeneration) and BMDX94 (Multipass Transgeneration).

Multivariate outliers (mistaken data values 5) would be revealed by using FLAG on the new variable calculated for each case n as the double summation,

$$\Sigma_i \Sigma_j (x_{ni} - \bar{x}_i)(x_{nj} - \bar{x}_j)\hat{\sigma}^{ij}, \tag{2}$$

after first inverting the sample covariance matrix $[\hat{\sigma}^{ij}] = [\hat{\sigma}_{ij}]^{-1}$,

$$\hat{\sigma}_{ij} = \Sigma_{n=1}^{N} (x_{ni} - \bar{x}_i)(x_{nk} - \bar{x}_j)/f(N). \tag{3}$$

This suggestion is made in the spirit of the more complicated Dixon (1969) program, BMDX74 (Identification of Outliers).

Least squares, or linear-regression computations have been less useful in the analysis of medical data than in other data-analysis applications. But if accurate least squares operations are available as transgeneration operations, the bulk of modern data analysis techniques become available. Joiner et al (1970) presents a strong case for this approach with engineering data.

SUBSAMPLE HISTOGRAMS OR PLOTS (SHOP)

As a possible next step, to investigate the dependence of a dichotomized response variable on predictor variables, one may look at histograms (and scatter plots) of various single predictors (and paired predictors) for the two subsamples defined by the response. For example, an age effect would appear as a difference in shape between the two age histograms: one of the nonfailure patients, the other of the failure patients. A joint age-by-tumor-size effect would appear as a difference in concentration pattern between the two age-by-tumor-size scatter plots.

One can make such comparisons of response-specific histograms (and plots) of predictor variables within subsamples defined by joint intervals on further sets of predictors. For example, a joint age-tumor-size effect would appear as a difference in shape between the two age histograms (failure and nonfailure) within at least one subsample defined by tumor size, say greater than 1.5 inches.

A difficulty with scatter plots is that both predictor variables ought to range over a practical approximation to a continuum. Even then, two or more histograms can often substitute for a scatter plot. A difficulty with both histograms and scatter plots is that a multiple-choice variable with many unordered values (for example, name of hospital in a cooperative trial of 16 hospitals) cannot serve well as an axis. The subsamples defined by the multiple-choice values must be graphed separately (for example, histograms of the response variable, one for each hospital).

The computer program, Subsample Histograms Or Plots (SHOP) (Jacobson, 1971), a modification of Dixon's (1970) BMD05D, gives printer-output, whole-sample and various-subsample histograms and scatter plots of selected variables and pairs of variables. This program allows an option to partition a print page into as many as four separate histograms with a single count scale for ease of comparison, as in figure 2. There also ought to be an option to obtain shared-axis, or superimposed histograms, distinguished by print symbols. The actual counts should be printed as well as plotted, and the marginal counts too. Then much of the function of the usual contingency table program could be performed by SHOP'ing.

Armitage et al (1969, table 6) noted that in the subsample of patients who had not previously had mastectomies (because they were first diagnosed at an advanced stage) there was an apparent differential nonfailure rate: 23% (8/35) for the adrenalectomized patients, and 66% (19/29) for the hypophysectomized patients. The rates for the corrected and updated data are similar, 23% (9/39) and 62% (20/32). The SHOP output for this nonmastectomy subsample with the four histograms of menopausal state when first seen (figure 2) shows an apparently more extreme choice-of-operation effect for the premenopausal women, 17% (2/12) and 73% (11/15).

The question naturally arises whether these apparent effects are real. One type of answer is obtained by calculating from these data subsets, $\underset{\sim}{D}$, the weighted likelihood ratios (Dickey and Lientz, 1970), also called Bayes factors (Good, 1950, 1965), for no-effect null hypotheses, H, versus their parameter-set complements, $\overline{H}$,

$$L_D(H, \overline{H}) = p(\underset{\sim}{D} | H) / p(\underset{\sim}{D} | \overline{H}), \qquad (4)$$

where, for various low-information prior distributions conditional on H and on $\overline{H}$,

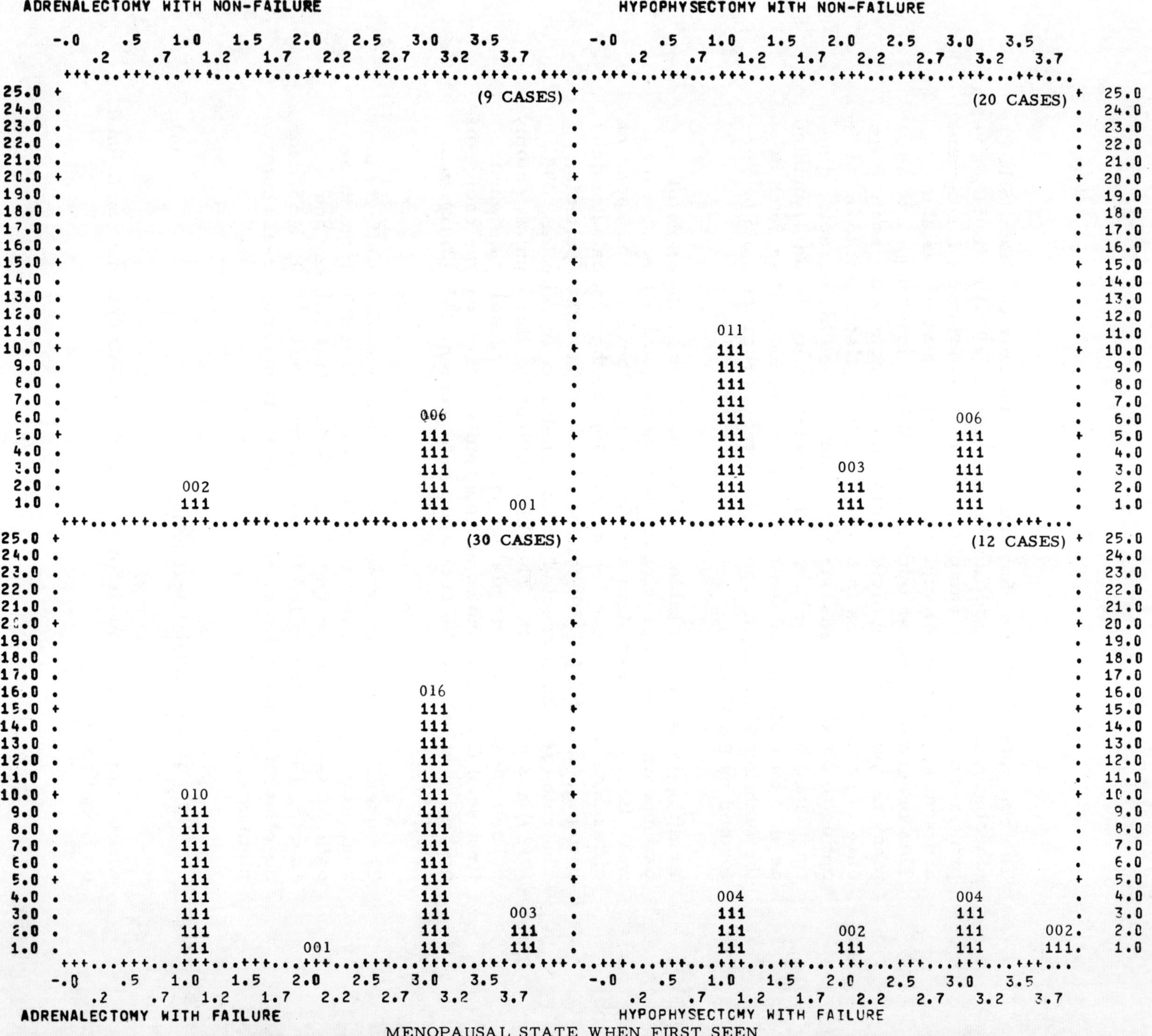

MENOPAUSAL STATE WHEN FIRST SEEN

1 - PREMENOPAUSAL, 2 - MENOPAUSAL, 3 - POSTMENOPAUSAL, 4 - HYSTERECTOMY

$$p(\underset{\sim}{D} \mid H) = E_{\pi \mid H}\, p(\underset{\sim}{D} \mid \underset{\sim}{\pi}), \tag{5}$$

$$p(\underset{\sim}{D} \mid \bar{H}) = E_{\pi \mid \bar{H}}\, p(\underset{\sim}{D} \mid \underset{\sim}{\pi}), \tag{6}$$

($\underset{\sim}{D}$ not random, nor conditioned on, in the expectations), where $\underset{\sim}{\pi}$ represents the parameters of the statistical models $p(\underset{\sim}{D} \mid \underset{\sim}{\pi})$. The weighted likelihood ratio is interpreted as the ratio of the posterior odds for H to the prior odds, no matter what the (finite nonzero) prior odds,

$$P(H \mid \underset{\sim}{D}) / P(\bar{H} \mid \underset{\sim}{D}) = L_D(H, \bar{H}) \cdot P(H)/P(\bar{H}). \tag{7}$$

Hence the weighted likelihood ratio reports the inferential content of the data $\underset{\sim}{D}$ for H.

The weighted likelihood ratio is a function of the prior distribution, and so should be reported for several choices of prior distribution under $\bar{H}$. The prior distributions under the singular subset H are assumed to be induced as the limit of conditional distributions under $\bar{H}$ (Savage's density principle, Dickey and Lientz, 1970).

The weighted likelihood ratios are given here for three contexts:

1. 2 x 2 4-cell multinomial model, hypothesis of independence (Gunel and Dickey, 1972).

   $\underset{\sim}{D} = (n_{11}, n_{12}, n_{21}),\ n_{ij} > 0,\ i, j = 1, 2,\ n_{..} = N$ fixed.

   $\underset{\sim}{\pi} = (\pi_{11}, \pi_{12}, \pi_{21}),\ \pi_{ij} > 0,\ i, j = 1, 2,\ \pi_{..} = 1.$

   $\underset{\sim}{D} \mid \underset{\sim}{\pi} \sim$ Multinomial $(\underset{\sim}{\pi}, N)$.

   $p(\underset{\sim}{\pi} \mid \bar{H}) = [\Pi_{ij}\, \pi_{ij}^{a-1}]\, \Gamma(4a)/[\Gamma(a)]^4,\ a > 0,$ Dirichlet with flattening constant a.

H: $\pi_{11} = \pi_{1.}\pi_{.1}$.

$$L_D(H,\bar{H}) = \frac{[\Pi_i \Gamma(n_{i.}+2a-1)][\Pi_j \Gamma(n_{.j}+2a-1)]\Gamma(n_{..}+4a)}{[\Pi_{ij}\Gamma(n_{ij}+a)][\Gamma(n_{..}+4a-2)]^2}$$

$$\cdot \frac{[\Gamma(a)]^4[\Gamma(4a-2)]^2}{[\Gamma(2a-1)]^4\Gamma(4a)} . \qquad (8)$$

2. Two-binomial model, hypothesis of equal rates (Dickey and Lientz, 1970).

For i = 1, 2,

$D_i = n_{i1}$, $n_{i.}$ fixed,

$\pi_i = \pi_{i1}/\pi_{i.}$, $\underset{\sim}{\pi} = (\pi_1, \pi_2)$.

$D_i | \underset{\sim}{\pi} \sim$ Binomial $(\pi_i, n_{i.})$, independent for i = 1, 2.

$p(\pi_i | \bar{H}) = \pi_i^{a-1}(1-\pi_i)^{a-1}/B(a,a)$, $a > 0$, beta with parameters (a, a), independent for i = 1, 2.

H: $\pi_1 = \pi_2$.

$\underset{\sim}{D} = (D_1, D_2)$.

$$L_{\underset{\sim}{D}}(H,\bar{H}) = \frac{[\Pi_j \Gamma(n_{.j}+2a-1)][\Pi_i \Gamma(n_{i.}+2a)]}{[\Pi_{ij}\Gamma(n_{ij}+a)]\Gamma(n_{..}+4a-2)}$$

$$\cdot \frac{[\Gamma(a)]^4\Gamma(4a-2)}{[\Gamma(2a-1)]^2[\Gamma(2a)]^2} . \qquad (9)$$

3. One-binomial model, hypothesis of a fixed rate (Jeffreys, 1961, p. 256).

   $D = s$, fixed n.

   $D \mid \pi \sim \text{Binomial}(\pi, n)$.

   $p(\pi \mid \bar{H}) = \pi^{a-1}(1-\pi)^{a-1}/B(a, a)$, $a > 0$, beta with parameters (a, a).

   $H: \pi = \pi_H$, π_H fixed.

$$L_D(H, \bar{H}) = \pi_H^s (1-\pi_H)^{n-s} \cdot \frac{\Gamma(n+2a)}{\Gamma(s+a)\,\Gamma(n-s+a)} \cdot \frac{[\Gamma(a)]^2}{\Gamma(2a)}. \qquad (10)$$

See also Good (1950, 1955) for similar Bayes factors.

Table 1 reports the evidence for an operation effect in the mastectomy-negative subsample. Note that prior odds for no-effect are diminished by the data to posterior odds of about 1% of the prior odds and that the stronger evidence for this change comes from the adrenalectomy cases. Table 2 reports the evidence in the mastectomy-negative and premenopausal-when-first-seen subsubsample; here the diminution of odds is to about 3%, which is weaker evidence for the analogous inference. The weakening appears to follow from the reduction in sample size.

The reporting of these subsample weighted likelihood ratios is less vulnerable to spurious selection effects from the multiplicity of candidate subsamples than is the use of the usual subsample P values. Each prior (and hence posterior) odds ratio for no-real-effect in a subpopulation tends to be greater the more the number of such subpopulations under consideration; hence the more extreme must be each weighted likelihood ratio to attract attention.

Table 1. The weighted likelihood ratio for equal nonfailure rate for adrenalectomy and hypophysectomy, subsample of mastectomy-negative cases.

Model	H	Data $\underset{\sim}{D}$		$L_D(H, \bar{H})$		
1.	Independence	(n_{ij})		a = .75	1.00 (Uniform)	1.50
2 x 2		9	30			
Multinomial		20	12	.0055	.0088	.012
2.	$\pi_1 = \pi_2$	(n_{ij})				
Two-		9	30			
Binomial		20	12	.011	.013	.014
3.	$\pi = \frac{1}{2}$	s	n-s			
One-		9	30	.017	.015	.014
Binomial		20	12	2.03	1.74	1.42

Table 2. The weighted likelihood ratio for equal nonfailure rate for adrenalectomy and hypophysectomy, subsubsample of mastectomy-negative premenopausal-when-first-seen patients.

Model	H	Data $\underset{\sim}{D}$		$L_D(H,\bar{H})$		
1.	Independence	(n_{ij})		a = .75	1.00 (Uniform)	1.50
2 x 2		2	10			
Multinomial		11	4	.013	.023	.036
2.	$\pi_1 = \pi_2$	(n_{ij})				
Two-		2	10			
Binomial		11	4	.025	.033	.044
3.	$\pi = \frac{1}{2}$					
One-		2	10	.22	.21	.21
Binomial		11	4	.75	.67	.60

Still, the dependence in prior opinion over subcells is not accounted for by the use of such individual weighted likelihood ratios, which should be based on all the data to completely eliminate spurious selection effects. Thus, Bayesian full-data procedures are proposed in the next section.

In a choice-of-treatment clinical decision experiment (a comparative therapy trial) spurious selection effects are not as important as they would at first seem to be. After an initial trial, to any subpopulations for which there arise apparent effects, this apparently better treatment can then be assigned. Then each subsample success rate can be monitored sequentially against both of the treatment's full-sample success rates to see whether the apparent effect holds up. Many interesting questions arise about size of initial trial, choice of subpopulations, and rules for stopping.

SHOP IN FULL TOTALITY (SIFT)

SHOP, discussed in the previous section, is useful for looking into already suspected effects, but is hit-or-miss for the unanticipated. The number of subsamples defined by r merely dichotomous predictor variables, $j = 1, 2, \ldots, k$ at-a-time is

$$\Sigma_{j=1}^{k} \binom{r}{j} \cdot 2^j \ .$$

The maximum number k of predictor variables amenable to human communication and comprehension is about $k = 3$, for which, from just $r = 16$ possible dichotomous predictors, there are $4480 + 480 + 32 = 4992$ candidate subsamples, too many for complete SHOP'ing.

Consider the "null" model in which each j-dimensional predictor cell X, given a number n_X of cases, has a number s_X of "successes" binomially distributed with a common success rate $\pi_X \equiv \pi$, say $\pi = \hat{\pi} = s_{.}/n_{.}$. The standardized observations are $z_X = (s_X - \hat{\pi} n_X)/\hat{\sigma}(s_X)$, where $\hat{\sigma}^2(s_X) = \hat{\pi}(1 - \hat{\pi})/n_X$ and $n_X \neq 0$.

They can all be calculated, and displayed in various ways. For example, within intervals of n_X values, histograms of z_X can be made, even if the cells X overlap. At least the z_X values on the null-model tails, say $|z_X| > 2$, should be listed with their actual s_X, n_X, and X's.

With these procedures there is no necessity for the predictor variables to be dichotomies. But of course too many values for a predictor decimates the subsample sizes.

We note in passing that the dependence structure of an r-dimensional contingency table can be FIRE-analyzed similarly to the above by calculating for all the j = 2 and 3 dimensional cells X, the standardized observations

$$y_X = (n_X - \hat{E}n_X)/\hat{\sigma}(n_X),$$

where $\hat{E}n_X \neq 0$ and $\hat{\sigma}(n_X)$ are calculated from the one-dimensional marginals for the "independence" model, $\hat{E}n_X = \Pi n_{x_i}$ and

$$\hat{\sigma}^2(n_X) = [(\hat{E}n_X)/n_.] [1 - (\hat{E}n_X)/n_.] /n_. .$$

Let us return to the notion of a histogram of the z_X's having n_X values in a single grouping interval, but with nonoverlapping cells X. Suppose, as enough <u>real</u> data will always imply, that the "null" model $\pi_X \equiv \pi$ is rejected because the histogram has greater spread than a binomial distribution. In particular, consider the model of independent binomial s_X's, conditional on π_X, with fixed constant subsample size $n_X \equiv n$, and the π_X's themselves distributed according to an unknown, nonsubjective, prior distribution. The variability of the success rate π_X over distinct subpopulations X is certainly a worthy object of interest and is subject to statistical inference.

The reader can verify that the first and second order moments of π_X are related to the <u>marginal</u> moments of its estimate $\hat{\pi}_X = s_X/n_X$ by

$$E\pi_X = E\hat{\pi}_X, \tag{11}$$

$$E[\pi_X(1 - \pi_X)] = (n_X - 1)^{-1} n_X E[\hat{\pi}_X(1 - \hat{\pi}_X)], \tag{12}$$

$$\sigma^2(\pi_X) = (E\pi_X)(1 - E\pi_X) - E[\pi_X(1 - \pi_X)]. \tag{13}$$

Hence the moments of π_X can be estimated from usual estimates of the marginal moments of $\hat{\pi}_X$,

$$\text{Est. of } E\hat{\pi}_X = \Sigma_X \hat{\pi}_X / \Sigma_X 1, \tag{14}$$

$$\text{Est. of } E[\hat{\pi}_X(1 - \hat{\pi}_X)] = \Sigma_X \hat{\pi}_X(1 - \hat{\pi}_X)/\Sigma_X 1. \tag{15}$$

Bias in estimating $\sigma^2(\hat{\pi}_X)$ and $\sigma^2(\pi_X)$ seems unavoidable even though the cells X do not overlap.

For each grouping interval of n_X, the variability of π_X from cell to cell can be estimated, say after reconstructing the cells to yield the target values of n_X. The covariability of π_X's can then be studied in various ways, since, for example,

$$E\pi_X\pi_Y = \tfrac{1}{2}[E(\pi_X + \pi_Y)^2 - E\pi_X^2 - E\pi_Y^2].$$

One may wish to estimate the success rate π_X for a specific cell X on the basis of s_X, n_X, and also the data from other (non-intersecting) cells X, of not necessarily restricted sizes n_X. We give a Bayesian answer to this problem which helps prevent the "selection effect." (By selection effect we mean that, by chance, even under the null model, a few undeserving cells among many would have extreme estimated response rates.)

Suppose the unknown population of π_X values belongs (to a good approximation) to the 2-parameter beta family,

$$p(\pi_X | a,b) = \pi_X^{a-1}(1-\pi_X)^{b-1}/B(a,b), \quad a > 0, \ b > 0.$$

In a full Bayesian analysis, a prior distribution $p(a,b)$ is used to obtain a distribution of a,b posterior to the data $\underset{\sim}{D} = \{s_X, \text{ all } X\}$,

$$p(a,b | \underset{\sim}{D}) \propto [\Pi_X p(s_X | a,b)]\, p(a,b), \tag{16}$$

where

$$p(s_X | a,b) = \int_0^1 p(s_X | \pi_X) p(\pi_X | a,b)\, d\pi_X$$

$$= B(s_X + a, \ n_X - s_X + b)/B(a,b). \tag{17}$$

Then for any particular cell X,

$$p(\pi_X | \underset{\sim}{D}) = \int_0^\infty \int_0^\infty da\, db\; p(\pi_X | a,b,\underset{\sim}{D}) p(a,b | \underset{\sim}{D}), \tag{18}$$

where

$$p(\pi_X | a,b,\underset{\sim}{D}) = \frac{\pi_X^{s_X+a-1}(1-\pi_X)^{n_X-s_X+b-1}}{B(s_X+a, n_X-s_X+b)} \cdot \tag{19}$$

Hence the full Bayesian posterior moments

$$E[\pi_X^i (1-\pi_X)^j | \underset{\sim}{D}]$$

are two-dimensional integrals (with the weight functions $p(a,b | \underset{\sim}{D})$) (eq. 18) of the usual one-dimensional posterior beta moments,

$$E[\pi_X^i (1-\pi_X)^j | a,b,\underset{\sim}{D}] = \frac{B(s_X+a+i, n_X-s_X+b+j)}{B(s_X+a, n_X-s_X+b)} \cdot \tag{20}$$

For example, the full Bayesian estimate $\hat{\pi}_X = E(\pi_X \mid \underset{\sim}{D})$ is an integral of the usual one-dimensional Bayesian estimate

$$E(\pi_X \mid a, b, \underset{\sim}{D}) = (s_X + a)/(n_X + a + b). \tag{21}$$

If two-dimensional quadratures are unacceptable to the data analyst, and if he has enough cells of nearly constant size n_X, he can use equations 19-21 directly, after first estimating the parameters a, b of an unknown beta population as follows. Use equations 11-15 to estimate the beta moments, which satisfy

$$E\pi_X = a/(a+b) \tag{22}$$

$$E[\pi_X(1 - \pi_X)] = ab/[(a+b+1)(a+b)], \tag{23}$$

then solve

$$a = (E\pi_X) \cdot (a+b) \tag{24}$$

$$b = (1 - E\pi_X) \cdot (a+b) \tag{25}$$

where

$$a+b = E[\pi_X(1 - \pi_X)]/\{(E\pi_X)(1 - E\pi_X) - E[\pi_X(1 - \pi_X)]\}. \tag{26}$$

Strictly speaking, the sample averages (equations 14 and 15) should be summations over cells other than the one X of interest for equations 19-21. We suspect that in practice this distinction is not necessary. These procedures are related to ideas in Dickey (1968a, 1968b, 1969).

DISCRIMINANT ANALYSES

Patient subpopulations may exist that differ in regard to the response variable outcomes yet cannot be described naturally in terms of the observed predictor variables. (By described naturally we mean they cannot be defined by logical operations on three or less observables.) A mathematical description may then be useful; for example, definition by a threshold on a linear discriminant function of many variables, each of small importance. Or one may need a purely operational description; for example, a patient who is more similar, in terms of predictor values, to an observed favorable-outcome patient than to any unfavorable-outcome patient, may tend to have a favorable outcome. The problems of finding nonnaturally-describable subpopulations are the problems of discriminant analysis in a broad sense.

There have been three types of approaches to discriminant analysis in medicine:

- by linear (or quadratic) discriminant functions;
- by Bayes' theorem, assuming independently distributed predictor variables within the response-variable-defined subpopulations; and
- by use of a distance measure in the space of joint predictor-variable outcomes, and the notion that close points tend to have similar response outcomes.

Each type can be viewed as an application of Bayes' theorem.

<u>Linear Discriminant Functions</u>

Armitage et al (1969) used Walker and Duncan's (1967) iterative maximum-likelihood fits to Cox's model (1966),

$$P(\text{Success} \mid x_1, \ldots, x_r) = e^{2Y}/(1+e^{2Y}), \tag{27}$$

in which Y is a linear function of the predictors,

$$Y = \beta_0 + \beta_1 x_1 + \beta_2 x_2 + \ldots + \beta_r x_r. \tag{28}$$

Alternatively, one could fit the usual Fisher's (1936) linear discriminant function for the two groups, with estimated coefficients,

$$\hat{\alpha}_i = \Sigma_{j=1}^{r} \hat{\sigma}^{ij}(\bar{x}_{iS} - \bar{x}_{iF}),\ i = 1, 2, \ldots, r, \tag{29}$$

$$\hat{\alpha}_0 = -\tfrac{1}{2}\Sigma_{i=1}^{r} \hat{\alpha}_i (\bar{x}_{iS} + \bar{x}_{iF}) + \log_e (N_S/N_F), \tag{30}$$

as in Truett et al (1967). The use of Fisher's discriminant is formally based on the assumption of multivariate normal distributions with a single covariance matrix for the distributions of $x_1, x_2, \ldots, x_r$ under the two groups. Essentially, one would have $\hat{\alpha}_1, \ldots, \hat{\alpha}_r$ determined by a fit to the model,

$$p(x_1, \ldots, x_r \mid \text{Success}) = (2\pi)^{-\frac{1}{2}r} [\det(\sigma^{ij})]^{\frac{1}{2}} \cdot \exp[-\tfrac{1}{2}\Sigma\Sigma\sigma^{ij}(x_i - \mu_{iS})(x_j - \mu_{jS})], \tag{31}$$

where

$$\alpha_i = \Sigma_{j=1}^{r} \sigma^{ij}(\mu_{iS} - \mu_{iF}),\ i = 1, 2, \ldots, r, \tag{32}$$

and similarly for $p(x_1, \ldots, x_r \mid \text{Failure})$.

Given a joint distribution of $x_1, x_2, \ldots, x_r$ and success-failure equations 27, 28, 31 and 32 hold simultaneously with

$$2Y = \alpha_0 + \alpha_1 x_1 + \alpha_2 x_2 + \ldots + \alpha_r x_r, \tag{33}$$

and

$$\alpha_0 = -\tfrac{1}{2}\Sigma_{i=1}^{r} \alpha_i(\mu_{iNF} + \mu_{iF}) + \log_e[P(S)/P(F)], \tag{34}$$

which yields $P(S | x_1, \ldots, x_r) = \frac{1}{2}$ when $2Y = 0$.

Fisher's linear discriminant function is easier to obtain in practice than the estimated linear function in Cox's model, but its formal justification requires multivariate normality; however, medical symptom variables usually do not even approximately range over a continuum, and are even less jointly normal. Cornfield (1971), Truett et al (1967), and Gilbert (1968) suggest that in practice, Cox's discriminant is not an appreciable improvement over Fisher's discriminant.

Sarfaty and Tallis (1970) apply an interesting predictive extension of Fisher's discriminant analysis to the Guy's Hospital data.

The Independent-Symptoms Assumption

Apparently beginning with Ledley and Lusted (1959) physicians (for example, Warner, 1961; Lusted, 1968; and Jacquez, 1964, 1970) have been interested in using Bayes' theorem for medical diagnosis, and have done so with some success. This theorem is expressed as

$$P(\text{Disease} | \text{Symptoms}) = P(\text{Symptoms} | \text{Disease}) \cdot P(\text{Disease})/P(\text{Symptoms}),$$

where P(Symptoms) is the sum of the numerator over the possible Diseases. In order to estimate P(Symptoms | Disease) where Symptoms $= X = (x_1, x_2, \ldots, x_r)$, it is usually assumed roughshod that

$$P(x_1, x_2, \ldots, x_r | \text{Disease}) = \Pi_i P(x_i | \text{Disease}),$$

since it is easier to estimate the margins than the joint probabilities of individual discrete predictor variables.

Proximity in Symptom Space

Fix and Hodges (1951) proposed nonparametric discrimination on the space of joint predictor-variable values X by multidimensional grouping of the data within a ball B(X) centered on the X of a case to be predicted. The radii differ from ball to ball as the minima to include at least a given number m of cases in each ball. The given minimum number of cases $m \leq n_{B(X)}$ is chosen to stabilize the estimates.

$$\hat{\pi}_X = \hat{\pi}_{B(X)} = s_{B(X)}/n_{B(X)}, \tag{35}$$

with variances

$$\pi_{B(X)}(1 - \pi_{B(X)})/n_{B(X)}.$$

The special Fix-Hodges procedure with $m = 1$ yields predictions of response, but not probability estimates $\hat{\pi}_B$. This "nearest-neighbor rule" (Cover, 1968, but not Hills' procedure, 1967 of the same name) predicts the same response as that experienced by the nearest observed case to the case of interest. The computer time requirements for $m = 1$ are much less than for the stable probability estimates because the varying radius for the latter requires continual sorting of the cases by their distances from cases in question.

In another answer to this computer time problem, Dickey (1968b) suggested choosing a constant radius by comparing various radii for the predictive performance of their corresponding probabilities.

One difficulty with the use of a constant radius is that some balls may be empty or nearly empty, thus occasionally yielding no probabilities or misleading probabilities. Dickey (1968b) proposed, more generally, the smoothed estimate,

$$\hat{\pi}_X = \Sigma_Y w_{Y-X} \cdot s_Y / \Sigma_Y w_{Y-X} \cdot n_Y, \tag{36}$$

where w_Z is a moot weight function, for example the normal density kernel,

$$w_Z = w_{\sigma^2, Z} = \exp(-\tfrac{1}{2}\sigma^{-2}\|Z\|^2), \tag{37}$$

or the indicator for a ball of radius ρ,

$$w_Z = w_{\rho^2, Z} = \begin{cases} 1 \text{ if } \|Z\|^2 \leq \rho^2 \\ 0 \text{ if otherwise.} \end{cases} \tag{38}$$

Since in practice most cells Y are empty, the summations in equation 36 are best performed over cases. Of course, the denominator is a sum of numerators over success and failure. In fact, the estimates in equations 35 and 36 are a special application of Bayes' formula for two response classes whose prior probabilities are estimated by the whole-sample frequencies,

$$s_{.}/n_{.} \ (n_{.} = N) \text{ and } (n_{.} - s_{.})/n_{.};$$

and the conditional cell probabilities are given by the smoothed estimates,

$$\hat{p}(X \mid \text{Success}) = \Sigma_Y w_{Y-X} \cdot (s_Y/s_{.})/\Sigma_Z w_Z. \tag{39}$$

The use of a smoothed-histogram estimate of a probability mass or density function (equation 39) is reminiscent of work begun by Whittle (1958) and Parzen (1962) for one-dimensional X, and by Cacoullos (1966) and Dickey (1968a) in higher dimensions. Its use in Bayes' theorem for discrimination seems to have been first proposed by Van Ryzin (1965, 1966) and Dickey (1968b).

Discrimination by equation 36 is of interest for the lack of underlying assumptions. But there are important disadvantages. The distance measure, say norm Y-X , is quite arbitrary and is expensive to choose by trial and error. (In practice, the authors like to divide each X-coordinate by its standard deviation, then use the usual Euclidean rooted sum of squared differences.) The related arbitrariness of the smoothing weight function w_{Y-X} can, in theory, be reduced by considerations akin to the suggestions made in the previous section (cf. Dickey, 1968a and 1969). Specht (1971) shows that it is not necessary to keep the complete history of cases on file but only joint sample moments. In contrast to the usual linear discriminant function, the lack of a closed form for the prediction function hinders the search for important versus noise variables.

Performance Scores

Given a set of estimated probabilities of success

$$\hat{\pi}_X,\ X_1, X_2, \ldots, X_N,$$

and corresponding actual response-outcomes

$$\delta_1, \delta_2, \ldots, \delta_N$$

with $\delta = 1$ for the response-outcome success and $\delta = 0$ for failure, we predict success $\hat{\delta} = 1$ if $\hat{\pi}_X$ exceeds some threshold, say $\frac{1}{2}$, and failure $\hat{\delta} = 0$ otherwise. Then the fraction-correct scores can be calculated,

$$N(\text{SUCC}) = \Sigma_1^N \delta_n \tag{40}$$

$$N(\text{FAIL}) = \Sigma_1^N (1 - \delta_n) = N - N(\text{SUCC}) \tag{41}$$

$$\text{NPREDS}(\text{SUCC}) = \Sigma_1^N \hat{\delta}_n \tag{42}$$

$$\text{NPREDS}(\text{FAIL}) = \Sigma_1^N (1 - \hat{\delta}_n) = N - \text{NPREDS}(\text{SUCC}) \tag{43}$$

$$\text{NCORRECT(SUCC)} = \Sigma_1^N \hat{\delta}_n \delta_n \tag{44}$$

$$\text{NCORRECT(FAIL)} = \Sigma_1^N (1 - \hat{\delta}_n)(1 - \delta_n) \tag{45}$$

$$\text{NCORRECT} = \text{NCORRECT(SUCC)} + \text{NCORRECT(FAIL)}; \tag{46}$$

yielding NCORRECT/N; and for k = SUCC, FAIL; NCORRECT(k)/N(k) and NCORRECT(k)/NPREDS(k).

One may wish not only that the sign of $\hat{\pi}_X - \frac{1}{2}$ be valid but also that, of the times $\hat{\pi}_X$ takes a given value, the actual success rate be nearly $\hat{\pi}_X$. Namely, for a given partition of the unit interval [0, 1] into subintervals

$$I = [0, 1/J), [1/J, 2/J), \ldots, [1 - 2/J, 1 - 1/J), [1 - 1/J, 1],$$

one desires small values of

$$\text{VALID} = N^{-1}\Sigma_I N_I[N_I(\text{SUCC})/N_I - P_I]^2, \tag{47}$$

where

$$P_I = \tfrac{1}{2}(0 + 1/J), \tfrac{1}{2}(1/J + 2/J), \ldots, \tfrac{1}{2}(1 - 1/J + 1)$$

and N_I refers to those estimates $\hat{\pi}_X$ falling in I.

But even if VALID = 0, it would still be desirable, for many of the subintervals, to have I correspond to a set of data cases nearly all success or nearly all failure. Namely, $\hat{\pi}_X$ should cut the data to yield small values of

$$\text{SHARP} = N^{-1}\Sigma_I \Sigma_{n|I}[\delta_n - N_I(\text{SUCC})/N_I]^2$$

$$= N^{-1}\Sigma_I N_I[N_I(\text{SUCC})/N_I][N_I(\text{FAIL})/N_I]. \tag{48}$$

(The second equality was pointed out to us by Prof. Marcello Pagano.)

Define the total sum of squares,

$$\text{APPBAR} = N^{-1}\Sigma_I \Sigma_{n|I} [\delta_n - P_I]^2 = \text{SHARP} + \text{VALID}. \tag{49}$$

(The second equality has appeared, in essence, in the literature on probabilistic weather forecasting, Sanders, 1958).

APPBAR is an I-grouped approximation to Brier and Allen's (1951) PBAR score,

$$\text{PBAR} = N^{-1}\Sigma_1^N (\delta_n - \hat{\pi}_{X_n})^2, \tag{50}$$

the average squared error of $\hat{\pi}_X$ as a predictor of δ.

See also Dickey (1968b) and Hills (1967) for other measures of discriminatory power.

Performance with the Guy's Hospital Data

The computer programs described in Walrath, Goldman and Dickey (1971) compute the above predictive performance scores for

- the smoothed probability estimates in equation 36 (and their obvious Bayes' theorem generalizations to several response classes with user-specified prior class probabilities) with iteration on a parameter α of the weight function w, e.g., $\alpha = \rho^2$ (equation 38), σ^2 (equation 37);
- the Fix-Hodges nearest-neighbor procedure ($m = 1$); and
- the logistic-form probability estimate (equation 27) for given linear discriminant function coefficients β_j.

Figure 3 depicts sample output for the first procedure. The program option is used to omit missing-value cases (5 cases omitted).

Table 3 summarizes the performance of the three discriminant-analysis procedures above for the Guy's Hospital data. The sets of variables used for table 3 are the nine variables of figure 3; the first six of these nine variables; and the single variable, Bulbrook's linear discriminant recalculated by equation 1. For the last procedure, we used the values of $\beta_0, \ldots, \beta_6$ given in Armitage et al (1969), and did not group together simple and radical mastectomies. Optimal scores for the performance measures are underlined.

Note that the smoothed probability estimators performed about as well as the linear discriminant. This appears to be due primarily to a lack of discriminatory information in the data, for when the response variable is included as an extra predictor variable, the nonparametric (6+1) seven-variable procedures give about 95% correct predictions. The (9+1) ten-variable versions then give about 83%; this 12% reduction in performance is caused by the noise of only three additional variables, illustrating the need for isolating important predictor variables.

Figure 3. Smoothed discrimination probability estimation output.

```
115 ADRENALECTOMY PATIENTS

NON-FAILURE(CLASS 1) AND FAILURE(CLASS 2) PREDICTED USING 9 VARIABLES

    TYPE OF MASTECTOMY - NONE (NEGATIVE) 0

                       - RADICAL OR SIMPLE (POSITIVE) 1

    LOG(F/S TO ABLATION TIME)

    LESION (TWO VARIABLES) -

        BREAST     1  1

        OSSEOUS   -0 -2

        VISCERAL  -1  1

    FOUR STEROIDS -

        17OHCS-17HYDROXICORTICOSTEROIDS

        A-ANDROSTERONE

        E-AETIOCHANOLONE

        D-DEHYDROEPIANDROSTERONE

    AGE AT MASTECTOMY


NO. OF CASES-  115

N - NO. OF CASES USED-  110

ALPHAL-        12.0500        ALPHAM-        15.0500        THICK-

DISEASE CLASS    1        TV(K) =    1.00        SV(K) =    1.00

DISEASE CLASS    2        TV(K) =    1.00 .      SV(K) =    1.00

    K               N(K)            P(K)

    1                54             .491

    2                56             .509
```

```
 ALPHA        PBAR         NCORRECT       /N                 /NPREDS            NKPATSEACHRANGE/NPATSEACHRANGE
---------  K --------     -------------  -----------------  ----------------   --------------------------------
           --

  13.250   .23331          70            110 (  .64)        110 (  .64)
        1      .23331           28            54 (  .52)           42 (  .67)
                                                                                [ 0.00,  .10)   1/  1  ( 1.00)
                                                                                [  .10,  .20)   0/  1  ( 0.00)
                                                                                [  .20,  .30)   0/  2  ( 0.00)
                                                                                [  .30,  .40)   0/  3  ( 0.00)
                                                                                [  .40,  .50)  25/ 61  (  .41)
                                                                                [  .50,  .60)  20/ 33  (  .61)
                                                                                [  .60,  .70)   3/  4  (  .75)
                                                                                [  .70,  .80)   3/  3  ( 1.00)
                                                                                [  .80,  .90)   0/  0  (-0.00)
                                                                                [  .90, 1.00]   2/  2  ( 1.00)
                                                                                    VALID =    .01684
                                                                                    SHARP =    .21257
                                                                                   APPBAR =    .22941
        2      .23331           42            56 (  .75)           68 (  .62)
                                                                                [ 0.00,  .10)   0/  2  ( 0.00)
                                                                                [  .10,  .20)   0/  0  (-0.00)
                                                                                [  .20,  .30)   0/  3  ( 0.00)
                                                                                [  .30,  .40)   1/  4  (  .25)
                                                                                [  .40,  .50)  13/ 33  (  .39)
                                                                                [  .50,  .60)  36/ 61  (  .59)
                                                                                [  .60,  .70)   3/  3  ( 1.00)
                                                                                [  .70,  .80)   2/  2  ( 1.00)
                                                                                [  .80,  .90)   1/  1  ( 1.00)
                                                                                [  .90, 1.00]   0/  1  ( 0.00)
                                                                                    VALID =    .01684
                                                                                    SHARP =    .21257
                                                                                   APPBAR =    .22941

  13.350   .23119          74            110 (  .67)        110 (  .67)
        1      .23119           30            54 (  .56)           42 (  .71)
                                                                                [ 0.00,  .10)   1/  1  ( 1.00)
                                                                                [  .10,  .20)   0/  1  ( 0.00)
                                                                                [  .20,  .30)   0/  2  ( 0.00)
                                                                                [  .30,  .40)   1/  4  (  .25)
                                                                                [  .40,  .50)  22/ 60  (  .37)
                                                                                [  .50,  .60)  22/ 34  (  .65)
                                                                                [  .60,  .70)   3/  3  ( 1.00)
                                                                                [  .70,  .80)   3/  3  ( 1.00)
                                                                                [  .80,  .90)   0/  0  (-0.00)
                                                                                [  .90, 1.00]   2/  2  ( 1.00)
                                                                                    VALID =    .02170
                                                                                    SHARP =    .20407
                                                                                   APPBAR =    .22577
        2      .23119           44            56 (  .79)           68 (  .65)
                                                                                [ 0.00,  .10)   0/  2  ( 0.00)
                                                                                [  .10,  .20)   0/  0  (-0.00)
                                                                                [  .20,  .30)   0/  3  ( 0.00)
                                                                                [  .30,  .40)   0/  3  ( 0.00)
                                                                                [  .40,  .50)  12/ 34  (  .35)
                                                                                [  .50,  .60)  38/ 60  (  .63)
                                                                                [  .60,  .70)   3/  4  (  .75)
                                                                                [  .70,  .80)   2/  2  ( 1.00)
                                                                                [  .80,  .90)   1/  1  ( 1.00)
                                                                                [  .90, 1.00]   0/  1  ( 0.00)
```

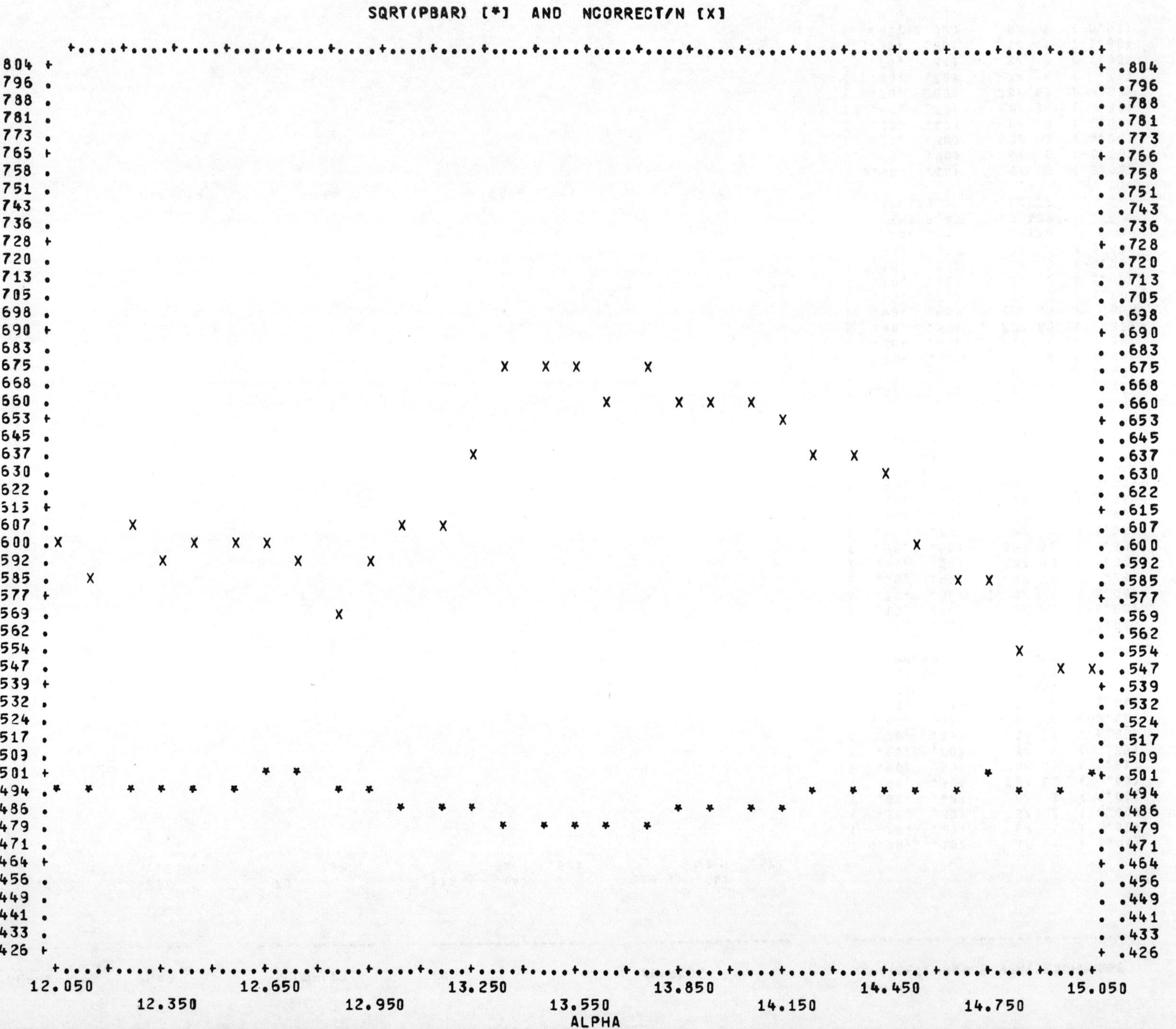
SQRT(PBAR) [*] AND NCORRECT/N [X]
.804
.796
.788
.781
.773
.765
.758
.751
.743
.736
.728
.720
.713
.705
.698
.690
.683
.675
.668
.660
.653
.645
.637
.630
.622
.615
.607
.600
.592
.585
.577
.569
.562
.554
.547
.539
.532
.524
.517
.509
.501
.494
.486
.479
.471
.464
.456
.449
.441
.433
.426
12.050
12.350
12.650
12.950
13.250
13.550
13.850
14.150
14.450
14.750
15.050
ALPHA

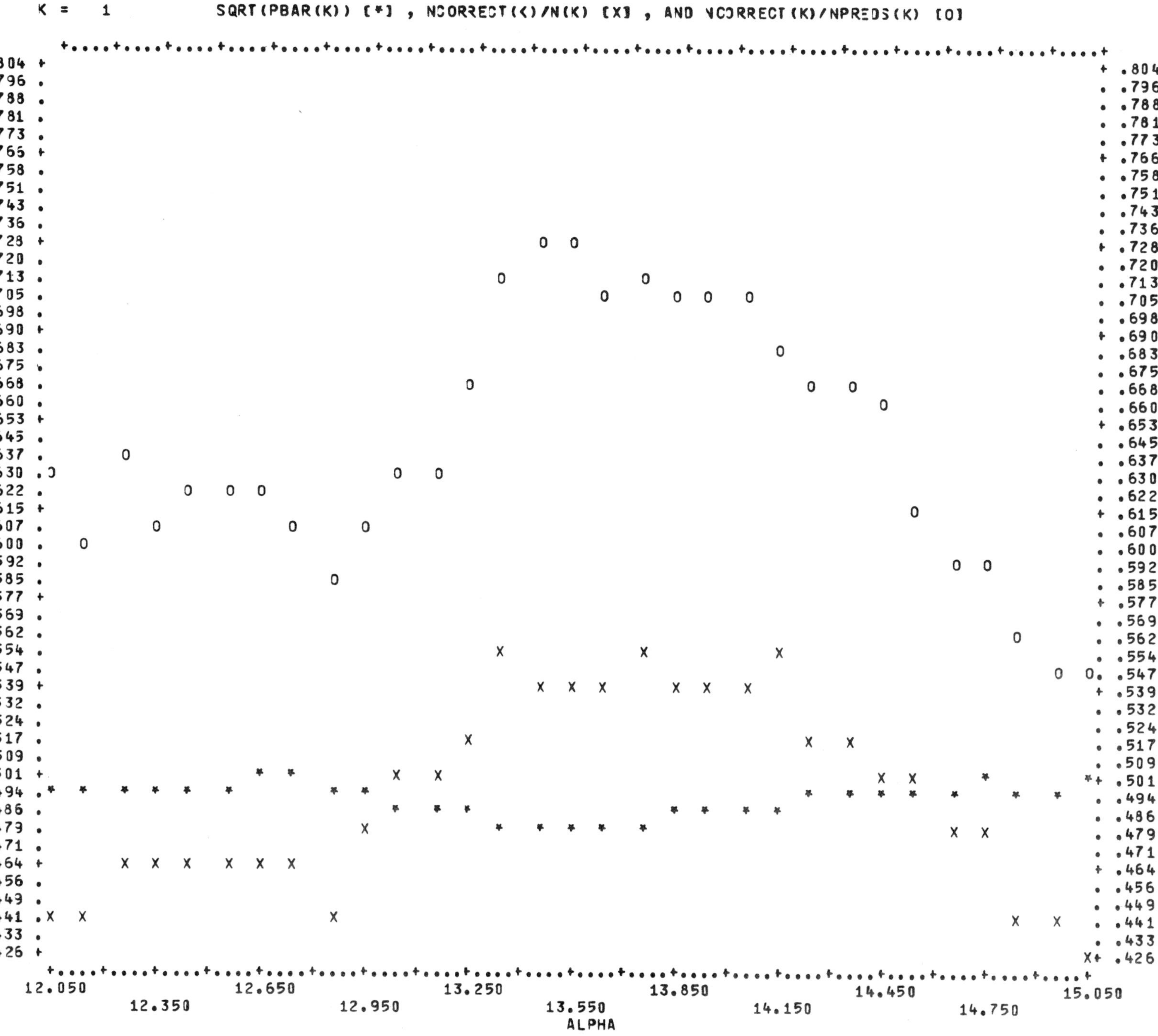
K = 1
SQRT(PBAR(K)) [*] , NCORRECT(K)/N(K) [X] , AND NCORRECT(K)/NPREDS(K) [O]
12.050
12.350
12.650
12.950
13.250
13.550
13.850
14.150
14.450
14.750
15.050
ALPHA

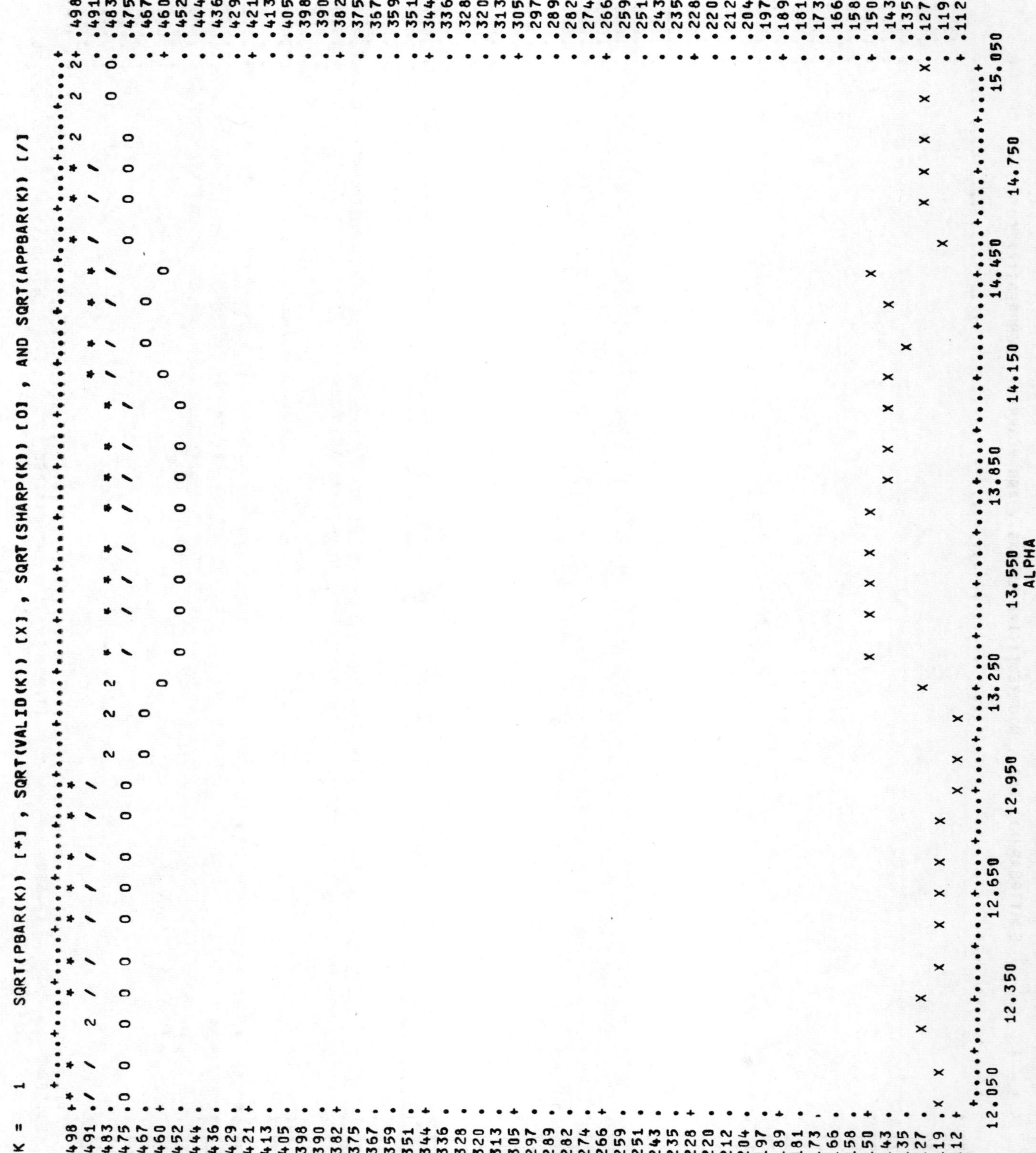
K = 1 SQRT(PBAR(K)) [*] , SQRT(VALID(K)) [X] , SQRT(SHARP(K)) [O] , AND SQRT(APPBAR(K)) [/]
.498 .491 .483 .475 .467 .460 .452 .444 .436 .429 .421 .413 .405 .398 .390 .382 .375 .367 .359 .351 .344 .336 .328 .320 .313 .305 .297 .289 .282 .274 .266 .259 .251 .243 .235 .228 .220 .212 .204 .197 .189 .181 .173 .166 .158 .150 .143 .135 .127 .119 .112
12.050 12.350 12.650 12.950 13.250 13.550 13.850 14.150 14.450 14.750 15.050
ALPHA

Table 3. Performance of discriminant analysis procedures.

Discrimation Procedure	Number of Predictor Variables	Hypophysectomy (95 cases)				Adrenalectomy (115 cases)			
		α	NCORRECT/N x 100%	PBAR	VALID	α	NCORRECT/N x 100%	PBAR	VALID
Ball Smoothing	1	.327	71	.21327	.02284	1.20	54	.24795	.00754
		.599	72	.21389	.02166	1.38	57	.25015	.00683
		1.25	68	.21888	.01380	7.62	53	.25032	.00060
	6	1.09	69	.25905	.04755	1.51	68	.23525	.04573
		1.56	65	.23515	.01825	1.85	63	.22822	.02718
		9.00	55	.24152	.00258	19.0	55	.24506	.00009
	9	4.03	74	.27161	.08919	13.3	67	.23394	.02439
		8.00	60	.23229	.00318	13.4	66	.23365	.02166
						19.0	55	.23839	.00432
Normal-Kernel Smoothing	1	.348	69	.21931	.01531	3.05	54	.25305	.00067
		.580	72	.22065	.03063	3.07	54	.25034	.00067
		5.80	55	.24020	.00009	3.27	53	.25027	.00049
	6	.230	58	.29076	.06031	.170	64	.27044	.05443
		3.50	54	.24222	.00991	1.31	59	.23983	.01985
		4.30	57	.24253	.00264	15.1	53	.24799	.00049
	9	.695	61	.27373	.05857	2.60	56	.24271	.00286
		7.10	55	.24580	.00132	5.10	57	.24529	.00079
		8.10	54	.24566	.00363	7.60	61	.24690	.01165
Fix-Hodges Nearest Neighbor	1		56				41		
	6		57				58		
	9		53				48		
Logistic Function of Linear Discriminant	6*		63	.22669	.04909		65	.21169	.00751

*In this case the variable, Type of Mastectomy, takes the values 0(none), 1(simple), 2(radical).

DATA SET DESCRIPTION

DATA - ADVANCED BREAST CANCER DATA (J. DICKEY)

CASES AND VARIABLES

210 CASES (PATIENTS)
50 VARIABLES - 43 MEASUREMENT, 3 CATEGORICAL, 4 IDENTIFICATION

BRIEF DESCRIPTION

VAR	NAME
***	****
1	PATIENT NUMBER
2	AGE AT MASTECTOMY OR WHEN FIRST SEEN (AGEM)
3	MONTH OF MASTECTOMY OR WHEN FIRST SEEN (MONTHM)
4	YEAR OF MASTECTOMY OR WHEN FIRST SEEN (YEARM)
5	TYPE OF MASTECTOMY (TYPE)
6	MENOPAUSAL STATE AT MASTECTOMY OR WHEN FIRST SEEN (MPSTATEM)
7	NUMBER OF YEARS POSTMENOPAUSAL (POSTM)
8	MAXIMUM DIMENSION OF TUMOR (TUMOR)
9	PATHOLOGICAL NODES (PATHNODES)
10	GRADE (GRADE)
11	AGE AT RECURRENCE (AGER)
12	MONTH OF FIRST RECURRENCE (MONTHR)
13	YEAR OF FIRST RECURRENCE (YEARR)
14	MENOPAUSAL STATE AT RECURRENCE (MPSTATER)
15	NUMBER OF YEARS POSTMENOPAUSAL AT RECURRENCE (POSTR)
16	TIME FROM MASTECTOMY OR WHEN FIRST SEEN TO RECURRENCE (FREEPERIOD)
17	RESULTS OF ADMINISTRATION OF ADROSTEROIDS (ANDROS)
18	RESULTS OF ADMINISTRATION OF OESTROGENS (OESTRO)
19	AGE AT ABLATION (AGEA)
20	DAY OF ABLATION (DAYA)
21	MONTH OF ABLATION (MONTHA)
22	YEAR OF ABLATION (YEARA)
23	MENOPAUSAL STATE AT ABLATION (MPSTATEA)
24	NUMBER OF YEARS POSTMENOPAUSAL AT ABLATION (POSTA)
25	TYPE OF OPERATION (OPERATION)
26	TIME FROM MASTECTOMY OR WHEN FIRST SEEN TO ABLATION (MTIME)
27	TIME FROM RECURRENCE TO ABLATION (RTIME)
28	SITE OF PREDOMINANT LESION (LESION)
29	CLINICAL RESULT OF ABLATION (ABRESULT)
30	MEAN CLINICAL VALUE ASSESSED AFTER THREE MONTHS (MCV3)
31	AREA UNDER MEAN CLINICAL VALUE CURVE (AREA)
32	TIME FROM MASTECTOMY OR WHEN FIRST SEEN TO DEATH (SURVIVAL)
33	TIME FROM ABLATION TO DEATH (ABTODEATH)
34	DAY LAST REPORTED ALIVE OR OF DEATH (DAYD)
35	MONTH LAST REPORTED ALIVE OR OF DEATH (MONTHD)
36	YEAR LAST REPORTED ALIVE OR OF DEATH (YEARD)
37	CARD NUMBER
38	PATIENT NUMBER
39	17-HYDROXYCORTICOSTEROIDS, URINE SAMPLE (17OHCS)
40	17 OHCSN, BLOOD SAMPLE (17OHCSN)
41	DEHYDROEPIANDROSTERONE, URINE SAMPLE (D)
42	DN, BLOOD SAMPLE (DN)
43	ANDROSTERONE, URINE SAMPLE (A)
44	AN, BLOOD SAMPLE (AN)
45	AETIOCHOLANOLONE, URINE SAMPLE (E)
46	EN, BLOOD SAMPLE (EN)
47	DISCRIMINANT FUNCTION (DISC)
48	DISCN (DISCN)

49 SURVIVAL STATUS (STATUS)
50 CARD NUMBER

PURPOSE OF STUDY

THE DATA CONCERN 210 WOMEN WITH ADVANCED BREAST CANCER WHO ATTENDED THE BREAST CLINIC, GUY'S HOSPITAL, LONDON AND RECEIVED EITHER HYPOPHYSECTOMY OR BILATERAL ADRENALECTOMY WITH OOPHORECTOMY BETWEEN JANUARY 1958 AND DECEMBER 1965. THE PURPOSE OF THE STUDY IS TO EVALUATE SURVIVAL.

REFERENCE

ARMITAGE, P., C. K. MCPHERSON, AND J. C. COPAS. STATISTICAL STUDIES OF PROGNOSIS IN ADVANCED BREAST CANCER. J. CHRON. DIS. 22, 1969, 343-360. (NEW CASES HAVE BEEN ADDED TO THE DATA SINCE THIS PUBLICATION.)

DETAILED DESCRIPTION

VAR	COL	FORMAT	DESCRIPTION
***	***	******	***********
1	1-3	XXX.	PATIENT NUMBER
2	4-5	XX.	AGE AT MASTECTOMY OR WHEN FIRST SEEN (YEARS)
3	6-7	XX.	MONTH OF MASTECTOMY OR WHEN FIRST SEEN
4	8-9	XX.	YEAR OF MASTECTOMY OR WHEN FIRST SEEN
5	10	X.	TYPE OF MASTECTOMY
			0 = NONE
			1 = SIMPLE
			2 = RADICAL
6	11	X.	MENOPAUSAL STATE AT MASTECTOMY OR WHEN FIRST SEEN
			1 = PREMENOPAUSAL
			2 = MENOPAUSAL
			3 = POSTMENOPAUSAL
			4 = HYSTERECTOMY
			BLANK = UNKNOWN
7	12-13	XX.	NUMBER OF YEARS POSTMENOPAUSAL
			0 = NOT APPLICABLE
			BLANK = UNKNOWN
8	14-15	X.X	MAXIMUM DIMENSION OF TUMOR (INCHES)
			0 = UNKNOWN
9	16	X.	PATHOLOGICAL NODES
			0 = UNKNOWN
			1 = NEGATIVE
			2 = POSITIVE
10	17	X.	GRADE
			0 = UNKNOWN
			1 = GRADE 1
			2 = GRADE 2
			3 = GRADE 3
11	18-19	XX.	AGE AT RECURRENCE (YEARS)
			0 = NOT APPLICABLE
12	20-21	XX.	MONTH OF FIRST RECURRENCE
			0 = NOT APPLICABLE
13	22-23	XX.	YEAR OF FIRST RECURRENCE
			0 = NOT APPLICABLE
14	24	X.	MENOPAUSAL STATE AT RECURRENCE
			0 = NOT APPLICABLE
			1 = PREMENOPAUSAL
			2 = MENOPAUSAL
			3 = POSTMENOPAUSAL
			4 = HYSTERECTOMY
			BLANK = UNKNOWN

```
15  25-26    XX.   NUMBER OF YEARS POSTMENOPAUSAL AT RECURRENCE
                     0 = NOT APPLICABLE
                     BLANK = UNKNOWN
16  27-29    XXX.  TIME FROM MASTECTOMY OR WHEN FIRST SEEN TO
                   RECURRENCE (MONTHS)
                     0 = NOT APPLICABLE
17     30    X.    RESULTS OF ADMINISTRATION OF ANDROSTEROIDS
                     0 = NOT APPLICABLE
                     1 = SUCCESS
                     2 = INTERMEDIATE
                     3 = FAILURE
                     4 = RESULTS UNKNOWN
18     31    X.    RESULTS OF ADMINISTRATION OF OESTROGENS
                     0 = NOT APPLICABLE
                     1 = SUCCESS
                     2 = INTERMEDIATE
                     3 = FAILURE
                     4 = RESULTS UNKNOWN
19  32-33    XX.   AGE AT ABLATION (YEARS)
20  34-35    XX.   DAY OF ABLATION
21  36-37    XX.   MONTH OF ABLATION
22  38-39    XX.   YEAR OF ABLATION
23     40    X.    MENOPAUSAL STATE AT ABLATION
                     1 = PREMENOPAUSAL
                     2 = MENOPAUSAL
                     3 = POSTMENOPAUSAL
                     4 = HYSTERECTOMY
                     BLANK = UNKNOWN
24  41-42    XX.   NUMBER OF YEARS POSTMENOPAUSAL AT ABLATION
                     0 = NOT APPLICABLE
                     BLANK = UNKNOWN
25     43    X.    TYPE OF OPERATION
                     1 = ADRENALECTOMY WITH OOPHORECTOMY
                     2 = HYPOPHYSECTOMY
26  44-46    XXX.  TIME FROM MASTECTOMY OR WHEN FIRST SEEN
                   TO ABLATION (MONTHS)
27  47-49    XXX.  TIME FROM RECURRENCE TO ABLATION (MONTHS)
                     0 = NOT APPLICABLE
28     50    X.    SITE OF PREDOMINANT LESION
                     1 = BREAST
                     2 = BONE
                     3 = VISCERAL
29     51    X.    CLINICAL RESULT OF ABLATION
                     1 = SUCCESS
                     2 = INTERMEDIATE
                     3 = FAILURE
30  52-54    XX.X  MEAN CLINICAL VALUE ASSESSED AFTER THREE
                   MONTHS
                      0 = DETERIORATION OF ALL LESIONS
                     ...
                     12 = IMPROVEMENT OF ALL LESIONS
31  55-57    XXX.  AREA UNDER MEAN CLINICAL VALUE CURVE
                     BLANK = UNKNOWN
32  58-60    XXX.  TIME FROM MASTECTOMY OR WHEN FIRST SEEN TO
                   DEATH (MONTHS)
                     BLANK = ALIVE
33  61-62    XX.   TIME FROM ABLATION TO DEATH (MONTHS)
                     BLANK = ALIVE
34  63-64    XX.   DAY LAST REPORTED ALIVE OR OF DEATH
35  65-66    XX.   MONTH LAST REPORTED ALIVE OR OF DEATH
36  67-68    XX.   YEAR LAST REPORTED ALIVE OR OF DEATH
37     80    X.    CARD NUMBER
                     1 = CARD 1

38   1-3     XXX.  PATIENT NUMBER
39   4-6     XX.X  17-HYDROXYCORTICOSTEROIDS (URINE SAMPLE)
40   7-9     X.XX  17 OHCSN (BLOOD SAMPLE)
```

```
                            BLANK = UNKNOWN
41   10-13   XXXX.      DEHYDROEPIANDROSTERONE (URINE SAMPLE)
                            BLANK = UNKNOWN
42   14-16    XXX.      DN (BLOOD SAMPLE)
                            BLANK = UNKNOWN
43   17-20   XXXX.      ANDROSTERONE (URINE SAMPLE)
                            BLANK = UNKNOWN
44   21-24   XXXX.      AN (BLOOD SAMPLE)
                            BLANK = UNKNOWN
45   25-28   XXXX.      AETIOCHOLANOLONE (URINE SAMPLE)
                            BLANK = UNKNOWN
46   29-32   XXXX.      EN (BLOOD SAMPLE)
                            BLANK = UNKNOWN
47   33-37  XXXXX.      DISCRIMINANT FUNCTION
                        (DF = 80 - 80(17 OHCSN) + E)
                            BLANK = UNKNOWN
48   38-42  XXXXX.      DISCN
                            BLANK = UNKNOWN
49      43      X.      SURVIVAL STATUS
                            1 = ALIVE
                            2 = DEAD, NO SPECIFIC CAUSE REPORTED
                            3 = ACCIDENTAL DEATH
                            4 = OPERATIONAL DEATH
50      80      X.      CARD NUMBER
                            2 = CARD 2

FORMAT IS (F3.0,3F2.0,2F1.0,F2.0,F2.1,2F1.0,3F2.0,F1.0,F2.0,
F3.0,2F1.0,4F2.0,F1.0,F2.0,F1.0,2F3.0,2F1.0,F3.1,2F3.0,4F2.0,
11X,F1.0/F3.0,F3.1,F3.2,F4.0,F3.0,4F4.0,2F5.0,F1.0,36X,F1.0)

N = 210
```

LOCATION OF DATA

CARD IMAGE - FS.C073.CARMI1

LISTING OF CARD 1 (THE FIRST TWO LINES ARE COLUMN GUIDES)

COLUMN 80, CARD NUMBER, CONTAINS A ''1.'' IT IS NOT PRINTED BELOW BECAUSE OF LACK OF SPACE.

```
0          1                2         3                   4             5              6
1234567890 1234567 890123 45678901 23456789 0123 456789 01 234567 89012 345678

0014809552 1000000 491056 10001330 50191257 3012 027014 23 056023 03205 050558
0023511472 1000000 380350 10002830 45301257 3071 121093 32 050048 13311 121258
0034011541 1002510 420756 10002030 43140158 1002 038018 13 000000 03801 310158
0042709562 1001520 280857 10001130 29310158 1001 016005 33 000000 01700 090258
0054911570 2004000 000000 00000030 49310158 3011 002000 13 014007 00705 180658
0065005552 3052520 520157 30702033 53190258 3081 033013 31 100168 05017 080759
0073410532 1001011 361155 10002530 38070258 3012 052027 31 120090 06008 171058
0085202452 3020010 600253 31009601 65280258 3151 156060 31 120575 20549 310362
0094809560 3032500 000000 00000000 49070358 3041 018000 33 000000 01801 230358
0106610512 3102011 720957 31607103 72040358 3162 077006 32 065128 09822 211259
0112606541 1001000 280556 10002320 30010458 1001 046023 11 096447 08944 221161
0124410570 1000800 000000 00000010 44010458 1002 005000 21 100134 02015 300659
0136909570 3181500 000000 00000003 69150458 3182 007000 22 093079 01710 100259
0145501562 3080020 560257 30901303 57070558 3101 028015 32 067042 03507 051258
0154505570 1003500 000000 00000010 46250458 1002 011000 21 105461 05139 100861
0164709562 3170020 481157 31801430 48160558 3181 020006 22 107061 02707 111258
0174511542 1001022 461255 10001330 48090558 2002 042029 22 067135 06624 210560
0185406551 3020000 550456 30301003 57300558 3051 035025 12 047121 06025 200660
0193109562 1000010 320158 10001630 33060658 3011 021005 32 105052 02908 100259
0205212552 4003020 550358 40002700 55100658 4002 030003 23 000002 04414 100859
0213712542 1000510 400158 10003730 41170658 1001 042005 23 040019 04806 251258
```

0224309442	1000000	570558	31016400	57020758	3101	166002	23	040008	17913	060859
0234610570	2000000	000000	00000030	47080758	3012	009000	31	110212	02920	060360
0243710392	1001000	520554	30217500	56080758	3061	225050	33	000008	22702	090958
0254611540	1004000	000000	00000000	50180758	3031	044000	22	060084	05712	010859
0264904572	1003023	491057	10000600	50140758	1001	015009	22	120056	02409	290459
0273202562	1001520	331057	10002030	34280758	1001	029009	13	000000	03001	200858
0286504582	3235023	650458	32300000	65050858	3231	004004	22	060086	02926	290960
0295005551	1000023	521157	20003030	53110858	3011	039009	22	090078	04809	110559
0303207580	1003500	000000	00000000	32050858	1002	001000	23	045038	01413	150959
0314404552	1000000	450756	10001500	47010858	3012	040025	13	000006	04203	301058
0324704542	1002020	500257	10003420	51300958	1002	053019	13	000016	05603	261258
0334509580	1008000	000000	00000000	45141058	1001	001000	13	000000	00100	231058
0344407552	1000000	470258	30103100	47121058	3021	039008	22	080047	04607	190559
0354008552	1000000	430658	10003400	43061058	1002	038004	31	110142		011167
0364608570	4003020	000000	00000030	47281058	4001	014000	13	000000	01400	281058
0374801562	1001520	490657	10001700	51271058	3012	033016	22	090175	07036	101161
0384306562	1000012	450458	10002230	45301258	3011	030008	23	000001	03201	030259
0394111552	1001210	430658	10003130	44300159	1001	038007	23	007010	04910	071259
0403908542	1002020	430358	10004320	44170359	1001	055012	11	090117	06712	150360
0414002542	1000000	410255	10001240	45130359	3041	061049	23	000005	06302	270559
0424204572	1000013	430558	10001300	44310359	2001	023010	22	105076	03208	101259
0434201572	1000002	420657	10000530	44090459	1002	027022	23	015011	03306	011059
0444407572	1001220	450958	10001420	45240459	1001	021007	11	110040	02605	140959
0454202542	1003022	470259	10006030	47010559	1001	063003	21	104157	09028	190861
0464804562	1000010	510159	30203330	51110559	3022	037004	33	000002	03902	020759
0474712560	1001020	000000	00000003	49200559	3012	029000	21	100235	05223	240461
0485507522	3071011	600557	31205802	62050659	3142	083025	11	120130	12341	211062
0495603592	3082010	560559	30800200	56300659	3081	003001	23	000000	00603	260959
0505401570	3052500	000000	00000031	56170759	3071	030000	13	039018	04616	131160
0516812580	3242500	000000	00000001	68160759	3252	007000	13	047018	01508	260360
0525310572	3032010	540359	30401703	54070859	3041	022005	22	060164	04220	080461
0534002591	1004003	400559	10000300	40230759	1002	005002	22	090191	03226	071061
0546208580	3040000	000000	00000000	63100859	3051	012000	13	000000	01200	150859
0554602570	4000000	000000	00000020	48210859	4001	030000	33	000000	03100	030959
0565705590	4003800	000000	00000003	57070859	4002	003000	33	000000	00401	020959
0575007580	1000000	000000	00000020	51210859	1002	013000	23	000003	01502	151059
0584303560	1004000	000000	00000020	46170959	3022	042000	12	075144	07836	170962
0594803580	1001500	000000	00000000	50290959	1002	018000	11	120154	03921	170661
0604502572	1001521	470559	10002700	48031159	1001	033006	31	120972		011167
0614905590	2003200	000000	00000000	49051159	3012	006000	23	074023	01004	200360
0623012571	1000000	300158	10000130	32201159	1001	023022	22	105074	03310	250960
0635306550	3030000	000000	00000000	57241159	3071	053000	11	120249	09340	180363
0645912522	3072010	631256	31104801	66111259	3141	084036	31	120147	10016	050461
0655108580	3033000	000000	00000002	52141259	3042	016000	22	086170	04630	170662
0664309552	4003023	470959	40004800	47220160	4001	052004	32	120093	06311	211260
0674501590	3014000	000000	00000030	46110360	3011	014000	12	090050	03117	220861
0685904592	3071000	601159	30800703	60220360	3082	011004	23	060023	02513	040561
0695007590	1004000	000000	00000000	50190560	1002	010000	32	079090	02010	250361
0704212561	1000020	430957	10001000	46240660	1001	042033	13	000000	04503	250960
0715206600	3031200	000000	00000000	52040860	3032	002000	13	018012	01615	271061
0723708572	1002020	390759	10002300	39220760	1001	035012	22	090148	05419	230262
0735903570	3143000	000000	00000000	63160860	3182	041000	32	077051	04908	220461
0744302580	1004000	000000	00000030	45281060	1001	032000	12	093065	04715	220162
0756203600	3124000	000000	00000000	63181060	3131	007000	23	037060	02013	171161
0764411600	1002800	000000	00000000	44231260	1002	001000	11	110325	03231	200763
0774107602	1002020	411160	20000500	41060161	2002	006002	23	000002	01711	081261
0784211562	1001023	461260	10004900	46310161	1002	050001	33	040029	05908	061061
0795612592	3062022	570161	30701300	57030361	3072	015002	11	113149	06248	190265
0805911600	3101200	000000	00000003	59110461	3102	005000	21	116158	02217	270962
0814509562	3011822	480160	30404010	49060361	3052	054014	33	000000	05400	070361
0824207602	1003020	420261	10000700	42110461	1002	009002	13	000003	01102	200661
0834307591	1001000	441060	10001500	45140461	1002	021006	22	060067	03414	310562
0845005592	2001023	520361	30102200	53020561	3012	024002	31	120129	03814	050762
0854307600	1003000	000000	00000030	44130661	3011	011000	33	000000	01201	010761
0864909602	2001020	490561	20000800	49080861	2002	011003	21	075339	04736	180864
0874302572	1000000	471260	30304630	47010861	3042	054008	31	098	05704	291161
0885603562	3022023	581258	30403130	60010861	3062	065034	11	120166	08925	240863

0895608592	3080000	571060	30901403	58250861	3101	024010	13	003003	02803	041261
0904402592	1000020	451260	10002200	46150961	1001	031009	22	074087	04312	060962
0915607580	3101200	000000	00000000	60260961	3141	038000	23	015019	04911	250862
0925512600	3020020	000000	00000030	55010961	3021	009000	23	000001	01001	061061
0935411580	3033000	000000	00000010	57220961	3072	034000	11	120098	05015	020163
0945408610	3021800	000000	00000000	55131061	3032	002000	33	000002	00301	211161
0955901582	3091523	610160	31102400	63131061	3131	045021	12	060114	08338	061264
0964902552	4000000	530360	40006130	55311061	4001	080019	23	060024	08808	200662
0973209610	1001000	000000	00000000	32131061	1002	001000	21	100307	04443	260565
0985806600	3100000	000000	00000003	59011261	3111	018000	13	075049	02709	060962
0996112602	3092020	621061	31001000	62291161	3101	011001	13	000011	01402	050262
1004908590	1001500	000000	00000000	51191261	1001	028000	13	030016	03406	090662
1014310561	1002000	480961	10005900	48300162	1001	063004	33	000012	06603	240462
1024407602	1000020	450461	10000900	45020162	1002	018009	11	120536		011167
1034805582	1000500	500859	10001530	52260262	3011	045030	13	000003	05005	200762
1043906600	1004000	000000	00000010	40160262	3021	020000	13	000004	02505	020762
1055011582	1001222	531061	10003500	53200262	1002	039004	21	120585		011167
1064704592	1001510	490861	10002820	49230262	1001	034006	11	120100	04813	050463
1074302552	1000010	500162	30408300	50060462	3041	086003	13	052028	09509	130163
1084107610	1003500	000000	00000000	41060362	1002	008000	13	000003	01407	280962
1095703602	4002022	591161	40002030	59090362	4001	024004	22	060042	03006	200962
1106103610	3062500	000000	00000000	62270362	3071	012000	13	000001	01301	300462
1115312582	3010000	550660	30301820	57250462	3051	040022	23	036020	04707	191162
1124408552	1000520	480959	10004920	50100562	1002	081032	32	060028	08706	031162
1135512592	3030512	570362	30502700	57240462	3052	028001	21	112693		011167
1145011581	1002800	531061	30203500	53170562	3022	042007	12	120156	08138	010865
1156203620	4001500	000000	00000000	62210562	4001	002000	11	100322	04543	171265
1165606582	3102020	600162	31404333	60040662	3152	048005	22	112039	05205	281062
1174405592	1002022	470262	10003300	47140662	1002	037004	23	000006	05013	290763
1183701612	1003023	370961	10000820	38180962	1002	020012	32	030050	04626	221164
1195801580	3040520	000000	00000000	62070962	3081	056000	13	000000	05600	180962
1204703612	1001022	480262	10001130	48250962	1001	018007	21	083216	03920	020664
1215405620	3055000	000000	00000000	55220962	3062	004000	12	057025	01309	180663
1225912572	3081510	640562	31305303	64051062	3132	058005	23	015010	06104	230163
1234308602	1000000	440162	10001700	45261062	1001	026009	23	000014	02802	191262
1245703620	4004000	000000	00000003	57121162	4002	008000	13	018012	01810	070963
1254705602	4000000	490961	40001630	50301162	4001	030014	23	020027	04414	250164
1264608620	2004000	000000	00000000	46111262	2002	004000	21	116438	04036	261265
1273411610	1000000	000000	00000030	35181262	3011	013000	23	000000	01401	050163
1285512602	3081512	571162	31002300	57181262	3101	024001	11	120739		011167
1293201572	1000000	380562	30506430	39181262	3052	071007	11	120233	09119	030864
1304512601	4000000	471162	40002300	47180163	4002	025002	12	075094	03813	030264
1315308612	3030000	541062	30401400	55290163	3051	017003	33	026021	02407	200863
1324402622	1000020	450163	10001100	45010363	1002	013002	31	120114	02917	280764
1332602611	1000000	270362	10001330	28020463	1002	026013	11	120383	07145	120167
1344110582	3040010	450163	30805103	45190363	3082	053002	33	060021	07219	141064
1356010610	3103520	000000	00000000	62160463	3121	018000	23	015009	03517	150964
1365902630	3061000	000000	00000000	59080463	3061	002000	22	060040	01008	201263
1375012592	1001020	510361	10001620	54200563	1001	041026	23	060045	05109	030364
1384105552	1000000	430657	10002510	49230763	1001	098073	12	080082	11012	200764
1395202622	3030000	530263	30401200	53180663	3042	016004	13	000003	02509	260364
1404405630	1005500	000000	00000000	44160763	1002	002000	11	120203	01917	111264
1415309620	3082500	000000	00000000	54020863	3091	011000	12	105201	03525	280865
1422306621	1000000	231162	30100530	24150763	3021	013008	23	000006	01502	190963
1436006630	3122500	000000	00000000	60060863	3121	002000	21	110152	01514	290964
1444705630	1004800	000000	00000000	47090863	1001	003000	33	000000	00300	210863
1453806601	1001000	410363	10003300	41130863	3012	038005	13	000004	04103	071163
1463408622	1003022	350763	10001100	35230863	1001	012001	31	120131	02412	270864
1475704602	3071520	581261	30802002	60060963	3101	041021	11	120386	08532	060566
1483803630	1002200	000000	00000020	38101063	1002	007000	11	081141	02417	150365
1493801611	1000000	400163	10002400	41250363	1001	026002	11	108347	07133	181266
1504104630	4003200	000000	00000020	41121163	4001	007000	13	000000	00801	081263
1513902621	1000500	400863	10001820	41170764	1002	029011	22	067037	03607	280265
1523201602	1001510	350263	10003730	35051263	1001	047010	12	086072	06821	040965
1534904622	1002020	500763	20001530	51280164	2001	021006	23	030014	03009	211064
1545803622	3081413	601063	31001900	60280164	3102	022003	23	000005	03007	050964
1554711582	1002022	480959	20001000	53140264	3011	063053	31	120268	09633	221166

```
1562608632 1001520 261263 30100430 26210264 3012 006002 13 000000 00701 230364
1572506612 1000000 260662 10001230 28190364 1001 033021 11 120593       011167
1584005612 1002012 420163 10002010 43100364 1002 034014 22 030282 06026 230566
1595609612 4000520 571163 40002600 57240364 4002 030004 33 000000 03101 140464
1606206612 3201510 640463 32202202 65010464 3231 034012 31 120138 05521 050166
1615401630 3093000 000000 00000003 60130664 3101 017000 13 013009 03316 051065
1625411632 3112021 540264 31100300 55310764 3122 008005 23 030047 02517 301265
1635305561 2000000 570161 30405600 61310764 3082 098042 21 096200 12022 300566
1643703632 1001510 380664 10001504 38221064 1002 019004 22 069131 03818 070566
1654301642 1001520 440964 10000800 44071064 1001 009001 33 000003 01001 221164
1665808630 3052500 000000 00000003 59231064 3061 014000 13 000004 01601 021264
1674512622 1000523 460763 20000730 47191164 2002 023016 33 000000 02401 111264
1684410640 2002000 000000 00000000 44291264 2002 002000 23 000000 00301 140165
1695003640 1003000 000000 00000030 50181264 1001 009000 13 000003 01102 160265
1704603582 4001020 491160 40003220 53271164 4002 080048 12 105361       011167
1714108601 1004000 440763 30303620 45141264 3042 052017 21 087233       011167
1725807611 3031520 610864 30603703 62120165 3072 042005 13 000001 04806 250765
1734302630 1003001 000000 00000020 45010165 3011 023000 13 000007 02603 140465
1744905612 1000010 520664 20003630 53150265 2001 045008 32 094065 05813 250366
1754909632 1000510 490364 10000630 50030265 3011 017011 13 000000 01701 200265
1764210561 1002000 471061 10006000 48230265 1002 063003 33 000012 10602 240465
1774006612 1000020 410662 10001240 44050365 3031 045033 22 095043 05006 290865
1785610630 3094000 000000 00000002 57160365 3101 017000 13 000001 03417 240866
1794806502 4000000 600661 40013203 64060465 4001 178046 12 082067 21234 220268
1804610622 1000000 470763 10000910 48140465 1002 030021 33 037049 03930 170967
1814201591 1000000 450262 10003710 48200465 1001 075038 33 000000 07601 150565
1825512600 3022020 000000 00000002 60200465 3071 052000 11 120211       011167
1834210640 1002200 000000 00000030 43210565 1002 007000 22 037066 01912 310566
1843504630 1002223 000000 00000030 37190565 1001 025000 13 000006 03313 210666
1854102592 2001522 420260 20001210 47020665 2002 076064 23 030175 10328 010168
1864510622 1000000 460663 30100820 48110665 3032 032024 23 000003 03402 050865
1874412640 1002000 000000 00000030 44300665 1001 006000 13 004011 01407 040266
1884102632 1000020 431264 30202200 43030865 3032 030008 33 002005 03505 210366
1895608640 3054200 000000 00000003 57251065 3061 014000 33 010009 03622 150867
1905101640 2004000 000000 00000020 53051165 2002 022000 22 077037 02407 100666
1914908612 4002020 530565 40004510 53031265 4002 052007 33 000000 05200 071265
1924310582 1000510 500365 30107730 51211265 3011 086009 31 120    08604 150466
1934103421 1000000 540355 30215601 64211265 3121 285129 11 120    28319 070867
1944203622 1002522 441063 10001920 45171265 3021 045026 32 000    04610 141166
1954201652 1001822 430965 10000830 43140266 1001 013005 31 120113 02310 311266
1965807650 3045500 000000 00000020 58240266 3052 008000 23 000000 01001 130466
1975211582 3021513 561162 30604803 60210366 3101 088040 11 113319       011267
1985005622 2000021 530365 30303420 54120466 3041 047013 33 060023 05306 021066
1995303652 3050020 541165 30500803 55180466 3061 013005 33 000002 01502 130666
2005207632    2520 530764    01220 55100566    1 038026 22 057          010168
2012704631 1000000 301065 10003030 30190466 1002 036006 33 010006 03903 250766
2025501660 3070000 000000 00000003 55060766 3071 006000 13 000003 01610 130567
2035903612 3100000 621264 31304533 64210766 3152 064019 23 033017 07308 150467
2042702660 1000000 000000 00000000 27050866 2002 001000 23 015008 00504 261266
2054305652 1000000 440566 10001200 44060966 2002 016004 32 043026 02004 250167
2064409642 1000000 460566 10002000 46071066 2002 025005 22 107041 02904 090267
2076309660 3080000 000000 00000034 63211066 3081 001000 23 057239       010168
2085812632 3130020 601265 31502403 61301266 3161 036012 33 000004 04002 010367
2094505622 1001023 490766 30305020 50310167 3031 056006 33 030019 06307 230867
2103806512 1000000 480461 10011810 53200167 3061 192069 31 100041       010168
```

LISTING OF CARD 2 (THE FIRST TWO LINES ARE COLUMN GUIDES)

```
0           1           2              3              4              ...   8
123 456 789 0123 456 7890 1234 5678 9012 34567 89012 3             ...   0

001 068 646 0089 148 0290 1000 0350 1023 -0114 +0846 2                   2
002 109 697 0174 204 0442 1227 0412 1245 -0380 &1006 2                   2
003 146 718 0156 232 0391 1334 0940 1346 -0148 +1070 2                   2
004 115 885 0054 562 0263 2377 0208 2317 -0632 &1518 2                   2
005 068 658 0539 158 2150 1040 1269 1062 &0805 &0878 2                   2
```

006	070	619	0125	122	0504	0881	0772	0910	&0292	&0750	2		2
007	021	775	0090	321	0319	1641	0475	1637	+0405	+1230	2		2
008	011	518	0045	056	0185	0537	0238	0569	&0230	&0366	2		2
009	059	658		158	0020	1040	0020	1062	−0372	&0878	2		2
010	029	386	0062	038	0083	0417	0288	0447	+0136	+0142	2		2
011	040	871	0197	535	0347	2280	0911	2239	&0771	&1486	2		2
012	046	708	0232	218	0990	1279	1084	1291	+0796	+1038	2		2
013	020	488	0136	043	0483	0454	0605	0486	+0525	+0238	2		2
014	052	581	0115	094	0083	0748	0322	0778	−0014	&0622	2		2
015	107	687	0577	191	2378	1175	1539	1197	+0763	+0974	2		2
016	029	667		168	0075	1084	0089	1107	−0063	&0910	2		2
017	081	667	0164	168	0591	1084	0236	1107	−0332	+0910	2		2
018	079	581	0244	094	0483	0748	0442	0778	−0110	&0622	2		2
019	183	834	0170	442	0543	2018	0644	1991	−0740	&1390	2		2
020	157	602	0111	107	0535	0813	0382	0841	−0794	+0686	2		2
021	069	740	0200	264	0796	1445	0595	1452	&0123	&1134	2		2
022	088	581	0116	094	0502	0748	0396	0778	−0228	&0622	2		2
023	105	676	0184	180	0768	1132	0762	1148	+0002	+0942	2		2
024	113	592	0194	100	0359	0776	0590	0809	−0234	&0654	2		2
025	040	646	0070	148	0175	1000	0138	1023	−0102	&0846	2		2
026	108	646	0291	148	0770	1000	1360	1023	&0576	&0846	2		2
027	060	820	0179	417	0119	1936	0800	1910	&0400	&1358	2		2
028	112	518	0181	056	0506	0537	0618	0569	−0198	&0366	2		2
029	174	619	0254	122	1272	0881	0972	0910	−0340	&0750	2		2
030	093	847	0068	473	0336	2099	0256	2065	−0408	+1422	2		2
031	130	676	0352	180	0547	1132	0603	1148	−0357	+0942	2		2
032	089	638		139	0167	0957	0099	0984	−0533	+0814	2		2
033	231	697	0248	204	1388	1227	0616	1245	−1152	&1006	2		2
034	074	676	0112	180	0422	1132	0303	1148	−0109	&0942	2		2
035	058	718	0162	232	0457	1334	0845	1346	+0462	+1070	1		2
036	080	676	0072	180	0288	1132	0398	1148	−0162	&0942	4		2
037	078	638	0344	139	1055	0957	1045	0984	+0501	+0814	2		2
038	135	697	0024	204	0402	1227	0411	1245	−0589	&1006	2		2
039	155	708	0215	218	0424	1279	0510	1291	−0650	&1038	2		2
040	115	708	0393	218	1077	1279	1015	1291	&0175	&1038	2		2
041	128	697	0246	204	0391	1227	0795	1245	−0149	&1006	2		2
042	187	708	0243	218	2087	1279	0899	1291	−0517	&1038	2		2
043	094	708	0244	218	0366	1279	0527	1291	−0154	+1038	2		2
044	072	697	0246	204	0657	1227	0900	1245	&0404	&1006	3		2
045	085	676	0260	180	2250	1132	0720	1148	&0120	&0942	2		2
046	086	638	0240	139	1093	0957	0547	0984	−0061	+0814	2		2
047	077	658	0322	158	0357	1040	0680	1062	+0144	+0878	2		2
048	084	542	0124	068	0286	0607	0271	0640	−0321	+0462	2		2
049	060	592	0339	100	0622	0776	0897	0809	&0497	&0654	2		2
050	074	592	0189	100	0472	0776	0567	0809	&0055	&0654	2		2
051	061	496	0167	046	0271	0473	0209	0506	−0099	+0270	2		2
052	166	610	0046	115	0718	0845	0313	0873	−0935	&0718	2		2
053	108	750	0085	282	0627	1514	0973	1514	+0187	+1166	2		2
054	044	535	0119	064	0249	0582	0402	0617	&0130	&0430	2		2
055	077	667	0082	168	0195	1084	0440	1107	−0096	&0910	2		2
056	067	581		094	0119	0748	0119	0778	−0337	+0622	2		2
057	273	646	0248	148	1014	1000	1064	1023	−1040	+0846	2		2
058	087	687	0092	191	0462	1175	0753	1197	+0217	+0974	2		2
059	117	646	0225	148	0942	1000	1381	1023	+0525	+0846	2		2
060	152	667	0171	168	0741	1084	0686	1107	−0450	&0910	1		2
061	163	658	0052	158	1031	1040	0372	1062	−0850	+0878	2		2
062	086	847	0288	473	0692	2099	0686	2065	&0078	&1422	2		2
063	119	581	0196	094	0371	0748	0503	0778	−0369	&0622	2		2
064	078	511	0106	053	0218	0513	0328	0548	−0216	&0334	2		2
065	128	667	0115	168	0458	1084	0850	1107	−0094	+0910	2		2
066	152	676	0442	180	1985	1132	1874	1148	&0738	&0942	2		2
067	064	687		191	0471	1175	0228	1197	−0204	&0974	2		2
068	060	559	1481	078	0477	0661	0892	0692	+0492	+0526	2		2
069	033	646	0207	148	0649	1000	0790	1023	−0194	+0846	2		2
070	118	687	0130	191	1296	1175	0711	1197	−0153	&0974	2		2
071	071	628	0080	130	0363	0920	0571	0944	+0020	+0782	2		2
072	150	762	0096	295	1405	1574	0795	1567	−0325	&1198	2		2

073	062	535	0082	064	0359	0582	0706	0617	+0290	+0430	2		2
074	070	697	0040	204	0498	1227	0211	1245	-0269	&1006	2		2
075	082	535	0108	064	0179	0582	0399	0617	-0177	&0430	2		2
076	132	708	0141	218	1265	1279	1180	1291	+0204	+1038	2		2
077	055	740	0111	264	0646	1445	0612	1452	+0252	+1134	2		2
078	016	730	0030	248	0122	1393	0069	1396	+0021	+1102	2		2
079	123	581	0304	094	1580	0748	1570	0778	+0664	+0622	2		2
080	085	566	0806	083	0451	0687	0884	0718	+0284	+0558	2		2
081	088	566	0469	083	2090	0687	1751	0718	+1127	+0558	4		2
082	081	730	0282	248	1090	1393	1395	1396	+0823	+1102	2		2
083	082	697	0181	204	1515	1227	1395	1245	+0815	+1006	2		2
084	068	619	0141	122	0896	0881	0978	0910	+0518	+0750	2		2
085	151	708	0044	218	1170	1279	0260	1291	-0865	&1038	2		2
086	066	658	0854	158	1217	1040	1104	1062	+0656	+0878	2		2
087	084	676	0041	180	0968	1132	0628	1148	+0036	+0942	3		2
088	135	559	0149	078	1250	0661	1530	0692	+0530	+0526	2		2
089	090	575		088	0247	0716	0546	0748	-0090	&0590	2		2
090	126	687	0042	191	0367	1175	0140	1197	-0784	&0974	2		2
091	044	959	0067	078	0149	0661	0145	0692	-0127	&0526	2		2
092	066	602	0054	107	0100	0813	0330	0841	-0115	&0686	2		2
093	080	628	0075	130	0524	0920	0856	0944	+0294	+0782	2		2
094	060	602	0070	107	0434	0813	0492	0841	+0037	+0686	2		2
095	082	535	0314	064	0783	0582	0289	0617	-0287	&0430	2		2
096	071	602	0055	107	0370	0813	0335	0841	-0135	&0686	2		2
097	133	847	0300	473	2080	2099	2475	2065	+1495	+1422	2		2
098	046	566	0064	083	0162	0687	0160	0718	-0124	&0558	2		2
099	107	542	0054	068	0626	0607	0768	0640	-0008	&0462	2		2
100	036	638	0074	139	0414	0957	0177	0984	-0032	&0814	2		2
101	094	676	0152	180	0336	1132	0544	1148	-0124	&0942	2		2
102	125	697	0623	204	2250	1227	2050	1245	+1134	+1006	1		2
103	130	628	0097	130	1070	0920	0800	0944	-0160	&0782	2		2
104	184	750	0135	282	0708	1514	0696	1514	-0694	&1166	2		2
105	091	619	0142	122	0761	0881	0768	0910	+0121	+0750	1		2
106	083	658	0047	158	0554	1040	0351	1062	-0231	&0878	2		2
107	111	646	0178	148	0610	1000	0634	1023	-0187	&0846	2		2
108	102	740	0288	264	0844	1445	1098	1452	+0362	+1134	2		2
109	124	566	0173	083	0925	0687	0663	0718	-0397	&0558	2		2
110	088	542	0082	068	0216	0607	0278	0640	-0346	&0462	2		2
111	088	581	0168	094	0325	0748	0349	0778	-0275	&0622	2		2
112	096	646	0222	148	1541	1000	1152	1023	+0464	+0846	2		2
113	162	581	0341	094	2664	0748	2052	0778	+0836	+0622	1		2
114	114	619	0194	122	1258	0881	1055	0910	+0221	+0750	2		2
115	173	542	0140	068	0602	0607	1059	0640	-0247	&0462	2		2
116	044	559	0229	078	0646	0661	0743	0692	+0472	+0526	2		2
117	100	676	0157	180	2075	1132	1268	1148	+0548	+0942	2		2
118	095	775	0262	321	1206	1641	0755	1637	+0074	+1230	2		2
119	103	542	0193	068	0706	0607	0688	0640	-0073	&0462	2		2
120	070	667	0101	168	0242	1084	0215	1107	-0262	&0910	2		2
121	151	602	0616	107	2541	0813	2361	0841	+1233	+0686	2		2
122	092	526	0103	060	0248	0559	0915	0592	+0259	+0398	2		2
123	180	697	0345	204	0889	1227	1089	1245	-0271	&1006	2		2
124	076	581	0174	094	0668	0748	1507	0778	+0976	+0622	2		2
125	147	646	0097	148	0348	1000	0164	1023	-0929	&0846	2		2
126	167	687	0337	191	1782	1175	1686	1197	+0430	+0974	2		2
127	123	809	0098	389	1575	1858	0728	1837	-0176	&1326	2		2
128	065	581	0096	094	0276	0748	0430	0778	-0011	&0622	1		2
129	102	762	0119	295	1749	1574	1072	1567	+0333	+1198	2		2
130	065	676	0090	180	0445	1132	0527	1148	+0087	+0942	2		2
131	069	602	0042	107	0679	0813	0449	0841	-0021	&0686	2		2
132	135	697	0369	204	1387	1227	1360	1245	+0362	+1006	2		2
133	112	912	0218	665	3128	2582	1435	2512	+0619	+1582	2		2
134	086	697	0244	204	0701	1227	0937	1245	+0331	+1006	2		2
135	078	542	0129	068	0291	0607	0472	0640	-0072	&0462	2		2
136	116	566	0048	083	0188	0687	0454	0718	-0397	&0558	2		2
137	091	610	0130	115	1480	0845	1216	0873	&0568	&0718	2		2
138	115	658	0188	158	1496	1040	1000	1062	&0162	&0878	2		2
139	075	619	0096	122	0213	0881	0277	0910	-0241	+0750	2		2

140	099	708	0065	218	0450	1279	0244	1291	-0468	+1038	2	2
141	033	610	0103	115	0551	0845	0539	0873	&0353	&0718	2	2
142	080	953	0089	790	0736	2931	0942	2818	&0382	&1678	2	2
143	078	667	0149	168	0642	1084	0748	1107	&0204	&0910	2	2
144	018	676	0025	180	0225	1132	0169	1148	&0105	&0942	2	2
145	086	740	0048	264	0387	1445	0498	1452	-0110	+1134	2	2
146	085	809	1195	389	2480	1858	2110	1837	&1510	&1326	2	2
147	097	559	0474	078	0902	0661	1346	0692	&0650	&0526	2	2
148	086	775	0073	321	0954	1641	0340	1637	-0271	+1230	2	2
149	096	740	1067	264	1006	1445	0698	1452	&0010	&1134	2	2
150	124	740	0326	264	0949	1445	1025	1452	&0112	&1134	2	2
151	065	730	0029	248	0383	1393	0139	1396	-0301	+1102	2	2
152	127	809	0412	389	1208	1858	1486	1837	&0550	&1326	2	2
153	091	638		139		0957	1026	0984	&0378	&0814	2	2
154	096	559	0083	078	0344	0661	0269	0692	-0419	+0526	2	2
155	087	619	0182	122	1209	0881	1213	0910	+0597	+0750	2	2
156	124	925	0196	692	1134	2692	0704	2612	-0208	+1614	2	2
157	064	897	0125	610	1173	2483	0830	2415	&0410	&1550	1	2
158	078	718	0084	232	0538	1334	0388	1346	-0156	+1070	2	2
159	127	581		094	0324	0748	0487	0778	-0449	+0622	2	2
160	058	518	0167	056	0620	0537	0923	0569	&0539	&0366	2	2
161	112	559	0245	078	1495	0661	2558	0692	&1742	&0526	2	2
162	094	602	0085	107	0846	0813	0538	0841	-0232	+0686	2	2
163	069	550	0170	073	0319	0631	0392	0667	-0080	+0494	2	2
164	117	775	0148	321	1266	1641	0524	1637	-0332	+1230	2	2
165	064	708		218		1279	0681	1291	&0251	&1038	2	2
166	136	566	0337	083	1192	0687	1654	0718	&0648	&0558	2	2
167	078	676	0044	180	0219	1132	0196	1148	-0346	+0942	2	2
168	092	708	0052	218	0127	1279	0182	1291	-0474	+1038	2	2
169	039	646	0674	148	0430	1000	1098	1023	&0865	&0846	2	2
170	097	619	0138	122	0681	0881	0553	0910	-0161	+0750	1	2
171	134	697	0121	204	1389	1227	0908	1245	-0084	+1006	1	2
172	114	542	0220	068	0325	0607	0518	0640	-0313	+0462	2	2
173	203	697	0419	204	2780	1227	1562	1245	&0028	&1006	2	2
174	128	619	0426	122	1024	0881	1408	0910	&0455	&0750	2	2
175	074	646	0196	148	0480	1000	1267	1023	&0747	&0846	2	2
176	064	667		168		1084		1107		+0910	2	2
177	145	708	0195	218	4434	1279	1962	1291	&0882	&1038	2	2
178	054	581	1246	094	0542	0748	1015	0778	&0667	&0622	2	2
179	075	526	0172	060	1165	0559	1974	0592	&1457	&0398	2	2
180	099	676		180	0396	1132	0303	1148	-0593	+0942	2	2
181	058	667	0037	168	0202	1084	0392	1107	&0008	&0910	2	2
182	078	559	0216	078	1722	0661	1774	0692	&1228	&0526	1	2
183	168	718		232	1085	1334	0421	1346	-0846	+1070	2	2
184	116	785		343	1723	1690	1988	1698	&1140	&1262	2	2
185	098	676	0134	180	0139	1132	0499	1148	-0203	+0942	2	2
186	124	667	0229	168	1255	1084	0603	1107	-0309	+0910	2	2
187	059	708	0187	218	0515	1279	2540	1291	&2148	&1038	2	2
188	105	718	0042	232	0322	1334	0381	1346	-0382	+1070	2	2
189	079	581	1404	094	0728	0748	0883	0778	&0328	&0622	2	2
190	242	619	0478	122	2008	0881	1382	0910	-0476	+0750	2	2
191	131	619	0327	122		0881	0684	0910	-0286	+0750	2	2
192	093	638	0447	139	2285	0957	1482	0984	&0820	&0814	2	2
193	071	526	0975	060	0457	0559	0548	0592	&0062	&0398	2	2
194	095	697	1193	204	0878	1227	1813	1245	&1235	&1006	2	2
195	094		0327		1450		1446		+0772		2	2
196	062		0106		0198		0140		-0277		2	2
197	099		0816		0761		1309		+0598		1	2
198	098		0159		0663		0975		+0271		2	2
199	075		1427		0719		0910		+0392		2	2
200	082		0222		1363		2358		+1784		1	2
201	133		0177		0471		0452		-0535		2	2
202	114		2062		1095		1750		+0915		2	2
203	053				0180		0125		-0217		2	2
204	080		0079		0334		0308		-0249		2	2
205	105		0092		0416		0416		-0346		2	2
206	103		0051		0379		0413		-0329		2	2

207	115	0123	1451	0911	+0074	1	2
208	098	0165	0265	0788	+0085	2	2
209	113	0551	0883	0975	+0151	2	2
210	073	0074	0492	0517	+0010	1	2

REFERENCES

Armitage, P., C.K. McPherson and J.B. Copas (1969). Statistical studies of prognosis in advanced breast cancer. J. Chron. Dis. 22, 343-360.

Atkins, H., R.D. Bulbrook, M.A. Falconer, J.L. Hayward, K.S. Maclean, and P.H. Schurr (1968). Ten years' experience of steroid assays in the management of breast cancer, a review. Lancet, Dec. 14, 1255-1260.

Brier, G.W. and R.A. Allen (1951). Verification of weather forecasts. In Compendium of Meteorology, T.F. Malone, Ed., Boston, Amer. Meteorol. Soc., 841-848.

Bulbrook, R.D., F.C. Greenwood and J.L. Hayward (1960). Selection of breast cancer patients for adrenalectomy or hypophysectomy by determination of urinary 17-hydroxycorticosteroid and aetiocholanolone. Lancet 1, 1154.

Cacoullos, T. (1966). Estimation of a multivariate density. Ann. Inst. Statist. Math. Tokyo 18, 179-189.

Cornfield, J. (1971). Private conversation.

Cover, T.M. (1968). Estimation by the nearest-neighbor rule. I.E.E.E. Trans. Infor. Theory IT-14, 50-55.

Cox, D.R. (1966). Some procedures connected with the logistic qualitative response curve. In Research Papers in Statistics: Essays in Honour of J. Neyman's 70th Birthday, F.N. David, Ed., London, Wiley.

Dickey, J.M. (1968a). Smoothed estimates for multinomial cell probabilities. Ann. Math. Statist. 39, 561-566.

Dickey, J.M. (1968b). Estimation of disease probabilities conditioned on symptom variables. Math. Biosci. 3, 249-265.

Dickey, J.M. (1969). Smoothing by cheating. Ann. Math. Statist. 40, 1477-1482.

Dickey, J.M. and B.P. Lientz (1970). The weighted likelihood ratio, sharp hypotheses on chances, the order of a Markov chain. Ann. Math. Statist. 41, 214-226.

Dixon, W.J. (1969). BMD Biomedical Computer Programs, X-Series Supplement. Berkeley and Los Angeles, Univ. of California Press.

Dixon, W.J. (1970). BMD Biomedical Computer Programs. 2nd ed., 3rd printing, revised. Berkeley and Los Angeles, Univ. of California Press.

Fisher, R.A. (1936). The use of multiple measurements in taxonomic problems. Ann. Eugen. 7, 179-188.

Fix, E. and J.L. Hodges, Jr. (1951). Discriminatory analysis, nonparametric discrimination. USAF School of Aviation Med., Randolph Field, Texas, Project 21-49-004, Report 4, Contract AF41(128)-31, February, 1951.

Gilbert, E.S. (1968). On discrimination using qualitative variables. J. Amer. Statist. Assoc. 63, 1399-1412.

Goldman, R., J. Walrath, E. Jacobson, and J. Dickey (1971). First Look at Graphs. Research Report 41, revised. Statistics Dept., State Univ. of New York at Buffalo.

Good, I.J. (1950). Probability and the Weighting of Evidence. New York, Hafner.

Good, I.J. (1965). The Estimation of Probabilities. Cambridge, M.I.T. Press.

Gunel, E. and J. Dickey (1972). Bayes factors for independence in contingency tables. (In preparation).

Hills, M. (1967). Discrimination and allocation with discrete data. Appl. Statist. 16, 237-250.

Jacobson, E. (1971). SHOP: a computer program for printer plots and histograms of sub-classes. Research Report 52, Statistics Dept., State Univ. of New York at Buffalo.

Jacquez, J. A. (1964). The diagnostic process. In Computer Diagnosis and Diagrammatic Methods, J. A. Jacquez, Ed., Springfield, Illinois, Charles C Thomas.

Jeffreys, H. (1961). Theory of Probability, 3rd ed., Oxford, Clarendon Press.

Joiner, B. L., J. R. Rosenblatt and J. W. Dean (1970). OMNITAB - and an example in data analysis. (Preliminary draft for discussion - not for publication), Revised 9/21.

Ledley, R. S. and L. B. Lusted (1959). Reasoning foundations of medical diagnosis. Science 130(3366), 9-21.

Lusted, L. B. (1968). Introduction to Medical Decision Making. Springfield, Illinois, Charles C Thomas.

Parzen, E. (1962). On estimation of a probability density function and mode. Ann. Math. Statist. 33, 1065-1076.

Radhakrishna, S. (1964). Discrimination analysis in medicine. Statistician 14, 147-167.

Sanders, F. (1958). The evaluation of subjective probability forecasts. Scientific Report No. 5, Contract AFCRC-TN-58-465. Cambridge, Mass. Inst. of Tech.

Sarfaty, G. and M. Tallis (1970). Probability of a woman with advanced breast cancer responding to adrenalectomy or hypophysectomy. Lancet, Oct. 3, 685-687.

Specht, D. F. (1971). Series estimation of a probability density function. Technometrics 13, 409-424.

Truett, J., J. Cornfield and W. Kannel (1967). A multivariate analysis of the risk of coronary heart disease in Framingham. J. Chron. Dis. 20, 511-524.

Tukey, J. W. (1962). The future of data analysis. Ann. Math. Statist. 33, 1-67.

Tukey, J. W. (1970). Exploratory Data Analysis (Limited Preliminary Edition), Vols. I, II, III. Reading, Mass., Addison-Wesley.

Van Ryzin, J. (1965). Non-parametric Bayesian decision procedures for (pattern) classification with stochastic learning. Trans. Fourth Prague Conf. Information Theory, Statistical Decision Functions and Random Processes.

Van Ryzin, J. (1966). Bayes risk consistency of classification procedures using density estimation. Sankhya A 28, 261-270.

Walker, S.H. and D. B. Duncan (1967). Estimation of the probability of an event as a function of several independent variables. Biometrika 54, 167-179.

Walrath, J., R. Goldman and J. Dickey (1971). Nonparametric discriminators: computer programs for research in computer assisted medical diagnosis. Research Report 48, revised. Statistics Dept., State Univ. of New York at Buffalo.

Warner, H.R., A.F. Toronto, L.G. Veasey and R. Stephenson (1961). A mathematical approach to medical diagnosis: application to congenital heart disease. J. Amer. Med. Assoc. 177, 177-183.

Whittle, P. (1958). On the smoothing of probability density functions. J. Roy. Statist. Soc., B 20, 334-343.

DISCUSSION OF THE DICKEY AND WALRATH CHAPTER

R. Elashoff

The Atkins et al (1968) Study

The problem

Atkins et al (1968) review the experience gained over 10 years concerning the clinical usefulness of urinary androgen and corticosteroid metabolites in the treatment of advanced breast cancer. Specifically, they investigate these questions:

1. Do these steroids predict short-term remission in patients receiving one of the two ablative treatments, adrenalectomy or hypophysectomy? Do steroid levels indicate which one of these ablative treatments is best?

2. Do steroid levels interact with other clinical variables relative to remission in patients receiving one of the two ablative treatments?

3. Can we identify patients who are unresponsive to either type of ablative treatment? Can we identify patients who are responsive to ablative treatment?

4. Is there a basis for choice between the ablative treatments?

The study design, the data and previous results

The Atkins et al (1968) paper reviews previous work done by the authors to investigate effects of steroid levels on remission in breast cancer, and comparisons between two ablative treatments, adrenalectomy and hypophysectomy. The study design and some of the previous results are given below.

Ablative treatment was used presumably because the standard therapies — mastectomy and hormone treatment — were not leading to satisfactory results. In the Atkins report, 206 patients were reviewed; subsequently, information on four additional patients was added. Among the 210 breast cancer cases, 116 had previously had a radical mastectomy and 23 a simple mastectomy. The remaining 71 cases were first diagnosed at an advanced stage where, presumably, mastectomy was inapplicable. Of the 210 cases, 132 received some form of hormone therapy as a palliative. Ablative treatment, hypophysectomy or bilateral adrenalectomy with oophorectomy, was then used to possibly prolong life; 95 patients received hypophysectomy and 115 adrenalectomy. During the first six months, 107 patients obtained at least partial remission of symptoms — 52 of the 95 hypophysectomy patients and 55 of the 115 adrenalectomy patients. One hundred and three patients obtained no remission of symptoms.

The original 206 patients were treated over a period of ten years during which time other clinical investigations were carried out and dovetailed into the hormone-assay trials.

All but 82 patients received previous hormone therapy. Premenopausal women and those within five years of the menopause were given androgens. No patients were included who had been previously castrated. The 82 patients who had ablative procedures without previous hormone therapy were selected at random. They were part of a trial in which the response to adrenalectomy or hypophysectomy of patients having no previous hormone therapy was compared with the response of patients who had the operations after hormones had been tried (Atkins et al, 1966).

In the first 65 patients of the series of 206, the decision on whether to perform an adrenalectomy or a hypophysectomy was made by random sample. Subsequently, in order to test the accuracy of steroid assays in prediction, the choice of operation was made according to the value of the discriminant described by Bulbrook et al (1960). This discriminant is computed from the twenty-four-hour urine levels of 17-OHCS and aetiocholanolone; it can be positive or negative. Atkins et al (1960) had previously shown that hypophysectomy gave slightly better results than adrenalectomy. Patients with a positive discriminant, and hence a good chance of response, were allocated to hypophysectomy. It was believed that patients with negative discriminants had little or no chance of responding to ablation; they were recommended for adrenalectomy. Sixty-seven patients were included in this trial. The remission rate from hypophysectomy in discriminant positive patients was 46%, and from adrenalectomy in discriminant negative patients, 9% (Atkins et al, 1964).

A reverse trial was then carried out in which patients with negative discriminants were sent for hypophysectomy and patients with positive discriminants were sent for adrenalectomy. Seventy-four patients were included in this trial, and it was found that patients with negative discriminants also had a low response rate to hypophysectomy (Atkins et al, 1968).

Thus, of the 206 patients reviewed by the authors, the type of operation was chosen by random sample for 65 patients and by the value of the discriminant function for the remaining 141 patients.

All patients were admitted to the hospital for a preoperative assessment, including a complete cataloguing of all lesions; photographic and radiographic records were made wherever possible. The fitness of each patient for operation was assessed independently by a physician. During their stay in the hospital, a three-day urine sample was collected. After operation the patients were examined at four weekly intervals and the measurements of all lesions were compared with the measurements obtained preoperatively. The nonfailure or failure of the operation was decided at least six months after the operation. The criteria of success were those described by Bulbrook et al (1960) and, in more detail, by Hayward (1966). Remission is defined as nonfailure.

Results

Through the use of two-, three- and four-way contingency tables, Atkins and his co-workers suggest these findings: First, relative to the question (1) at the beginning of this comment, patients with negative discriminant values have a lower remission rate than those with positive discriminant values; however, see remarks below. Also, a difference is suggested in remission rate and survival for those receiving hypophysectomy over adrenalectomy among patients with positive discriminant values.

Second, free period and menopausal state modify the predictive value of the discriminant function to forecast remission. Also, patients without previous mastectomy do not respond as well to adrenalectomy as they do to hypophysectomy, using remission as the response.

Third, patients with negative discriminants, who had gone through the menopausal state, and who had a fast progression of the disease, seem to perform poorly to the ablative treatments.

Fourth, hypophysectomy is better than adrenalectomy during the first year after surgery. However, this advantage relates to patients with no mastectomy and positive discriminant values.

The Reanalysis by Armitage, McPherson and Copas (1969)

Armitage and his colleagues used a mathematically more advanced analytic approach than Atkins et al to shed light on the four clinical questions defined at the beginning of this comment. A statistical reanalysis of a study is generally attempted to obtain new results, qualify old results, discover errors or omissions or to introduce an alternative statistical methodology. As indicated below, the Armitage reanalysis satisfies more than one of these criteria.

Methodological approach

Armitage et al not only study remission, they also investigate survival times and a clinical scoring system. But since Dickey and Walrath work with data primarily related to the response variable failure or nonfailure of symptoms to remiss, only the remission analysis of the Armitage group will be reported here.

To answer the first three clinical questions listed above, the following approach was adopted by the Armitage group:

1. Compute informal preliminary analyses to reduce the number of variables possibly predictive of remission to a manageable number. Throw out variables not observed on a modest number of individuals. Many of these preliminary analyses were carried out using a program written and described by Hills (1967) for stepwise discrimination with binary variables.

2. Assume that the probability of remission (nonfailure) is related to the K predictor variables found from step 1. This relation is given by

$$P = \Pr(\text{nonfailure}) = [1+\exp(-\beta_o - \Sigma\beta_i X_i - \Sigma\beta_{i_o} X_i X_{2_j})]^{-1} \; ; (1)$$

where X_i = value of predictor variable i; and β_o, β_i, β_{ij} are parameters whose values are to be determined.

3. Estimate the β_o, β_i, β_{ij} by the Fletcher-Powell method. Fit equation (1) many times using subsets of the K variables and different assumptions about the β_{ij}; for example, set some or all of the $\beta_{ij} = 0$.

4. Compute significance tests for the null hypotheses

$$H_o : \beta_{i_1} = 0, \beta_{i_2} = 0 \ldots \beta_{i_\ell} = 0; \; H_o : \beta_{ij} = 0;$$

where ℓ = number of variables in one of the subsets of the K variables. After carrying out these tests, inferences will be made about the value of steroids to predict remission; and about how steroid levels intersect with clinical variables in the prediction of remission.

5. Compute P = Pr(nonfailure) for various values of $X_1, X_2, \ldots, X_K$. Also obtain from the patient data the observed proportion of nonfailures among patients with P values between 0.0 and 0.1, between 0.1 and 0.2, etc. and between 0.9 and 1.0. These computations might enable one to identify a group of patients unresponsive to ablative treatment, and a group of patients responsive to ablative treatment.

6. The analysis to choose between ablative treatments is to use a standard life table approach based upon the 65 individuals in the first trial.

Some problems exist concerning inferences that might be drawn from steps 1 - 5 taken as a whole. All these problems are mentioned by the authors. First, when variables are chosen and subsequent analysis performed on these variables using the same patient data, the extent to which conclusions would hold up on another sample of patients is not known. For example, the significance tests described in step 4 would not follow the distribution theory used by the authors; and the computed P values would be biased estimates of the true P values.

The next problem concerns choice of the model relating P to $X_1, X_2, \ldots, X_K$. Equation (1) might not be the correct form of the relation. Truett and Cornfield (1967) indicated that equation (1) will give reasonable results even if the relation between P and the X_i is not given by equation (1). Perhaps, but examples are not difficult to construct where this robustness does not hold.

Still another difficulty might exist. The estimators used for P, β_i, β_{ij} are reasonable provided that there is negligible measurement error in the X_i. (Some statements about the steroid assay errors would have been welcome.)

Finally, the patients are selected in some particular way and treated by surgeons. Would the results reported in this paper hold for any other group of patients and surgeons?

Results

The authors report these conclusions: First, concerning question 1, "steroid values seem to provide the best means of prediction (of remission) for the data as a whole." Addition of other variables does not significantly improve the association.

Second, the results relating to question 2 accord well with Atkins' conclusions.

Third, there is an interaction between a history of mastectomy and the choice of ablative procedure. Prognosis is improved when the progress of the disease is slow; prognosis is worse when ablation occurs a few years after menopause. Thus, generally, Atkins' results related to question 3 are confirmed.

Fourth, hypophysectomy is better than adrenalectomy during the first year after ablation. However, the advantage of hypophysectomy is apparent only for patients without previous mastectomy. This advantage is reversed for patients with a mastectomy and a negative Bulbrook discriminant function.

Essentially, then, the results found by Armitage's group accord well with those of Atkins' group. A possible bonus provided by Armitage et al lies in their determination of a patient's P value. Ideally the observed proportion of remissions (nonfailures) would increase as P increases. Indeed, it would be convenient to have only two P values, 0 and 1, to be determined from the patients — those with P = 0 would be failures, and those with P = 1 would be nonfailures. Armitage et al, using 19 variables, find that the data clearly do not accord with this ideal.

The proportion of nonfailures in table 1 appears to increase as P increases. But the extent of this monotonic relation and, indeed, its correctness are difficult to judge with these data. Sampling fluctuations and selection error make it hazardous to render a judgment of the usefulness of the P function approach to predict remission for a different sample of patients.

Table 1. The prediction of P.

Predicted probability	Patients	Proportion nonfailure
0	3	0
0.1 -	21	0.24
0.2 -	25	0.16
0.3 -	14	0.36
0.4 -	16	0.31
0.5 -	25	0.64
0.6 -	28	0.57
0.7 -	23	0.87
0.8 -	10	0.90
0.9 -	21	0.90

Relation to other prognostic studies

Several prognostic studies for breast cancer patients have appeared using different data than those described in this paper (see for example, Cutler et al, 1969 and Zippin et al, 1967). The Armitage group does mention previous work on prognostic factors in breast cancer, but unfortunately, the most interesting factors from previous studies were incompletely measured in the Armitage study. This is not surprising. Different studies do not necessarily record the same data, the quality of the data is variable and the patient groups are sometimes quite different. Thus, replicating prognostic factors from study to study is difficult — although not impossible. Since 1968, no definitive conclusions have appeared in the literature concerning the usefulness of steroids to predict survival.

The Dickey and Walrath Reanalysis of the Data Reviewed by Atkins et al (1968) and Armitage et al (1969)

The value of steroids in predicting prognosis for survival in advanced breast cancer cases must be regarded as uncertain at this time. The same applies to the use of ablative treatment. However, the Atkins' conclusions are sufficiently suggestive to warrant attention and follow-up investigations.

Another reanalysis of the same data would have doubtful value for studying prognostic significance of the variance factors already studied. Indeed, the Dickey and Walrath paper brings out no new findings and does not further qualify suggestive inferences already made.

The Dickey and Walrath paper should be viewed in another light. The authors illustrate a useful set of computer programs to edit data for nonsense values, outliers, etc., by using the cancer data set. They find errors in the data set analyzed by the previous authors (although none serious enough to invalidate the suggestive inferences described by the previous authors). Dickey and Walrath also advance a Bayesian (or weighted) likelihood approach to examine whether certain apparent effects are real or not. The approach is innovative and essentially reviews some of Dickey's previous papers. Finally, they illustrate their nonparametric Bayes techniques for classification, and for determining the probability of nonfailure (remission) for each patient. These techniques are included in useful computer programs written by or under the direction of the authors.

In summary, the Dickey-Walrath paper is not a careful data analysis. Rather the data are used to illustrate statistical techniques developed in previous papers by the authors, and present a good data edit program.

Some methodological considerations

Two methodological issues are considered in applying classification techniques to data: the Dickey-Walrath nonparametric approach and the multiple logistic function. The Dickey-Walrath nonparametric approach seems promising for computing a patient's probability of remission. Their techniques might very well be superior to the use of the multiple logistic function to estimate P = probability of remission if the sample size is sufficiently large, the number of variables is small and selection effects are negligible.

It is useful to compute the classification errors resulting from the several classification techniques described by Dickey and Walrath. A small number of classification errors essentially means that the predictor variables separate the failures and nonfailures well. Suppose a single multiple logistic function is used to compute P and an individual is classified as nonfailure if $P \geq \frac{1}{2}$ and as failure otherwise. If the nine variables given by Armitage are used, we have the results shown in table 2.

Table 2. Classification of cases by one multiple logistic.

Actual group of patients	$P > \frac{1}{2}$ Nonfailure	$P < \frac{1}{2}$ Failure	Total
Nonfailure	65	42	107
Failure	29	74	103
Total	94	116	210

Sum along diagonal = 139

This table shows 66% correctly classified and a conditional probability of nonfailure being predicted of 69%. The nine predictor variables include:

- mastectomy type (none, simple, or radical);

- time from first seen or mastectomy to ablation (in months);
- site of predominant lesion (breast, bone, viscera);
- 17-hydroxycorticosteroids level;
- androsterone level;
- aetiocholanolone level;
- dehydroepiandrosterone level;
- age at mastectomy or when first seen; and
- type of operative treatment (hypophysectomy or bilateral adrenalectomy and oophorectomy).

When failure or nonfailure is predicted for each operative treatment separately, using the nine variables in each multiple logistic function, we have the results shown in table 3 (after Dickey).

Table 3. Type of operative treatment.

	Hypophysectomy	Adrenalectomy	Total
Number correctly classified	63	65	128
Number incorrectly classified	32	50	82
Total	95	115	210

Clearly, the percent correctly classified is not treatment dependent.

A stepwise linear discriminant function analysis can be applied to the same breast cancer data using the appropriate BMD program.

The first variable entered into the discrimination was the dose of aetiocholanolone, which resulted in correct classification in 59½ % of the 210 cases — hardly better than flipping a coin. The maximum number of cases correctly classified was 67% which occurred when only two more variables were used: the time to ablation, and dose of 17-hydroxycorticosteroids. The results at this point in the analysis are shown in table 4 below.

Table 4. Classification of cases by discriminant function.

Actual cases	Nonfailure	Failure	Total
Nonfailure	59	48	107
Failure	22	81	103
Total	81	129	210

Total of diagonal entries = 140

The choice of these three variables agrees with the best three variables chosen by the Armitage et al (1969) procedure — and was made with a lot less computation. The percent correctly classified is (140/210) x 100 = 67%.

For ball smoothing, Dickey and Walrath use a nonparametric Bayesian technique to classify the data, using the nine variables chosen by Armitage and carrying out the classification separately for the hypophysectomy and adrenalectomy cases. They choose the best neighborhood and use the reclassification estimator.

Table 5. Ball smoothing classification.

	Hypophysectomy	Adrenalectomy
Correctly classified	74	67
Incorrectly classified	21	48
Total	95	115

The percent correctly classified is (141/210) x 100 = 67%. Normal kernel smoothing produces somewhat worse results. Dickey and Walrath also used a Fix-Hodges nearest-neighbor rule which produced the worst results of any method used.

Other classification and/or clustering techniques might be tried. However, the message from the data is clear: the prognostic variables are not much help in predicting survival except to isolate a group of cases clearly not responsive to treatment.

Summary

The Atkins et al (1968), Armitage et al (1969) and data description sections of Dickey et al in this volume really constitute a single data analysis. Surely the latter two studies should have appeared as one data analysis.

The Dickey and Walrath work cannot stand alone as a data analysis of the breast cancer data under consideration. It should properly be regarded as an illustration of useful computer programs built around ideas of W. J. Dixon (as incorporated in the BMD series) and J. Dickey (as found in his several papers on Bayesian statistical inference).

This Conference on Data Analysis, involving problems from clinical medicine to rainfall prediction, may have really been a conference illustrating novel methods of data analysis. Clearly, the UCLA Conference was much more methodological than problem-oriented. This conclusion is not to be regarded as a criticism of the Conference. Indeed, readers of these Proceedings will form a broadly-based group of statisticians, many times more broadly-based than if the Proceedings had dealt exclusively with clinical biometric data analyses for prognosis in advanced breast cancer.

REFERENCES

Atkins, H., R.D. Bulbrook, M.A. Falconer, J.L. Hayward, K.S. Maclean and P.H. Schurr (1968). Ten years' experience of steroid assays in the management of breast cancer, a review. Lancet, Dec. 14, 1255-1260.

Atkins, H., M.A. Falconer, J.L. Hayward, et al (1966). The timing of adrenalectomy and of hypophysectomy in the treatment of advanced breast cancer. Lancet, April 16, 827-830.

Atkins, H., R.D. Bulbrook, M.A. Falconer, J.L. Hayward, K.S. Maclean and P.H. Schurr (1964). Urinary steroid estimations in the prediction of response to adrenalectomy or hypophysectomy. Lancet, Nov. 28, 1133-1136.

Atkins, H.J., M.A. Falconer, J.L. Hayward, K.S. Maclean, P.H. Schurr and P. Armitage (1960). Adrenalectomy and hypophysectomy for advanced cancer of the breast. Lancet 1, 1148-1153.

Armitage, P., C.K. McPherson and J.B. Copas (1969). Statistical studies of prognosis in advanced breast cancer. J. Chron. Dis. 22, 343-360.

Bulbrook, R.D., F.C. Greenwood and J.L. Hayward (1960). Selection of breast cancer patients for adrenalectomy or hypophysectomy by determination of urinary 17-hydroxycorticosteroid and aetiocholanolone. Lancet 1, 1154.

Cutler, S. J., M. M. Black, G. H. Fridell, et al (1966). Prognostic factors in cancer of the female breast. II. Reproducibility of histopathologic classification. Cancer 19, 75-82.

Hayward, J. L. (1966). Assessment of response to treatment at Guy's Hospital Breast Clinic. In Clinical Evaluation in Breast Cancer, New York, Academic Press.

Hills, M. (1967). Discrimination and allocation with discrete data. Appl. Statist. 16, 237.

Truett, J., J. Cornfield and W. Kannel (1967). A multivariate analysis of the risk of coronary heart disease in Framingham. J. Chron. Dis. 20, 511-524.

Zippin, C. and N. Petrakis (1971). Identification of high risk groups in breast cancer. Cancer 23, No. 6.

R. Kronmal

The Dickey-Walrath paper is a potpourri of statistical techniques and ideas including computer data screening, Bayesian analysis of contingency tables and a method for classifying cases into predefined groups. The focus of this critique is on the last section of the paper, that of classification, though I will discuss briefly some of Dr. Dickey's earlier ideas on computer data screening.

I would not wish to hazard a guess at the exact number of computer programs that have been written to screen data prior to statistical analysis, but I am sure that it is well over the hundred mark. Dickey and Walrath have added to this number with their "FIRE" and "SHOP" computer programs. Although these programs are interesting and innovative, their future will undoubtedly follow the same course as hundreds of other programs that have been written in the past. That is, they will be used for several years by the authors and their immediate associates and then, as both the computer facilities available and the personnel change, they will be forgotten and fall into total disuse. It seems to me (and I think this UCLA Conference on Statistical Computing is an opportune place to make the point) that if such programs as those developed by Dickey and Walrath are of general usefulness to the scientific community, we should not spend the time, effort and money to write them at installations all over the country; rather we would be much better served to devote our energies to convincing the few groups that have the ongoing capability of maintaining and distributing such programs to incorporate our ideas into their development. In particular, it seems to me that it would be natural to ask the organization that is so graciously providing us with the facilities for this meeting (i.e., UCLA Health Sciences Computing Facility) to attempt to incorporate such ideas into their highly successful BMD series of programs.

I read with a great deal of interest Dickey and Walrath's experience with the analysis of the breast cancer data for the purpose of classifying between failure and nonfailure classes, since in large part their experience paralleled our own with data from cardiovascular disease. In particular, examination of table 3 revealed that the linear models performed, for all practical purposes, as well as or better than the much more complicated methods based on nonparametric density estimation; and this was in spite of the fact that most of the variables available for classification clearly do not meet the assumptions of the linear models. We might also note that the nearest-neighbor model does quite poorly relative to the other model. The authors mention in the paper that the most commonly assumed model is that of Bayes' theorem with independence. However, this method was apparently not tried on the breast cancer data. Although it is obvious that the variables in the breast cancer data are nonindependent, it has been our experience that as the number of variables gets moderately large this model seems to perform as well as and in some cases considerably better than any of the other models we have examined. This, in my opinion, gives us pause to consider the possibility that the models we have proposed are clearly far from optimal and much research still needs to be done in this area.

Another important consideration in the comparison of various mathematical models for classification is how well the ultimate performance of the model reflects its ability to classify the cases on which the functions themselves are computed. It is well known that this "internal" check will usually give a much inflated estimate of the ability of the model to correctly classify the cases. In a similar study in coronary heart disease with more cases than were available to Dickey and Walrath, we found that not only was the drop off in performance extremely large, but in comparing various mathematical methods the order of performance given by the internal check had almost no correlation with the ordering found in classifying a new sample. In fact, in the particular Dickey-Walrath example, it appears from the relatively low frequencies of correct classification that there is little or no discriminatory power in the variables measured, and if a new sample were collected, flipping a coin would probably be as good a method of classification as the ones listed in this paper.

In conclusion the Dickey and Walrath paper provided us with many new ideas and gives indication of many problems still facing us in statistical computing.

J. Tukey

The general approach of borrowing strength for conditional probabilities almost has to be the right choice. Any comments must relate to alternate versions in detail. Two questions seem most important: (1) How should neighborhoods or distances be defined? (2) How should we make use of them once available?

Let us look at the second question first. At first glance, if we could order the other points with regard to their distance from the point we are considering, we might like to try some kind of sequential procedure. A little thought leaves us feeling that binomial sequential estimation is not too promising, but leaves us with an interest in related processes.

One approach that could be considered, though it is quite different from that proposed by the authors and only applicable to quite different data, would be to attempt a sequential trial to see which conditional probability among those for two or more therapies is ascertainably largest. Here a binomial sequential trial could make sense.

Another approach, more consistent with that already proposed, is the following: given observations y at distances r look at all results for $r \le r_o$, extrapolate a p value at r = 0, and assess the variance of this extrapolation; then choose (for each original point) the r_o that minimizes this variance subject to certain a priori restrictions (such as minimum number of points and/or a maximum radius). One extrapolation formula of some merit would be a straight line relating "ave y" to "$r^{2/3}$".

Returning to the first question, how to measure distance, it seems to me that one important consideration would be to adjust the distance measure to ensure, as far as possible, that constellations (of symptoms and history) that have quite different prognoses are not placed too near one another. The simplest way to assure this would seem to be an iterative process. Suppose that p_o is the initial estimate of a conditional probability and that A is a handy parameter (over which we are likely to want to explore), then with i, j representing cases, we could try

$$\{r_1(i,j)\}^2 = \{r_o(i,j)\}^2 + A \cdot \{p_o(i) - p_o(j)\}^2$$

iterating until the p_k settle down. It is important not to take A too large, since accidents might then lead to an increasing and irrationally complex split between successful and unsuccessful cases.

An alteration that might reduce this danger would be to let $p_o(i)$ be the values at i of a regression (linear in initial variables and stepwise into quadratics and cross products?) that attempts to fit the value of $p_o(i)$ and then to put

$$\{r_1(i, j)\}^2 = \{r_o(i, j)\}^2 + B \cdot \{p_o(i) - p_o(j)\}^2$$

Here any tearings apart would have to correspond to what can be described by something among the limited possibilities of the regression.

In all cases, I suppose one would do better to leave out the target point when estimating a conditional probability of it.

A third simple sort of modification, once a conditional probability pattern has been calculated, would be to try "bending" it to see if this offers improvement. A simple version would be to put

$$\text{logit } q = C + D \cdot \text{logit } p + E \cdot (\text{logit } p)^2$$

and try to adjust C, D and E to make q a better conditional probability estimate. (Use of a cube — instead of/or in addition to — a square might work better. If a square is used, its behavior might be "cut" to keep logit q a monotone function of logit p. For the formula given, this would make logit q constant for values of logit p beyond a certain one.)

CHAPTER 2

BLOC VOTING IN THE UNITED NATIONS

J. A. HARTIGAN
Department of Statistics, Yale University

This study is based on data from 50 roll call votes of 126 nations in the General Assembly of the United Nations in 1968. The full set of voting records is on file at the Health Sciences Computing Facility, University of California, Los Angeles, available on request. A Data Set Description at the end of this chapter contains a complete printout of the data and accompanying description. The 50 questions are listed on pages 3 and 4 of the Data Set Description with code names used in the text. Pages 5, 6 and 7 present the complete 50 X 126 voting record. A small subset of the voting record is presented in table 3.

Political scientists study the U.N. voting data to evaluate cohesion of existing coalitions of states, and to discover new coalitions. For example, Hovet (1960) considered regularly caucusing groups such as the Eastern European nations, the Benelux nations, the South American bloc, the Brazzaville African bloc, the North African bloc, etc. He also grouped the questions themselves by their committee of origin. Using these a priori groups as a base, he computed measures of cohesion such as the proportion of questions of a particular type on which the members of a bloc agree. Other studies of given groups are Ball (1951) and Riggs (1958).

Clustering and factor analysis have been used to identify groups of countries and groups of questions from the data themselves. For each pair of countries, Rieselbach (1960) computed the percentage of questions on which they agree, and then identified blocs of countries within which all pairs exceed a given threshold of agreement. A similar approach appears in Lijphart (1963). In the factor analysis approach of Alker (1964), each question is represented as a linear combination of hypothetical factors, which are named suggestively (and subjectively) as cold war factor, colonialist factor, etc. Then clusters of questions are identified as those highly weighted on the same factor. In Russett (1966) clusters of countries are identified as those highly weighted on factors such as "Western Community," "Afro-Asians," etc. A defect of the factor analytic approach is that the clusters must be picked by "eye" examination of the factor loading matrix. Factor analysis does violence to the data in supposing that the measurements are real rather than category valued.

The U.N. data were selected for this study as an archetype of a clustering problem. The principal characteristics of the data are:

- Clusters of countries, such as Mexico, Venezuela, Brazil, that vote alike across a wide variety of issues. Such blocs are evident in the data, and also in external evidence such as caucusing before votes. The formation of blocs or parties seems to be a consequence of parliamentary voting, so much voting data have this characteristic. Hadwen and Kaufman (1962) consider the mechanics of cohesion of blocs.

- Clusters of questions, such as the Hungarian amendments to the resolution expelling South Africa from UNCTAD, that are repeats of the same question due to parliamentary maneuvering. (Although many procedural questions are settled by a show of hands, so there is some selection in favor of "important," "substantive" questions for the roll calls.)

- Questions that are related in other ways than by the repetition mentioned above, such as the reversal between "China admission important" and "admit

China." (See table 2.)

- Responses categorized as YES, NO, ABSTAIN and ABSENT, that may reflect policy of the voting countries. There is a considerable risk in treating ABSTAIN and ABSENT as missing, or as halfway between YES and NO. Both ABSTAIN and ABSENT represent policy of blocs of countries on clusters of questions.

The clustering technique used in this study is aimed at data with these four characteristics. The basic unit is the bloc, or submatrix of the data, within which all values may be predicted from a single bloc value. Corresponding to each bloc is a cluster of countries and a cluster of questions. Thus clusters of countries are characterized by the clusters of questions on which they vote alike.

PRELIMINARY ANALYSES

In table 1, the YES, ABSTAIN, NO, ABSENT and UNKNOWN categories are counted for each question. These are coded 1, 2, 3, 5 and 0 in the data. Some questions, such as SOA9 "appeal for cash for South African education program," are almost universally agreed to. Other questions, such as SOA4 and SOA5, have almost identical response patterns over the four categories, suggesting that almost all countries voted the same on the two questions.

A question is passed if the YES votes form a majority of those voting YES or NO. (In this sense, ABSTAIN and ABSENT have equal effect.) The histograms in figure 1, of the counts of each type of vote over the 50 questions, suggest the existence of blocs of countries. Three modes appear in each of the histograms for YES, ABSTAIN, and NO. Some questions are passed almost unanimously, with 110-120 YES votes. Others are passed with about 80 YES votes, and yet others have only 20 or 40 YES votes. A suggested mechanism for generating these histograms postulates a number of blocs of 20 countries each. All countries of the same bloc vote together YES, ABSTAIN or NO; thus four

Table 1. Vote counts in the General Assembly, 1968.

Question	Vote* 1	2	3	5	0	Question	Vote* 1	2	3	5
KOR1	19	29	63	14	1	MOR1	114	3	0	9
UK1	86	19	9	11	1	INT1	68	29	16	13
CHI1	73	5	47	1		INT2	89	22	2	13
CHI2	44	23	58	1		SOA9	115	1	2	8
CHI3	63	29	32	2		GIB1	67	34	18	7
CHI4	30	27	67	2		NUC1	103	5	7	11
HUM1	58	36	7	25		NUC2	98	16	0	12
POR1	85	15	3	23		NUC3	75	30	9	12
SOA1	47	23	52	4		DIS1	108	7	0	11
SOA2	56	13	48	9		KOR2	73	28	15	10
SOA3	12	34	66	14		KOR3	71	20	25	10
SOA4	11	31	71	13		COL3	29	73	9	15
SOA5	11	29	73	13		COL4	41	66	4	15
SOA6	73	21	14	18		SEA1	112	7	0	7
SOA7	55	28	33	10		SEA2	85	25	9	7
SOA8	96	16	2	12		MON1	110	0	10	6
UK2	66	26	18	16		MON2	86	24	9	7
UK3	87	16	4	19		MON3	52	38	29	7
COL1	87	19	2	18		MON4	34	33	51	8
COL2	58	48	4	16		RUS1	36	33	53	4
POR2	58	42	10	16		RUS2	86	26	6	8
POR3	36	43	31	16		RUS3	93	22	6	5
POR4	82	25	7	12		RUS4	69	29	23	5
PAP1	72	24	19	11		RUS5	81	24	17	4
PAP2	61	17	37	11		RUS6	118	2	0	6

*1 = YES
2 = ABSTAIN
3 = NO
5 = ABSENT
0 = UNKNOWN

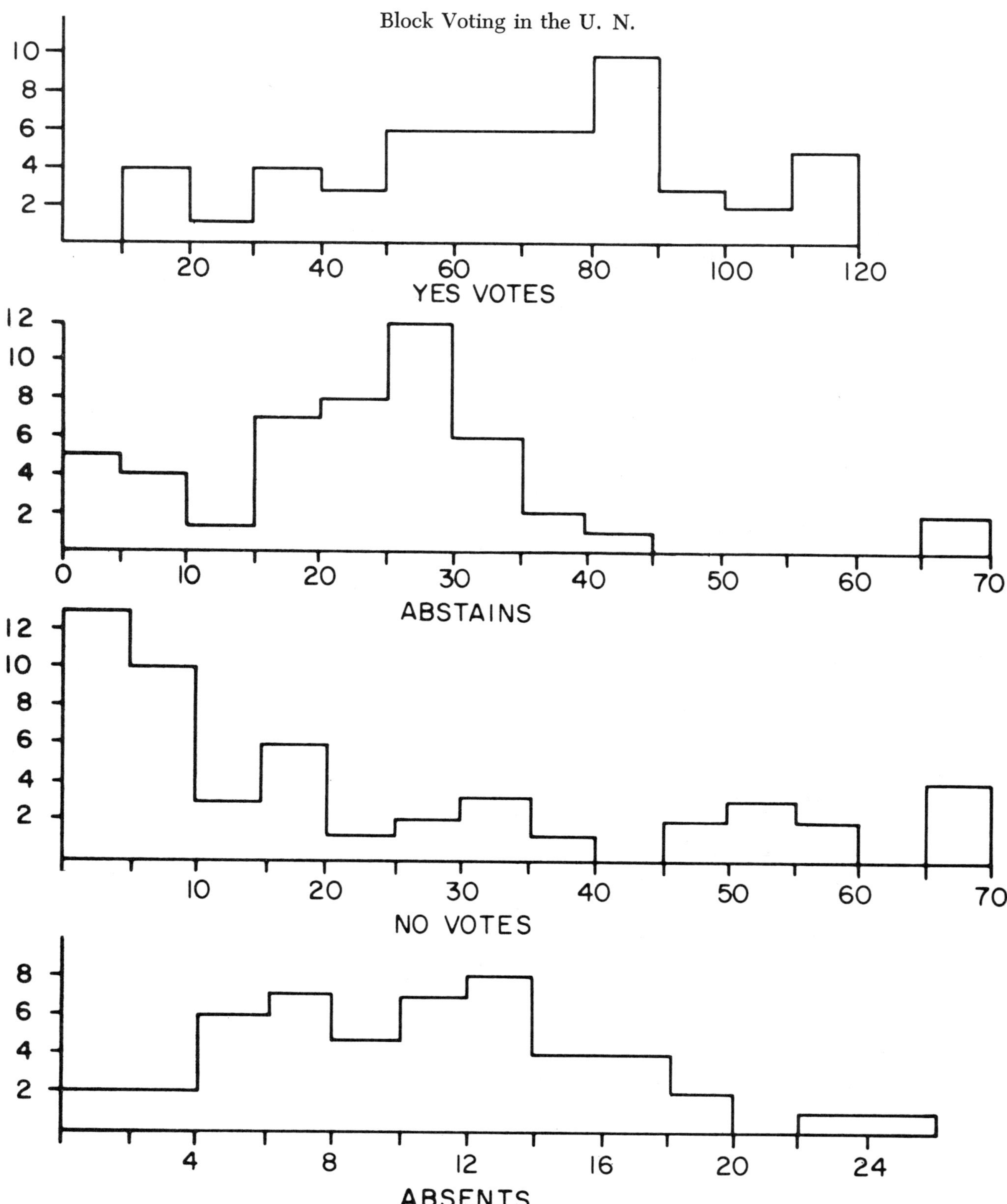

Figure 1. Histograms of counts of voting categories, over 50 questions.

blocs voting YES together set the mode in YES votes at 80. This model is crude and innacurate, but does show how blocs of countries explain modes in the histograms.

The ABSENTS are relatively few in number and do not exhibit the trimodality of the other three categories.

The next step is cross tabulation, or contingency tables, one for each pair of questions. A small sample of these tables (table 2) illustrates the types of relationships between questions that should be handled in the clustering. An alternative measure of relationship between questions is the correlation coefficient (treating ABSENT as ABSTAIN). It should be noted that highly related questions like CHI1 and CHI4, related by a shift, are not very highly correlated. A different measure of association is needed for shifts.

A PILOT STUDY

The clustering techniques used are illustrated on the small subset of U.N. data given in table 3.

Clusters of Countries

A standard algorithm (the weighted average algorithm) is used to cluster the countries. Each variable (question) is scaled to have mean zero and variance one, with ABSENT treated as missing values. This procedure is appropriate if variables are initially measured in different units, but could reasonably be omitted here. Its effect is to give variables that are nearly constant over the whole data set more weight in the clustering.

The algorithm finds the closest pair of countries where distance is Euclidean. For example,

USSR	1	1	1	1	3	1	2	3	1	3	2	2
USA	1	3	3	3	1	3	3	1	3	1	1	3

Table 2. Selected contingency tables, 1968 (absent votes ignored).

	CHI1			
CHI4	1	2	3	
1	27	2	1	30
2	23	3	1	27
3	23	0	44	67
	73	5	46	

Shift of CHI1 YES toward CHI4 NO.

	CHI3			
CHI4	1	2	3	
1	2	6	22	30
2	7	10	10	27
3	54	13	0	67
	63	29	32	

Reversal CHI3 YES to CHI4 NO.

	SOA4			
SOA5	1	2	3	
1	11	0	0	11
2	0	26	2	28
3	0	4	69	73
	11	30	71	

Identical questions.

	SOA4			
CHI1	1	2	3	
1	2	22	41	65
2	0	1	3	4
3	9	8	26	43
	11	31	70	

Weak relationship.

Key for votes: 1 = YES; 2 = ABSTAIN; 3 = NO.

Table 3. Clustering countries and issues: votes of 19 selected nations on 12 selected issues, General Assembly, 1968.

	1	2	3	4	5	6	7	8	9	10	11	12
USR	1	1	1	1	3	1	2	3	1	3	2	2
BGA	1	1	1	1	3	1	1	3	1	3	2	2
YUG	1	3	3	3	3	1	1	3	1	2	3	1
SYR	1	2	2	2	3	1	1	3	1	2	3	1
UAR	1	3	3	3	3	1	1	3	2	2	3	1
KEN	1	3	3	3	3	1	1	3	2	5	3	1
TAN	1	2	2	2	3	1	1	3	2	5	3	1
SEN	1	3	3	3	1	2	2	2	2	1	3	1
DAH	1	3	3	3	1	3	1	3	5	1	3	1
USA	1	3	3	3	1	3	3	1	3	1	1	3
UNK	1	3	3	3	1	1	3	2	3	1	1	3
FRA	1	3	3	3	3	1	2	3	3	1	1	3
SWE	1	3	3	3	3	1	2	3	3	1	1	3
NOR	1	3	3	3	3	1	3	2	3	1	1	3
ALA	1	3	3	3	1	3	1	3	3	1	1	3
NZ	1	3	3	3	1	3	1	1	3	1	1	3
MEX	1	2	2	2	1	3	3	1	3	1	1	1
VEN	1	2	2	2	1	3	3	1	3	1	2	1
BRA	1	2	2	2	1	3	3	1	3	1	1	3

Code for issues:

1. RUS6: Personal linguistic requirement in U.N. hiring.
2. SOA3: Add Hungarian preamble to South African expulsion from UNCTAD.
3. SOA4: Replace last paragraph of preamble by Hungarian amendment.
4. SOA5: Hungarian amendment of paragraphs 1 and 2 of South African expulsion.
5. CHI1: Declare the China admission question important.
6. CHI2: Recognize mainland China and expel Formosa.
7. CHI3: Make study commission on China important.
8. CHI4: Form study commission on China admission.
9. KOR1: Adopt USSR proposal to delete item on Korea.
10. KOR2: Call for eased tensions in Korea.
11. SOA2: Declare South African expulsion from UNCTAD important.
12. SOA7: Adopt South African expulsion.

Key for votes:

1 = YES
2 = ABSTAIN
3 = NO
5 = ABSENT

$$\begin{aligned} \text{DIST (USA,USSR)} = {} & (1-1)^2 + (1-3)^2 + (1-3)^2 + (1-3)^2 + (3-1)^2 \\ & + (1-3)^2 + (2-3)^2 + (3-1)^2 + (1-3)^2 + (3-1)^2 \\ & + (2-1)^2 + (2-3)^2 \end{aligned}$$

The closest pair of countries is United Arab Republic (UAR) and Kenya. These are amalgamated to form a cluster taking the average of UAR and Kenya on each variable. In later operations the two countries are ignored and the cluster is treated as a single object. Thus at each step two countries are amalgamated and the total number of clusters is reduced by one. The sequence of amalgamations can be reconstructed from the amalgamation distances on the output in table 4 (from the program BMDP2M). In the clustering, Russia (USSR) and Bulgaria are separated from all other countries. Within the other countries there is a cluster of West countries and African countries. The African cluster lies with the West because they both disagree with the USSR over the three Hungarian amendment proposals. This is a case where a single issue is given excessive weight in the clustering because of repetition, a really serious problem in clustering in general. If only one Hungarian question is included, the African cluster is grouped with the USSR and Bulgaria.

Clusters of Questions

A similar procedure clusters questions with 1- correlation used as a measure of distance. The clusters are given in table 5 (output from BMDP1M). There is a sad mixup of the various questions on China, on Korea, and on South Africa expulsion. Notice that KOR1 has correlation -.88 with KOR2. Countries vote oppositely on the two procedures. By changing YES to NO for KOR1, the correlation becomes .88. Thus the distance should be 1- absolute value of correlation. The variables may still be related by a shift which leaves the correlation low; a monotone measure like

$$P \{ (\text{KOR1 (I)} - \text{KOR1 (J)}) (\text{KOR2 (I)} - \text{KOR2 (J)}) \geq 0 \}$$

TABLE 4. AMALGAMATION DISTANCES

```
C N
  O      1 1 1 1 1 1 1 1 1 1
A .  1 2 0 6 5 2 3 4 1 9 8 7 8 9 4 7 5 6 3

S L  U B U N A F S N U B V M S D S T U K Y
  A  S U N E U R W O N R E E E A Y A A E U
E B  S L   W S A E R   A N X N H R N R N G
  E  R G S   T N D W K Z E I E O I Z   Y O
  L
```

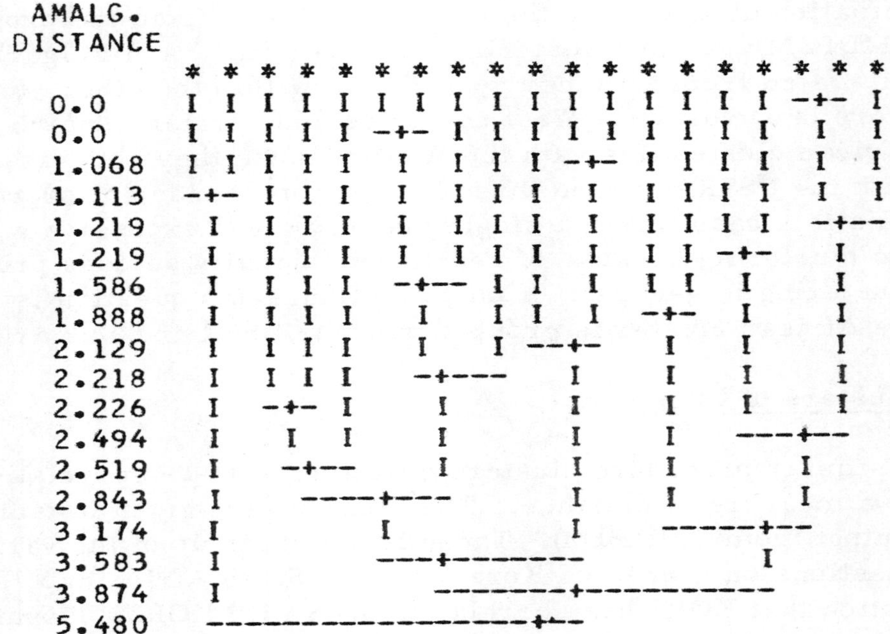

TABLE 5. TREE DIAGRAM OF CLUSTER BASED ON WEIGHTED AVERAGES ALGORITHM

```
   VARIABLE          DISTANCE IS GIVEN AS 50(1 - CORRELATION)
NAME     NO.
            ----------------------------------/
RUS6    ( 1) 50 50 50 50 50 50 50/50 50 50 50/
                                /           /
            -------------------/           /
SOA3    ( 2)  0/ 0/25 51 39 45/54 50 59 84/
              /  /           /           /
             /  /           /           /
SOA4    ( 3)/ 0/25 51 39 45/54 50 59 84/
              /           /           /
             /           /           /
SOA5    ( 4)/25 51 39 45/54 50 59 84/
                       /           /
            ----------/           /
KOR1    ( 9) 20/21/21/87 79 79 94/
              /  /  /           /
             /  /  /           /
CHI3    ( 7)/31/36/80 85 70 72/
              /  /           /
             /  /           /
SOA7    (12)/44/93 61 58 65/
              /           /
             /           /
CHI2    ( 6)/66 84 93 77/
                       /
            ----------/
SOA2    (11) 24 32 27/
                    /
            -------/
CHI4    ( 8) 14/24/
              /  /
             /  /
CHI1    ( 5)/18/
              /
             /
KOR2    (10)/
```

(where I and J are selected randomly from all countries) is more friendly to a shift.

The revised clusters in table 6 show RUS6 off by itself (everybody for); Hungarian questions (SOA3, SOA4 and SOA5) form a separate cluster; and the remainder form a large cluster.

Direct Clustering

Interpretation of the above clusters is complicated by their remoteness from the data. All that can be said is that "these objects are close in terms of this distance." Since the distance is often only slightly interesting, it is hard to be interested in the clusters.

It therefore seems appropriate to see cluster models that apply directly to the data so the clusters can be used to make statements about the data. One way to do this is to report means of variables for each cluster to get an idea of which variables are varying over which clusters. This technique could be used with any algorithm as a way of confronting the data and the clusters.

In the following, a cluster is considered to be any submatrix of the data that satisfies certain rules (which may vary from one data set to the next). Corresponding to this cluster of data values is a marginal cluster of rows and columns (rows and columns of the submatrix).

An example of direct clustering is shown in table 7. The rows and columns of data are reordered according to the trees on rows and columns already computed. Then submatrices of the data are considered whose marginal rows and columns conform to the clusters already obtained. The model within each submatrix is that variables are constant — no relation is supposed between variables.

For example, there is a Western neutral vote on China admission: the submatrix is France, Sweden, Norway, United Kingdom by CHI1 and CHI2. The value for CHI1, "China admission important,"

Table 6. Clusters of issues with distance = 100 (1 - ABS (correlation)). Based on 19 countries, amalgamating clusters in order of average distance between them.

RUS6	RUS6											
SOA3	100	SOA3										
SOA4	100	0	SOA4									
SOA5	100	0	0	SOA5								
SOA2	100	92	92	92	SOA2							
SOA7	100	78	78	78	14	SOA7						
KOR1	100	50	50	50	26	42	KOR1					
KOR2	100	32	32	32	54	70	12	KOR2				
CHI1	100	82	82	82	64	84	42	36	CHI1			
CHI2	100	90	90	90	68	88	42	46	14	CHI2		
CHI3	100	98	98	98	72	62	40	56	60	72	CHI3	
CHI4	100	100	100	100	48	84	42	48	28	32	30	CHI4

Table 7. Direct clustering of U. N. data. From initial marginal trees in tables 1 and 3. Variables are constant within clusters.

	RUS6	SOA3	SOA4	SOA5	SOA2	SOA7	KOR1	KOR2	CHI1	CHI2	CHI3	CHI4
USR	1	1	1	1	2	2	1	3	3	1	2	3
BGA	1	1	1	1	2	2	1	3	3	1	1	3
SEN	1	3	3	3	3	1	2	1	1	2	2	2
DAH	1	3	3	3	3	1	5	1	1	3	1	3
SYR	1	2	2	2	3	1	1	2	3	1	1	3
TAN	1	2	2	2	3	1	2	5	3	1	1	3
UAR	1	3	3	3	3	1	2	2	3	1	1	3
KEN	1	3	3	3	3	1	2	5	3	1	1	3
YUG	1	3	3	3	3	1	1	2	3	1	1	3
FRA	1	3	3	3	1	3	3	1	3	1	2	3
SWE	1	3	3	3	1	3	3	1	3	1	2	3
NOR	1	3	3	3	1	3	3	1	3	1	3	2
UNK	1	3	3	3	1	3	3	1	1	1	3	2
NZ	1	3	3	3	1	3	3	1	1	3	1	3
ALA	1	3	3	3	1	3	3	1	1	3	1	1
USA	1	3	3	3	1	3	3	1	1	3	3	1
BRA	1	2	2	2	1	3	3	1	1	3	3	1
VEN	1	2	2	2	2	1	3	1	1	3	3	1
MEX	1	2	2	2	1	1	3	1	1	3	3	1

Key for votes: 1 = YES; 2 = ABSTAIN; 3 = NO; 5 = ABSENT

```
TABLE 8.  DIRECT CLUSTERING OF DATA ON AGREEMENT
WITH USSR



KEY

0 = MISSING
1 = AGREE WITH USSR
2 = ABSTAIN
3 = DISAGREE WITH USSR


               00000000000000000000000000000000000000
               000000000000000000000 00000000000000
               00 000000000000000000 00000000000000
               00 00000000000 00000 00000000000000
               00 00000000 00 00000 00000000000000
              RUS6   CHI3   SOA2   CHI1   SOA4   SOA3
                 CHI4   SOA7   CHI2   KOR1   SOA5   KOR2
                 1  2  3  4  5  6  7  8  9 10 11 12
              =====================================
USSR         1H 1H 1  2  2H 2H 1  1H 1  1  1  1  1H
              H  H        H  H     H              H
BULG         2H 1H 1  1  2H 2H 1  1H 1  1  1  1  1H
              H  H        H  H     H==============H
SYRI         3H 1H 1  1  1H 3H 1  1H 1  2  2  2  2H
              H  H        H  H     H              H
TANZ         4H 1H 1  1  1H 3H 1  1H 2  2  2  2  0H
              H  H=================H==============H
YUGO         5H 1H 1  1  1  3H 1  1H 1  3  3  3  2H
              H  H           H     H              H
UAR          6H 1H 1  1  1  3H 1  1H 2  3  3  3  2H
              H  H           H     H              H
KENY         7H 1H 1  1  1  3H 1  1H 2  3  3  3  0H
              H  H           H     H              H
FRAN         8H 1H 1  2  3  1H 1  1H 3  3  3  3  3H
              H  H           H     H              H
SWED         9H 1H 1  2  3  1H 1  1H 3  3  3  3  3H
              H  H           H     H              H
NORW        10H 1H 2  3  3  1H 1  1H 3  3  3  3  3H
              H  H=================H              H
DAHO        11H 1H 1  1  1H 3H 3  3H 0  3  3  3  3H
              H  H        H  H     H              H
AUST        12H 1H 1  1  3H 1H 3  3H 3  3  3  3  3H
              H  H        H  H     H              H
SENE        13H 1H 2  2  1H 3H 2  3H 2  3  3  3  3H
              H  H========H  H     H              H
MEXI        14H 1H 3  3  1H 1H 3  3H 3  2  2  2  3H
              H  H        H  H     H              H
NEW         15H 1H 3  1  3H 1H 3  3H 3  3  3  3  3H
              H  H        H  H     H              H
VENE        16H 1H 3  3  1H 2H 3  3H 3  2  2  2  3H
              H  H        H  H     H              H
UN K        17H 1H 2  3  3H 1H 1  3H 3  3  3  3  3H
              H  H        H  H     H              H
UN S        18H 1H 3  3  3H 1H 3  3H 3  3  3  3  3H
              H  H        H  H     H              H
BRAZ        19H 1H 3  3  3H 1H 3  3H 3  2  2  2  3H
              H===================================H
```

is NO; the value for CHI2, "admit China," is YES.

There are many choices of clusters (submatrices of the data) that fit the data exactly. The choice in table 7 is intended to minimize the number of parameters (the total number of variables in each cluster) summed over all clusters. Note that a single value determines a variable within a cluster.

Homogenous Data

A simple direct clustering model, which is appropriate if all data values are comparable, is to assume that all values within a bloc are equal. This model is not appropriate here, since a NO on one question may be equivalent to a YES on another.

All variables are standardized on the USSR (perhaps the modal response would be better) so that if the USSR votes NO, all values of the variable are reversed.

The direct clustering algorithm described in Hartigan (1971) constructs clusters by successively splitting clusters into two, at each step minimizing the sum of squares within the cluster. This algorithm is applied to the U.N. data in table 8. ABSENTS are treated as missing.

The only unsatisfactory cluster is YUG, UAR, KEN FRA, SWE NOR by CHI4, CHI3, SOA7, SOA2 within which values are far from constant — there are definite clusters within this bloc, but they are not revealed by a single splitting step on either rows or columns.

THE FULL U.N. DATA

The steps of the pilot study are now repeated on the full U.N. data. In table 9, the countries are clustered using the weighted average algorithm. A few salient blocs are selected from the welter of suggested cluster — these are the western bloc, United States through Bolivia; an Afro-Asian bloc, Yugoslavia through Ethiopia; a South American bloc, Mexico through Paraguay; a

Soviet bloc, Romania through USSR. Many countries are not included in any of these blocs.

Table 10, in clustering the questions with distance equal to 1-correlation, shows similar but reversed questions are widely separated. For example, CHI2 and CHI4 are in widely different clusters. The salient blocs of questions (adjusting for reversals) are the East-West questions KOR1, CHI2, MON4 together with CHI4 through SOA1; the Russian language questions RUS4 through RUS2; the Imperialist questions POR3 through MOR1; some more Imperialist questions SOA3 through RUS6; and some disarmament questions NUC2 through SEA1.

In table 11, the data are standardized by agreement or disagreement with the United States. Thus all U.S. votes are YES or ABSTAIN. The country blocs agree fairly well with those of table 9, and the questions less well. There are about 7 blocs of countries and 5 clusters of questions in this analysis. A summary 7 X 5 table would describe a substantial fraction of votes in the U.N.

In table 12, the model within blocks is that variables are constant, but there is no relation between variables. This avoids the standardizing operation necessary in table 11. A typical bloc is the Ceylon, Singapore, Nepal, Burma by the East-West questions CH3 through KO3. Ceylon is taken as the leading country of this bloc; the values for Ceylon are printed, and the values for other countries are left blank if they agree with those of Ceylon. In this way, many values in the data matrix are left blank and it becomes possible to scan by eye.

The Eastern European bloc is the most monolithic with exact agreement on issues where they disagree with the majority. The most discordant country in the bloc is Romania.

TABLE 11. DIRECT CLUSTERING OF DATA ON AGREEMENT WITH USA

KEY

0 = MISSING
1 = AGREE WITH USA
2 = ABSTAIN
3 = DISAGREE WITH USA

		RUS6 1	MOR1 2	SOA9 3	SEA1 4	DIS1 5	SOA8 6	UK3 7	COL1 8	INT2 9	POR1 10	NUC2 11	NUC1 12	SEA2 13	COL2 14	GIB1 15	COL4 16	SOA4 17	NUC3 18	SOA5 19	SOA3 20	SOA6 21	COL3 22	KOR2 23	MON4 24	MON3 25	KOR1 26	KOR3 27	CHI1 28	PAP2 29	CHI2 30	SOA2 31	SOA1 32	SOA7 33	POR3 34	RUS1 35	RUS4 36	CHI3 37	CHI4 38	UK2 39	INT1 40	PAP1 41	RUS5 42	UK1 43	POR2 44	MON2 45	HUM1 46	POR4 47	RUS2 48	RUS3 49	MON1 50
IREL	1H	1	1	1	1	1	1	1	2	1	1	1	1	2H	2	2	2H	1	1	1	1	1H	2H	1	1	1	1	1	1	1	1H	1	1	1	1	1H	1	1	1H	2	0	2	1	2	3	1	2	2	1	1	3H
CHIN	2H	1	1	1	1	1	1	1	1	1	1	1	1	2H	2	1	2H	1	2	1	0	1H	2H	1	1	1	1	1	1	1	1H	1	1	2	1	1H	1	3	3H	2	2	2	1	3	2	2	3	3	2	1	1H
HOND	3H	1	1	1	1	1	1	0	1	1	0	1	1	1H	2	1	2H	1	1	1	1	3H	2H	1	1	0	1	1	1	2	1H	1	1	1	1	2H	1	2	3H	3	3	2	3	2	1	3	1	1	3	3	3H
GREE	4H	1	1	1	1	1	1	1	1	2	1	1	1	1H	2	1	2H	2	2	2	1	1H	2H	1	1	1	1	1	1	1	1H	1	1	1	1	2H	2	1	2H	3	1	2	2	3	2	3	2	2	3	3	3H
PARA	5H	1	1	1	1	1	1	1	1	1	1	1	1	1H	2	1	0H	2	1	2	2	0H	0H	1	1	2	1	1	1	2	1H	1	1	1	2	1H	1	3	3H	2	2	3	2	2	2	3	0	2	3	3	3H
BOLI	6H	1	1	1	1	1	1	1	1	1	1	1	1	1H	2	1	2H	1	1	1	1	0H	2H	1	2	1	0	1	1	1	1H	1	1	1	2	1H	1	1	2H	2	2	2	2	3	2	3	2	3	3	3	3H
LESO	7H	1	1	1	1	0	0	1	0	1	0	0	0	1H	0	3	2H	1	0	1	1	1H	2H	1	1	1	1	1	1	1	1H	1	1	1	0	2H	2	3	3H	2	3	2	2	0	0	3	0	0	1	2	3H
MALT	8H	1	0	0	1	1	0	0	0	0	0	1	1	1H	0	0	0H	2	1	2	2	0H	0H	1	1	1	0	1	1	0	1H	1	1	2	0	1H	1	1	2H	0	0	0	1	0	0	3	0	0	0	1	3H
ISRA	9H	1	1	1	1	1	1	1	1	1	1	1	1	2H	2	2	2H	0	2	0	0	0H	2H	1	1	1	1	1	1	1	1H	1	2	0	1	1H	1	1	1H	1	2	2	3	3	2	2	3	3	3	3	3H
SWAZ	10H	1	1	1	1	0	0	1	0	0	0	0	0	1H	0	2	2H	1	0	1	1	1H	1H	1	1	1	0	1	1	0	1H	1	1	1	2	0H	0	3	2H	2	2	0	2	0	0	3	0	2	0	0	3H
COST	11H	0	1	0	1	0	0	0	1	0	0	0	0	1H	2	1	2H	2	0	2	2	0H	2H	1	0	0	1	1	1	0	1H	1	1	0	2	2H	1	1	2H	2	0	0	3	3	2	0	2	1	3	3	3H
GAMB	12H	0	0	0	0	1	0	0	0	0	0	1	1	0H	0	0	1H	0	2	0	0	0H	1H	1	0	0	1	1	1	0	1H	0	0	0	0	0H	0	2	3H	0	0	0	0	0	0	0	0	0	0	0	0H
MALA	13H	1	1	1	1	1	1	1	1	1	1	1	1	1H	1	3	2H	3	1	3	3	2H	2H	1	1	1	1	1	1	1	1H	1	1	2	2	2H	2	2	2H	2	2	2	2	3	2	3	3	3	2	2	3H
LIBE	14H	1	1	1	1	1	1	1	1	1	0	1	1	1H	1	3	2H	1	1	1	1	1H	2H	1	1	1	1	1	1	1	1H	3	3	3	2	1H	2	2	2H	2	3	1	2	3	3	3	3	3	1	2	3H
THAI	15H	1	1	1	1	1	1	1	1	1	1	1	1	1H	2	2	2H	1	1	1	1	1H	2H	1	1	1	1	1	1	1	1H	3	3	3	2	2H	3	3	3H	2	2	2	3	3	2	3	2	3	2	3	3H
GABO	16H	0	1	1	1	1	1	0	0	1	0	1	1	1H	0	2	2H	1	1	1	1	1H	2H	1	2	1	2	1	1	1	1H	3	3	3	0	1H	1	2	2H	0	3	3	1	3	0	3	3	0	0	1	3H
MALD	17H	1	1	1	1	1	1	1	1	2	1	1	1	1H	2	3	2H	2	1	2	2	2H	2H	1	2	1	1	1	1	1	2H	1	2	2	2	2H	3	2	1H	2	2	1	3	3	2	3	3	2	2	3	3H
LAOS	18H	1	1	1	1	1	1	1	1	1	1	1	1	1H	1	2	2H	2	1	2	3	2H	2H	1	2	1	1	1	1	2	2H	1	1	2	2	2H	3	2	1H	2	2	2	3	3	2	3	2	3	3	3	3H
ALBA	19H	0	0	0	0	0	0	0	0	0	0	0	0	0H	0	0	0H	0	0	0	0	2H	0H	3	0	3	3	3	3	0	3H	3	3	3	0	0H	0	3	3H	0	0	0	0	0	0	0	0	0	0	0	0H
TURK	20H	1	1	1	1	1	1	1	1	1	1	1	1	1H	2	1	2H	2	1	2	2	2H	2H	1	1	2	1	1	1	1	1H	1	2	2	1	2	3	1	1H	3	3	3	3	3	1	2	2	2	3	3	3H
JAMA	21H	1	1	1	1	1	1	1	1	1	1	1	1	1H	1	3	2H	2	1	2	2	1H	2H	1	3	2	1	1	1	1	2H	3	3	3	2	2	2	1	1H	2	2	2	2	3	3	3	2	3	3	3	3H
VENE	22H	1	1	1	1	1	1	1	1	1	1	1	1	1H	2	1	2H	2	1	2	2	1H	2H	1	2	2	1	1	1	2	1H	2	2	3	1	2	2	1	1H	3	2	3	3	3	2	2	2	3	3	3	3H
NICA	23H	0	1	0	0	1	1	0	0	0	0	1	1	0H	0	1	0H	1	1	1	1	3H	0H	1	1	0	1	1	1	0	1H	1	1	1	0	1	1	1	2H	0	0	0	3	2	0	0	2	0	3	3	3H

TABLE 11, CONTINUED

SPAI	24H	1	1	1	1	1	1	1	1	1	2	1	1	1H	2	1	2H	0	1	0	0	0H	2H	2	3	2	1	1	1	2	1H	1	2	0	2	3	3	3	1H	3	3	3	3	2	2	3	2	2	3	3	3H000000000
	H													H			H					H	H								H								H												H000000000
ELSA	25H	1	1	1	1	1	1	1	1	1	1	1	1	1H	1	1	2H	1	1	1	1	3H	1H	1	1	1	1	1	1	1	1H	1	1	1	2	3	3	3	3H	3	3	3	3	2	3	2	1	3	3	3	3H000000000
	H													H			H					H	H								H								H												H000000000
COLO	26H	1	1	1	1	1	1	1	1	1	1	0	0	1H	2	1	2H	2	1	2	2	1H	2H	1	3	3	0	1	1	1	1H	1	1	2	2	2	3	1	1H	2	3	3	3	2	2	3	2	3	3	3	3H000000000
	H													H			H					H	H								H								H												H000000000
IVOR	27H	1	1	1	1	1	1	1	1	1	1	1	1	1H	2	1	1H	0	2	0	0	0H	1H	1	1	1	1	1	1	1	1H	0	2	0	2	2	2	3	3H	3	2	3	2	3	3	3	3	3	2	2	3H000000000
	H													H			H					H	H								H								H												H000000000
NIGE	28H	1	1	1	1	1	1	0	0	1	1	1	1	1H	0	2	1H	1	2	1	1	1H	1H	1	2	2	1	1	1	3	1H	3	3	3	0	2	2	3	3H	0	3	3	2	3	0	3	3	3	2	2	3H000000000
	H													H			H					H	H								H								H												H000000000
MAUR	29H	1	1	1	1	1	1	1	1	1	0	1	1	1H	1	3	0H	2	1	2	2	2H	0H	1	2	2	1	1	1	1	2H	1	1	2	2	1	2	2	1H	2	3	3	3	3	2	3	0	3	2	3	3H000000000
	H													H			H					H	H								H								H												H000000000
PANA	30H	1	1	1	1	1	1	1	1	0	0	1	1	1H	2	1	2H	1	1	1	1	1H	2H	1	1	1	1	1	1	1	1H	1	1	1	2	1	3	3	2H	3	0	2	3	3	2	3	2	3	3	3	3H000000000
	H													H			H					H	H								H								H												H000000000
TRIN	31H	1	1	1	1	1	1	1	1	2	1	1	1	1H	1	2	2H	1	1	1	1	1H	2H	1	3	2	1	1	1	1	2H	3	3	3	2	2	2	1	2H	3	2	2	2	3	3	3	0	3	3	3	3H000000000
	H													H			H					H	H								H								H												H000000000
CONG	32H	1	1	1	1	1	1	1	1	1	1	1	1	1H	1	1	2H	2	1	1	2	1H	2H	1	0	1	2	1	1	3	1H	3	3	3	3	1	2	3	3H	3	3	3	2	3	3	3	0	3	2	2	3H000000000
	H													H			H					H	H								H								H												H000000000
SING	33H	1	1	1	1	1	1	1	1	1	1	1	1	1H	1	3	2H	2	1	2	2	2H	2H	2	1	1	2	2	3	1	2H	2	2	2	2	3	3	2	3H	3	2	2	3	3	2	3	3	3	3	3	3H000000000
	H													H			H					H	H								H								H												H000000000
AFGA	34H	1	1	1	1	1	1	1	1	1	1	1	1	1H	2	1	2H	1	1	1	2	1H	2H	2	3	2	2	2	3	2	3H	1	3	3	2	2	3	3	3H	3	2	3	3	3	2	3	2	3	3	3	3H000000000
	H													H			H					H	H								H								H												H000000000
UP V	35H	1	1	1	1	0	1	1	1	1	1	0	0	2H	1	2	3H	1	0	1	1	1H	3H	1	2	2	2	2	1	3	1H	3	3	3	3	2	2	3	3H	3	3	3	2	3	3	3	3	3	2	2	3H000000000
	H													H			H					H	H								H								H												H000000000
ARGE	36H	1	1	1	1	1	1	1	1	1	1	1	1	1H	2	1	2H	2	1	2	2	1H	2H	1	3	3	1	1	1	1	1H	1	1	2	1	3	2	1	2H	3	3	3	3	3	2	3	2	2	3	3	3H000000000
	H													H			H					H	H								H								H												H000000000
GUAT	37H	1	1	1	1	1	1	1	1	0	1	1	1	1H	2	1	2H	3	1	3	2	1H	2H	1	3	3	1	1	1	2	1H	1	1	2	2	3	3	1	1H	3	3	3	3	3	1	3	2	3	3	3	3H000000000
	H													H			H					H	H								H								H												H000000000
DAHO	38H	1	1	1	1	1	1	1	1	1	1	1	1	1H	1	1	1H	1	1	1	1	2H	1H	1	2	2	0	1	1	2	1H	3	3	3	3	1	2	3	3H	3	3	2	3	3	3	3	3	3	2	2	3H000000000
	H													H			H					H	H								H								H												H000000000
NIGE	39H	1	0	1	1	1	1	0	0	2	1	1	1	1H	0	2	2H	1	1	1	1	1H	2H	2	3	3	2	2	3	2	2H	3	3	3	2	2	3	3	3H	0	2	3	3	3	0	3	3	2	3	3	3H000000000
	H													H=	==	==	==	==	==	==	==	==	==	==	==	==	==	==	==	==	=H=	==	==	==	==	==	==	==	=H=	==	==	==	==	==	==	==	==	==	==	==	=H000
MADA	40H	1	1	1	1	1	1	1	1	1	1	1	1	2H	1	2	1	0	1	0	0	0H	1	1	1	1	1	1	1	1	1H	0	0	0	2	1	2	3	2H	3	2	3	2	3	3	3	0	3	3	3	3H000000000
	H													H								H									H								H												H000000000
PHIL	41H	1	1	1	1	1	1	1	1	1	1	1	1	1H	1	1	2	2	2	1	2	1H	2	1	1	1	1	1	1	1	1H	1	3	3	1	1	2	3	3H	3	2	3	3	3	3	3	3	3	3	3	3H000000000
	H													H								H									H								H												H000000000
URUG	42H	1	1	1	1	0	1	0	0	1	1	0	0	1H	0	1	2	1	0	1	1	0H	2	1	0	0	1	1	1	1	1H	1	1	1	0	2	3	1	1H	0	3	3	3	0	0	3	2	0	3	3	3H000000000
	H													H								H									H								H												H000000000
TOGO	43H	1	1	1	1	1	1	0	1	1	1	1	1	1H	2	2	2	1	1	1	1	1H	2	1	1	1	1	1	1	2	1H	3	3	3	3	2	3	3	2H	3	3	3	3	3	3	3	3	3	3	3	3H000000000
	H													H								H									H								H												H000000000
IRAN	44H	1	1	1	1	1	1	1	1	1	1	1	1	1H	2	1	1	2	1	2	2	1H	1	1	2	1	0	1	1	1	2H	1	2	2	2	2	3	2	2H	2	2	3	3	3	3	3	3	3	3	3	3H000000000
	H													H								H									H								H												H000000000
HAIT	45H	1	1	1	1	1	1	0	0	1	1	1	1	1H	1	1	0	0	1	0	0	1H	0	1	2	2	0	1	1	1	1H	1	2	3	2	1	2	1	1H	0	3	2	3	3	3	3	2	3	3	3	3H000000000
	H													H								H									H								H												H000000000
CHIL	46H	1	1	1	1	0	1	1	0	1	1	1	1	1H	2	1	1	2	0	2	2	1H	2	1	2	2	1	1	1	1	1H	1	2	3	1	3	3	1	1H	0	3	3	3	3	2	3	3	3	3	3	3H000000000
	H													H								H									H								H												H000000000
CHAD	47H	1	1	1	1	1	0	1	1	1	1	1	1	1H	1	1	0	0	1	0	0	0H	0	0	2	2	1	0	1	1	1H	0	3	0	2	2	3	3	3H	3	2	3	3	3	3	3	3	3	3	3	3H000000000
	H													H								H									H								H												H000000000
DOMI	48H	1	1	1	1	1	1	1	1	1	0	1	1	1H	2	1	2	2	1	2	2	1H	2	1	1	1	3	1	1	1	1H	0	1	2	1	1	0	3	3H	3	3	3	0	3	2	3	0	3	0	0	3H000000000
	H													H								H									H								H												H000000000
PERU	49H	1	1	1	1	1	1	1	1	0	2	1	1	1H	2	1	2	2	1	2	2	2H	2	1	2	2	1	1	1	1	1H	1	1	2	2	3	3	3	2H	3	0	3	3	3	2	3	2	3	3	3	3H000000000
	H													H								H									H								H												H000000000
RWAN	50H	1	1	1	1	1	1	1	1	1	1	1	1	1H	1	1	1	1	2	1	1	1H	2	1	1	1	1	1	1	3	1H	3	3	3	2	1	2	3	3H	3	3	3	3	3	3	3	3	3	3	3	3H000000000
	H													H								H									H								H												H000000000
INDO	51H	1	1	1	1	1	1	1	1	1	1	1	1	1H	1	1	2	1	1	1	1	1H	2	1	1	1	2	2	0	1	0H	2	3	3	2	2	3	0	0H	3	3	3	3	3	3	3	3	3	3	3	3H000000000
	H													H								H									H								H												H000000000
LEBA	52H	1	1	1	1	1	1	1	1	1	1	1	1	1H	2	1	1	0	1	0	0	0H	2	2	1	1	1	2	1	1	2H	0	2	0	2	3	3	2	1H	3	3	3	3	3	2	3	3	3	3	3	3H000000000
	H													H								H									H								H												H000000000
CYPR	53H	1	1	1	1	1	1	1	1	1	1	1	1	1H	1	1	2	1	1	1	1	2H	2	1	2	2	1	1	1	1	2H	1	2	2	3	2	3	1	1H	3	3	3	3	3	3	3	3	3	3	3	3H000000000
	H													H								H									H								H												H000000000
SAUD	54H	1	1	1	1	1	1	1	1	1	1	1	1	1H	2	1	2	1	1	1	1	2H	2	0	1	1	2	0	1	3	1H	3	0	3	2	3	3	3	2H	3	2	3	3	3	2	3	3	3	3	3	3H000000000
	H													H								H									H								H												H000000000
SIER	55H	1	1	1	1	1	1	1	1	1	1	1	1	1H	2	3	1	1	1	1	1	1H	1	2	3	2	0	2	1	1	1H	3	3	3	2	2	3	2	1H	2	3	3	3	0	3	3	0	3	3	3	3H000000000
	H													H								H									H								H												H000000000
SENE	56H	1	1	1	1	1	1	1	1	1	1	1	1	1H	2	2	1	1	1	1	1	1H	2	1	2	2	2	1	1	2	2H	3	3	3	2	2	3	2	2H	2	3	3	3	3	2	3	3	3	3	3	3H000000000
	H													H								H									H								H												H000000000
GHAN	57H	2	1	1	1	1	1	1	1	1	1	1	1	1H	1	2	1	1	1	1	1	2H	1	2	1	1	1	2	3	2	2H	3	3	3	0	3	2	3	3H	3	3	3	2	3	3	3	3	3	2	3	3H000000000
	H													H								H									H								H												H000000000

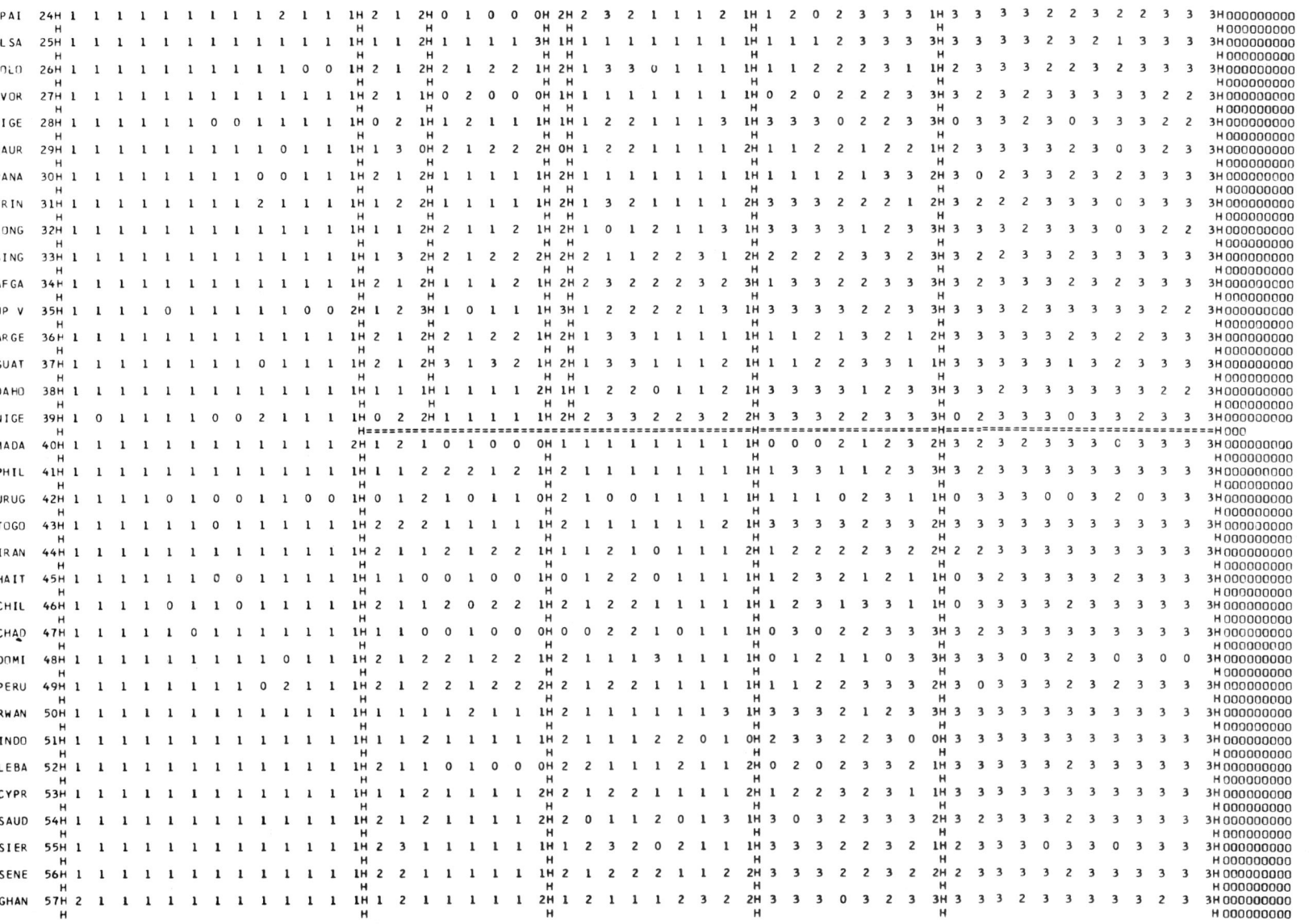

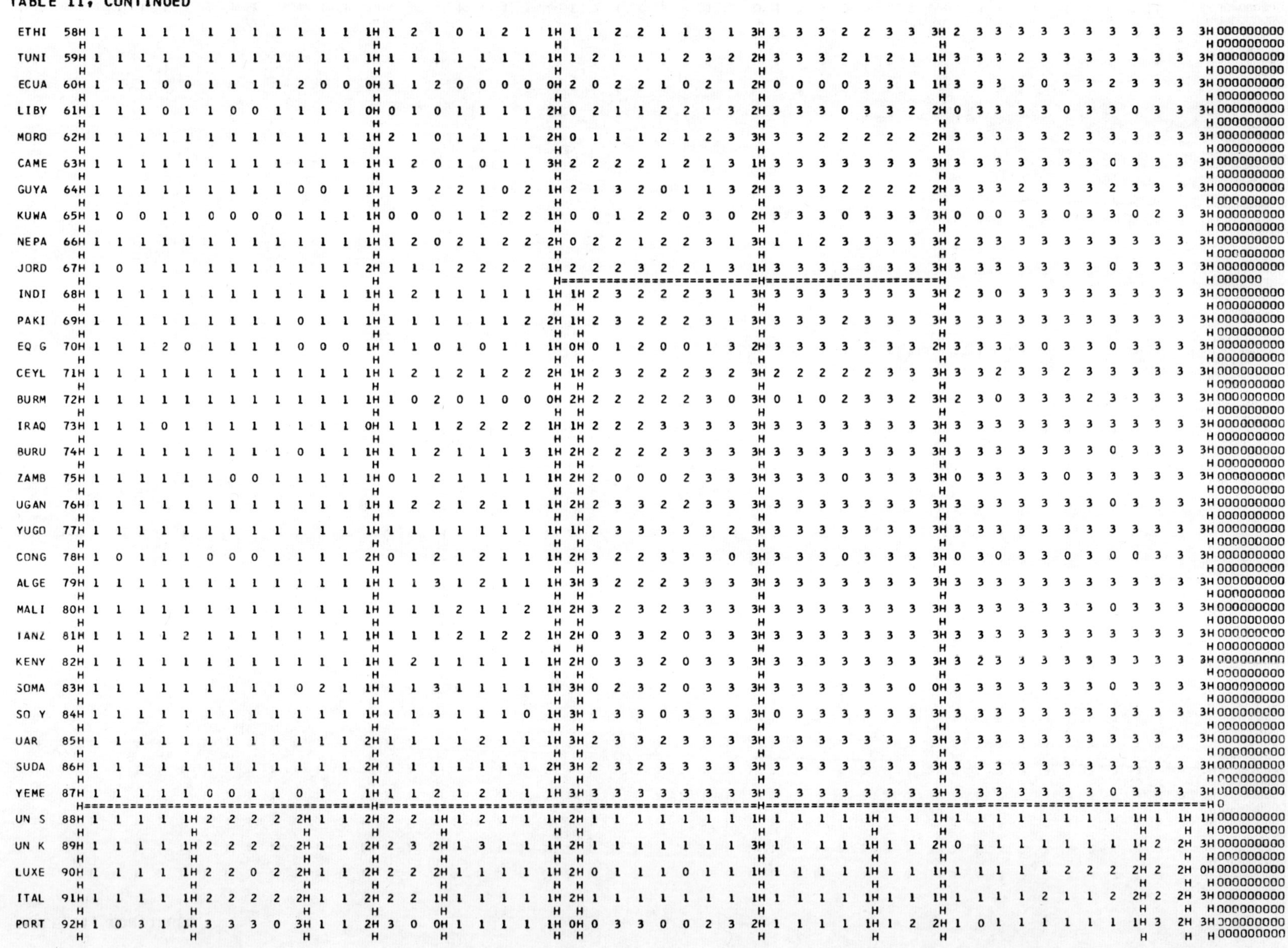

TABLE 11, CONTINUED

```
ETHI  58H 1  1  1  1  1  1  1  1  1  1  1  1  1H 1  2  1  0  1  2  1  1H 1  1  2  2  1  1  3  1  3H 3  3  3  2  2  3  3  3H 2  3  3  3  3  3  3  3  3  3  3  3H 000000000
        H                                      H                       H                          H                       H                                   H 000000000
TUNI  59H 1  1  1  1  1  1  1  1  1  1  1  1  1H 1  1  1  1  1  1  1  1H 1  2  1  1  1  2  3  2  2H 3  3  3  2  1  2  1  1H 3  3  3  2  3  3  3  3  3  3  3  3H 000000000
        H                                      H                       H                          H                       H                                   H 000000000
ECUA  60H 1  1  1  0  0  1  1  1  1  2  0  0  0H 1  1  2  0  0  0  0  0H 2  0  2  2  1  0  2  1  2H 0  0  0  0  3  3  1  1H 3  3  3  3  0  3  3  2  3  3  3  3H 000000000
        H                                      H                       H                          H                       H                                   H 000000000
LIBY  61H 1  1  1  0  1  1  0  0  1  1  1  1  0H 0  1  0  1  1  1  1  2H 0  2  1  1  2  2  1  3  2H 3  3  3  0  3  3  2  2H 0  3  3  3  3  0  3  3  0  3  3  3H 000000000
        H                                      H                       H                          H                       H                                   H 000000000
MORO  62H 1  1  1  1  1  1  1  1  1  1  1  1  1H 2  1  0  1  1  1  1  2H 0  1  1  1  2  1  2  3  3H 3  3  2  2  2  2  2  2H 3  3  3  3  3  2  3  3  3  3  3  3H 000000000
        H                                      H                       H                          H                       H                                   H 000000000
CAME  63H 1  1  1  1  1  1  1  1  1  1  1  1  1H 1  2  0  1  0  1  1  3H 2  2  2  2  1  2  1  3  1H 3  3  3  3  3  3  3  3H 3  3  3  3  3  3  3  0  3  3  3  3H 000000000
        H                                      H                       H                          H                       H                                   H 000000000
GUYA  64H 1  1  1  1  1  1  1  1  1  0  0  1  1H 1  3  2  2  1  0  2  1H 2  1  3  2  0  1  1  3  2H 3  3  3  2  2  2  2  2H 3  3  3  2  3  3  3  2  3  3  3  3H 000000000
        H                                      H                       H                          H                       H                                   H 000000000
KUWA  65H 1  0  0  1  1  0  0  0  0  1  1  1  1H 0  0  0  1  1  2  2  1H 0  0  1  2  2  0  3  0  2H 3  3  3  0  3  3  3  3H 0  0  0  3  3  0  3  3  0  2  3  3H 000000000
        H                                      H                       H                          H                       H                                   H 000000000
NEPA  66H 1  1  1  1  1  1  1  1  1  1  1  1  1H 1  2  0  2  1  2  2  2H 0  2  2  1  2  2  3  1  3H 1  1  2  3  3  3  3  3H 2  3  3  3  3  3  3  3  3  3  3  3H 000000000
        H                                      H                       H                          H                       H                                   H 000000000
JORD  67H 1  0  1  1  1  1  1  1  1  1  1  1  2H 1  1  1  2  2  2  2  1H 2  2  2  3  2  2  1  3  1H 3  3  3  3  3  3  3  3H 3  3  3  3  3  3  3  0  3  3  3  3H 000000000
        H                                      H                       H==========================H=======================H                                   H 000000
INDI  68H 1  1  1  1  1  1  1  1  1  1  1  1  1H 1  2  1  1  1  1  1  1H 1H 2  3  2  2  2  3  1  3H 3  3  3  3  3  3  3  3H 2  3  0  3  3  3  3  3  3  3  3  3H 000000000
        H                                      H                       H  H                       H                       H                                   H 000000000
PAKI  69H 1  1  1  1  1  1  1  1  1  0  1  1  1H 1  1  1  1  1  1  2  2H 1H 2  3  2  2  2  3  1  3H 3  3  3  2  3  3  3  3H 3  3  3  3  3  3  3  3  3  3  3  3H 000000000
        H                                      H                       H  H                       H                       H                                   H 000000000
EQ G  70H 1  1  1  2  0  1  1  1  1  0  0  0  1H 1  1  0  1  0  1  1  1H 0H 0  1  2  0  0  1  3  2H 3  3  3  3  3  3  3  2H 3  3  3  3  0  3  3  0  3  3  3  3H 000000000
        H                                      H                       H  H                       H                       H                                   H 000000000
CEYL  71H 1  1  1  1  1  1  1  1  1  1  1  1  1H 1  2  1  2  1  2  2  2H 1H 2  3  2  2  2  3  2  3H 2  2  2  2  2  3  3  3H 3  3  2  3  3  2  3  3  3  3  3  3H 000000000
        H                                      H                       H  H                       H                       H                                   H 000000000
BURM  72H 1  1  1  1  1  1  1  1  1  1  1  1  1H 1  0  2  0  1  0  0  0H 2H 2  2  2  2  2  3  0  3H 0  1  0  2  3  3  2  3H 2  3  0  3  3  3  2  3  3  3  3  3H 000000000
        H                                      H                       H  H                       H                       H                                   H 000000000
IRAQ  73H 1  1  1  0  1  1  1  1  1  1  1  1  0H 1  1  1  2  2  2  2  1H 1H 2  2  2  3  3  3  3  3H 3  3  3  3  3  3  3  3H 3  3  3  3  3  3  3  3  3  3  3  3H 000000000
        H                                      H                       H  H                       H                       H                                   H 000000000
BURU  74H 1  1  1  1  1  1  1  1  1  0  1  1  1H 1  1  2  1  1  1  3  1H 2H 2  2  2  2  3  3  3  3H 3  3  3  3  3  3  3  3H 3  3  3  3  3  3  3  0  3  3  3  3H 000000000
        H                                      H                       H  H                       H                       H                                   H 000000000
ZAMB  75H 1  1  1  1  1  1  0  0  1  1  1  1  1H 0  1  2  1  1  1  1  1H 2H 2  0  0  0  2  3  3  3H 3  3  3  0  3  3  3  3H 0  3  3  3  3  0  3  3  3  3  3  3H 000000000
        H                                      H                       H  H                       H                       H                                   H 000000000
UGAN  76H 1  1  1  1  1  1  1  1  1  1  1  1  1H 1  2  2  1  2  1  1  1H 2H 2  3  3  2  2  3  3  3H 3  3  3  3  3  3  3  3H 3  3  3  3  3  3  3  0  3  3  3  3H 000000000
        H                                      H                       H  H                       H                       H                                   H 000000000
YUGO  77H 1  1  1  1  1  1  1  1  1  1  1  1  1H 1  1  1  1  1  1  1  1H 1H 2  3  3  3  3  3  2  3H 3  3  3  3  3  3  3  3H 3  3  3  3  3  3  3  3  3  3  3  3H 000000000
        H                                      H                       H  H                       H                       H                                   H 000000000
CONG  78H 1  0  1  1  1  0  0  0  1  1  1  1  2H 0  1  2  1  2  1  1  1H 2H 3  2  2  3  3  3  0  3H 3  3  3  0  3  3  3  3H 0  3  0  3  3  0  3  0  0  3  3  3H 000000000
        H                                      H                       H  H                       H                       H                                   H 000000000
ALGE  79H 1  1  1  1  1  1  1  1  1  1  1  1  1H 1  1  3  1  2  1  1  1H 3H 3  2  2  2  3  3  3  3H 3  3  3  3  3  3  3  3H 3  3  3  3  3  3  3  3  3  3  3  3H 000000000
        H                                      H                       H  H                       H                       H                                   H 000000000
MALI  80H 1  1  1  1  1  1  1  1  1  1  1  1  1H 1  1  1  2  1  1  2  1H 2H 3  2  3  2  3  3  3  3H 3  3  3  3  3  3  3  3H 3  3  3  3  3  3  3  0  3  3  3  3H 000000000
        H                                      H                       H  H                       H                       H                                   H 000000000
TANZ  81H 1  1  1  1  2  1  1  1  1  1  1  1  1H 1  1  1  2  1  2  2  1H 2H 0  3  3  2  0  3  3  3H 3  3  3  3  3  3  3  3H 3  3  3  3  3  3  3  3  3  3  3  3H 000000000
        H                                      H                       H  H                       H                       H                                   H 000000000
KENY  82H 1  1  1  1  1  1  1  1  1  1  1  1  1H 1  2  1  1  1  1  1  1H 2H 0  3  3  2  0  3  3  3H 3  3  3  3  3  3  3  3H 3  2  3  3  3  3  3  3  3  3  3  3H 000000000
        H                                      H                       H  H                       H                       H                                   H 000000000
SOMA  83H 1  1  1  1  1  1  1  1  1  0  2  1  1H 1  1  3  1  1  1  1  1H 3H 0  2  3  2  0  3  3  3H 3  3  3  3  3  3  0  0H 3  3  3  3  3  3  3  0  3  3  3  3H 000000000
        H                                      H                       H  H                       H                       H                                   H 000000000
SO Y  84H 1  1  1  1  1  1  1  1  1  1  1  1  1H 1  1  3  1  1  1  0  1H 3H 1  3  3  0  3  3  3  3H 0  3  3  3  3  3  3  3H 3  3  3  3  3  3  3  3  3  3  3  3H 000000000
        H                                      H                       H  H                       H                       H                                   H 000000000
UAR   85H 1  1  1  1  1  1  1  1  1  1  1  1  2H 1  1  1  1  2  1  1  1H 3H 2  3  3  2  3  3  3  3H 3  3  3  3  3  3  3  3H 3  3  3  3  3  3  3  3  3  3  3  3H 000000000
        H                                      H                       H  H                       H                       H                                   H 000000000
SUDA  86H 1  1  1  1  1  1  1  1  1  1  1  1  2H 1  1  1  1  1  1  1  2H 3H 2  3  2  3  3  3  3  3H 3  3  3  3  3  3  3  3H 3  3  3  3  3  3  3  3  3  3  3  3H 000000000
        H                                      H                       H  H                       H                       H                                   H 000000000
YEME  87H 1  1  1  1  1  0  0  1  1  0  1  1  1H 1  1  2  1  2  1  1  1H 3H 3  3  3  3  3  3  3  3H 3  3  3  3  3  3  3  3H 3  3  3  3  3  3  3  0  3  3  3  3H 000000000
        H======================================H==================================================H===========================================================H 0
UN S  88H 1  1  1  1  1H 2  2  2  2  2H 1  1  2H 2  2  1H 1  2  1  1  1H 2H 1  1  1  1  1  1  1  1H 1  1  1  1  1H 1  1  1H 1  1  1  1  1  1  1  1  1H 1  1H 1H 000000000
        H              H              H        H        H              H  H                       H              H        H                          H     H     H 000000000
UN K  89H 1  1  1  1  1H 2  2  2  2  2H 1  1  2H 2  3  2H 1  3  1  1  1H 2H 1  1  1  1  1  1  1  3H 1  1  1  1  1H 1  1  2H 0  1  1  1  1  1  1  1  1H 2  2H 3H 000000000
        H              H              H        H        H              H  H                       H              H        H                          H     H     H 000000000
LUXE  90H 1  1  1  1  1H 2  2  0  2  2H 1  1  2H 2  2  2H 1  1  1  1  1H 2H 0  1  1  1  0  1  1  1H 1  1  1  1  1H 1  1  1H 1  1  1  1  1  2  2  2  2H 2  2H 0H 000000000
        H              H              H        H        H              H  H                       H              H        H                          H     H     H 000000000
ITAL  91H 1  1  1  1  1H 2  2  2  2  2H 1  1  2H 2  2  1H 1  1  1  1  1H 2H 1  1  1  1  1  1  1  1H 1  1  1  1  1H 1  1  1H 1  1  1  1  2  1  1  2  2H 2  2H 3H 000000000
        H              H              H        H        H              H  H                       H              H        H                          H     H     H 000000000
PORT  92H 1  0  3  1  1H 3  3  3  0  3H 1  1  2H 3  0  0H 1  1  1  1  1H 0H 0  3  3  0  0  2  3  2H 1  1  1  1  1H 1  2  2H 1  0  1  1  1  1  1  1  1H 3  2H 3H 000000000
        H              H              H        H        H              H  H                       H              H        H                          H     H     H 000000000
```

TABLE 11, CONTINUED

```
BELG  93H 1 1 1 1 1H 2 2 2 2 2H 1 1 2H 2 2 2H 1 2 1 1 1H 2H 1 1 1 1 1 1 1 1H 1 1 1 1 1H 1 1 1H 1 1 1 1 1 2 2 2 2H 2 2H 3H000000000
          H       H         H     H     H         H  H               H         H     H                    H     H  H000000000
AUST  94H 1 1 1 1 1H 2 2 2 2 2H 1 1 2H 2 3 1H 1 2 1 1 1H 1H 1 1 1 1 1 1 1 1H 1 1 1 1 1H 1 3 3H 1 1 1 1 1 2 1 1 2H 1 1H 3H000000000
          H       H         H     H     H         H  H               H         H     H                    H     H  H000000000
NETH  95H 1 1 1 1 1H 2 2 2 2 2H 1 1 1H 2 2 2H 1 1 1 1 1H 2H 1 1 1 1 1 1 1 2H 1 1 1 1 1H 1 1 1H 1 1 1 1 1 2 2 2 2H 2 2H 3H000000000
          H       H         H     H     H         H  H               H         H     H                    H     H  H000000000
SO A  96H 1 2 3 1 1H 3 3 3 3 3H 1 1 2H 3 0 2H 1 2 1 1 1H 2H 1 1 1 1 1 1 2 1H 1 1 1 1 1H 1 3 3H 1 1 1 1 1 1 1 1 1H 2 2H 3H000000000
          H       H         H     H     H         H  H               H         H     H                    H     H  H000000000
NEW   97H 1 1 1 1 1H 2 2 2 2 2H 1 1 2H 2 3 2H 1 2 1 1 1H 2H 1 1 1 1 1 1 1 1H 1 1 1 1 1H 1 3 1H 1 1 1 1 1 2 1 2 2H 2 2H 3H000000000
          H       H         H     H     H         H  H               H         H     H                    H     H  H000000000
JAPA  98H 2 1 1 1 1H 1 1 1 2 1H 1 1 1H 2 2 2H 1 1 1 1 1H 2H 1 1 1 1 1 1 1 1H 1 1 1 1 1H 1 1 1H 2 1 2 1 2 2 2 2 2H 2 2H 3H000000000
          H       H         H     H     H         H  H               H         H     H                    H     H  H000000000
ICEL  99H 1 1 1 1 1H 2 2 2 2 1H 1 1 1H 2 2 2H 1 1 1 1 1H 1H 1 1 1 1 1 1 1 2H 1 1 1 1 1H 2 1 1H 1 1 1 2 2 2 2 2 2H 2 2H 3H000000000
          H       H         H     H     H         H  H               H         H     H                    H     H  H000000000
NORW 100H 1 1 1 1 1H 2 2 2 2 1H 1 1 1H 2 2 2H 1 2 1 1 1H 2H 1 1 1 1 1 3 1 3H 1 1 1 1 1H 2 1 2H 1 1 1 2 2 2 2 2 2H 2 2H 3H000000000
          H       H         H     H     H         H  H               H         H     H                    H     H  H000000000
MALA 101H 1 1 1 1 2H 0 2 2 2 2H 2 2 2H 3 3 2H 1 2 1 1 1H 2H 1 1 1 1 1 1 1 1H 1 1 1 1 1H 1 3 3H 1 0 2 1 2 1 3 0 2H 1 2H 3H000000000
          H       H         H     H     H         H  H               H         H     H                    H     H  H000000000
DENM 102H 1 1 1 1 1H 2 2 2 2 1H 1 1 1H 2 3 2H 1 2 1 1 1H 2H 1 1 1 1 1 3 1 3H 1 1 1 1 1H 2 2 3H 1 1 1 2 2 2 2 2 2H 2 2H 3H000000000
          H       H         H     H     H         H  H               H         H     H                    H     H  H000000000
SWED 103H 1 1 1 1 1H 1 2 2 2 1H 1 1 1H 2 3 2H 1 1 1 1 0H 2H 1 1 1 1 1 3 1 3H 1 1 1 1 1H 2 2 3H 1 1 1 2 2 2 2 2 2H 2 3H 3H000000000
          H       H         H     H     H         H  H               H         H     H                    H     H  H000000000
BOTS 104H 0 0 0 0 0H 2 0 0 0 0H 0 0 0H 0 3 2H 1 0 1 1 1H 0H 1 0 0 0 1 1 1 1H 1 1 1 0 0H 0 3 3H 0 0 2 0 0 0 0 0 0H 0 0H 0H000000000
          H       H         H     H     H         H  H               H         H     H                    H     H  H000000000
AUST 105H 1 1 1 1 1H 2 2 2 2 1H 1 1 1H 2 2 2H 1 1 1 1 1H 2H 1 1 1 1 1 2 1 2H 1 1 1 2 2H 2 2 2H 1 2 1 2 2 2 0 2 2H 2 3H 3H000000000
          H       H         H     H     H         H  H               H         H     H                    H     H  H000000000
CANA 106H 1 1 1 1 1H 2 3 2 2 1H 1 1 2H 2 3 1H 1 2 1 1 1H 1H 1 1 1 1 1 1 1 2H 1 1 1 1 3H 3 1 2H 1 1 1 3 2 2 1 2 2H 3 3H 3H000000000
          H       H         H     H     H         H  H               H         H     H                    H     H  H000000000
FINL 107H 1 1 1 1 1H 2 2 2 2 1H 1 1 1H 2 2 2H 1 1 1 1 0H 2H 2 1 1 2 2 3 1 3H 1 1 1 1 2H 2 2 3H 1 2 1 2 2 2 2 2 2H 3 3H 3H000000000
          H       H         H     H     H         H  H               H         H     H                    H     H  H000000000
BRAZ 108H 1 1 1 1 1H 1 3 2 1 3H 1 1 1H 3 1 0H 2 1 2 2 2H 0H 1 3 3 1 1 1 1 1H 1 1 1 1 3H 3 1 1H 2 2 3 3 2 1 3 2 1H 3 3H 3H000000000
          H       H=================H=======================================H==================================================H00
CENA 109H 1 1 1 1 2H 1 1 1 1 0H 2 1 1H 2 2 2H 1 2 2 2 2H 1 1 2 2 1 1 1 1 1H 1 1 2 2 1 2 3 3H 3 2 2 2 3 2 3 3 3 2 2 3H000000000
          H       H         H     H     H         H                  H               H                                 H000000000
BARB 110H 1 1 1 1 1H 1 1 1 3 0H 1 1 1H 1 3 2H 2 1 1 2 0H 2 1 1 1 0 1 2 1 1H 3 3 3 2 2 1 2 1H 3 2 2 1 3 3 3 0 3 3 3 3H000000000
          H       H         H     H     H         H                  H               H                                 H000000000
MEXI 111H 1 2 1 1 1H 1 1 1 1 2H 1 1 1H 2 2 2H 2 1 2 2 1H 2 1 3 3 1 1 1 1 1H 1 2 3 2 3 3 1 1H 2 2 2 3 3 2 3 3 3 3 3 3H000000000
          H       H         H     H     H         H                  H               H                                 H000000000
FRAN 112H 1 2 2 1 2H 2 2 2 2 2H 2 1 2H 2 2 2H 1 2 1 1 1H 2 1 2 2 1 1 3 2 3H 1 1 1 1 3 3 2 3H 2 2 2 3 2 2 2 2 2 3 3 3H000000000
          H       H         H=====H=====H=========H==================H===============H=================================H00000
ROMA 113H 1 1 1 1 1H 1 1 1 1 1H 2 1 3H 1 1 1H 3 1 3 3 3H 1H 2 2 3 3 3 3 3 3H 2 2 2 3 3 3 3 3H 3 3 3 3 3 3 2 3 3H 3 3H 3H000000000
          H       H         H     H     H         H  H               H               H                    H     H  H000000000
MAUR 114H 1 1 1 1 2H 1 1 1 1 1H 2 2 1H 1 1 1H 1 2 1 1 2H 2H 3 2 2 2 3 3 3 3H 3 3 3 3 3 3 3 3H 3 3 3 3 3 3 3 3 3H 3 3H 3H000000000
          H       H         H     H     H         H  H               H               H                    H     H  H000000000
CAMB 115H 1 1 1 2 0H 1 1 1 1 1H 0 0 2H 1 1 2H 2 0 2 2 1H 2H 2 0 2 3 3 3 0 3H 3 3 3 3 3 3 3 3H 3 3 3 3 0 3 3 0 3H 3 3H 0H000000000
          H       H         H     H     H         H  H               H               H                    H     H  H000000000
UKRA 116H 1 1 1 2 1H 1 1 1 1 1H 2 3 3H 1 1 1H 3 3 3 3 3H 1H 2 3 3 3 3 3 3 3H 2 2 2 3 3 3 2 3H 3 3 3 3 3 3 2 3 3H 3 3H 1H000000000
          H       H         H     H     H         H  H               H               H                    H     H  H000000000
CUBA 117H 1 1 0 2 2H 0 0 2 0 2H 2 2 2H 0 1 2H 0 2 0 0 3H 2H 3 2 3 3 3 3 0 3H 3 3 3 0 3 3 3 3H 3 0 0 3 0 0 0 3 0H 3 3H 0H000000000
          H       H         H     H     H        .H  H               H               H                    H     H  H000000000
HUNG 118H 1 1 1 2 1H 1 1 1 1 1H 2 3 3H 1 1 1H 3 3 3 3 3H 1H 3 3 3 3 3 3 3 3H 2 2 2 3 3 3 3 3H 3 3 3 3 3 3 2 3 3H 3 3H 1H000000000
          H       H         H     H     H         H  H               H               H                    H     H  H000000000
BULG 119H 1 1 1 1 1H 1 1 1 1 1H 2 3 3H 1 1 1H 3 3 3 3 3H 1H 3 3 3 3 3 3 3 3H 2 2 2 3 3 3 3 3H 3 3 3 3 3 3 3 3 3H 3 3H 1H000000000
          H       H         H     H     H         H  H               H               H                    H     H  H000000000
USSR 120H 1 1 1 2 1H 1 1 1 1 1H 2 3 3H 1 1 1H 3 3 3 3 3H 1H 3 3 3 3 3 3 3 3H 2 2 2 3 3 3 2 3H 3 3 3 3 3 3 2 3 3H 3 3H 1H000000000
          H       H         H     H     H         H  H               H               H                    H     H  H000000000
GUIN 121H 1 1 1 1 2H 1 1 1 1 1H 2 2 2H 1 1 1H 1 2 1 1 0H 3H 3 3 2 2 3 3 3 3H 3 3 3 3 3 3 3 3H 3 3 3 3 3 3 3 3 3H 0 3H 3H000000000
          H       H         H     H     H         H  H               H               H                    H     H  H000000000
BSSR 122H 1 1 1 2 1H 1 1 1 1 1H 2 3 3H 1 1 1H 3 3 3 3 3H 1H 3 3 3 3 3 3 3 3H 2 2 2 3 3 3 2 3H 3 3 3 3 3 3 2 3 3H 3 3H 1H000000000
          H       H         H     H     H         H  H               H               H                    H     H  H000000000
POLA 123H 1 1 1 1 1H 1 1 1 1 1H 2 3 3H 1 1 1H 3 3 3 3 3H 1H 3 3 3 3 3 3 3 3H 2 2 2 3 3 3 2 3H 3 3 3 3 3 3 2 3 3H 3 3H 1H000000000
          H       H         H     H     H         H  H               H               H                    H     H  H000000000
CZEC 124H 1 1 1 1 1H 1 1 1 1 1H 2 3 3H 1 1 1H 3 3 3 3 3H 1H 3 3 3 3 3 3 3 3H 2 2 2 3 3 3 2 3H 3 3 3 3 3 3 2 3 3H 3 3H 1H000000000
          H       H         H     H     H         H  H               H               H                    H     H  H000000000
MONG 125H 1 1 1 1 1H 1 1 1 1 0H 2 1 3H 1 1 1H 3 3 3 3 3H 2H 3 3 3 3 3 3 3 3H 2 2 2 3 3 3 3 3H 3 3 3 3 3 3 2 3 3H 3 3H 1H000000000
          H       H         H     H     H         H  H               H               H                    H     H  H000000000
SYRI 126H 1 1 1 1 1H 1 1 1 1 1H 1 2 2H 1 1 1H 2 2 2 2 1H 3H 2 3 3 3 3 3 3 3H 3 3 3 3 3 3 3 3H 3 3 3 3 3 3 3 3 3H 3 3H 3H000000000
          H===================================================================================================================H
```

TABLE 12. DIRECT CLUSTERING OF U. N. VOTES

A CLUSTER IS A SUBMATRIX SURROUNDED BY DASHES. EACH QUESTION IDEALLY IS CONSTANT WITHIN EACH CLUSTER. A LEADING COUNTRY IS LISTED FOR EACH CLUSTER, AND VALUES IDENTICAL TO THIS COUNTRY WITHIN THE CLUSTER ARE PRINTED AS BLANKS. FOR EXAMPLE, SINGAPORE, CHI1, IS REFERRED TO THE CLUSTER LEADER, CEYLON, WHICH IS 3.

```
           MSRDS  PPCSSS  PSUCI GRRRRRIPUHUP CCCCPMMKK KMMSCCNNN SSSS
           ROUIE  OOOOOO  OOKON IUUUUUNOKUKA HHHHAOOOO OOOEOOUUU OOOO
           19611  322721  18312 112345142111 341223423 121243321 6543
         --------------------------------------------------------------
ALL      - 11111 -322311- 11111 131111111111 131311311 311122111 1333 -
         -        ------                                        -------
ARGENT   -           2              2  2 2   32   31            - 222--
URUGUA   -    5  -5553  -   55   2     5525  31   55         555-5333--
CHILI    -    5  -   1 2-    5          5    31   22       1 5  -    --
VENEZU   -       -   122-        2  2 2  2   31  222    2       -    --
BOLIVI   -       -2  3  -        1  322 22 2 32    2   5        -5333--
PERU     -       -2     - 2   5       5  2    2   22            -2   --
COLOMB   -       -2     -        2      222  31   31   5      55-    --
PANAMA   -       -2  3  - 5   5  1    5  2 2  2                 - 333--
ELSALV   -       -2113  -                32             2   1   -3333--
MEXICO   - 2     -2  1 2- 2     2     2 2  2 31   31            -    --
GUATAM   -       -23    -     5          2   31  231            - 11 --
NICARA   -  55 5 -5553  - 5 555  1  3 555225 32  55     5 555   -3333--
DOMINI   -       -    5 - 5      15555   5             1        -    --
HAITI    -       -2111 2-   55   1  2   52 2 31   22   5   55   - 555--
LAOS     -       -2 1   -       22    2 22 2 21 22 2            -2  1--
MAURIT   -       -2 1   - 5     312 2   25   21 2 22       55   -2   --
SPAIN    -       -2  5 2- 2            2 22   1  2212           -5555--
ISRAEL   -       -   5 2-       21  3 2 3  2 31         2 2  2  -5555--
LEBANO   -       -2  552-                    21 2   22     1    -5555--
ECUADO   -    55 -511555- 2              25  3122 2255-   5  555-5555--
TURKEY   -       - 3   2-        2     2 2   31   2     2       -2   --
IRAN     -       -21   2-        2    2 2    22 2  2   5   11   -    --
         -       -      -                   -----------         -    --
CEYLON   -       -2 1 22-       22         2-  3122122-2   11   -2   --
SINGAP   -       -2 1 22-       3     2    2-2  2113  -2        -2   --
NEPAL    -       -111   -       2       2   -    112  -2   55   -2   --
BURMA    -       -21155 -       5       2  5-2   5 2  -22       -5555--
         -       -      --     ------------------------         -    --
CHINA    -       -      -      - 123332 2  2-           232  2  - 335--
IRELAN   -       - 1 3  -    2 -2 3   52 22 -31         3 2     - 333--
JAPAN    -   2   -   3  -     2-2  2  32 22 -31         2       - 333--
MALDIV   -       -2    2-     2-32 111 2   3-21 2  2            -2   --
MALTA    - 55    -555   - 55555-5 5   555555-32  5     5   55   -5   --
MALAYS   -       -2 1   -      -32 222      -22                 -2111--
CENAFR   -    2  -2     - 5    -2  222  1   -     22        122 -2 3 --
IVORYC   -       -21 552-      - 2 222  1  1-              112  -5555--
GREECE   -       -   3  -     2- 211223212  -32              2  -   3--
PARAGU   -       -2  3  -      -  11 2 2 521-     22       55   -5   --
COSTA    -  555  -2  5  - 555 5- 211 153 2 5-32  555    5    555-5   --
HONDUR   -       - 3 3  - 5 5  - 211 113132 -2   25             -3333--
BRAZIL   -       - 333  - 3 32 - 31111 3 221-31   31       55   -2   --
         -       -      ---------------------                   -------
ITALY    -       - 3 3  --22222 212233323223-31         3 21          -
NETHER   -       -   3  --                3 -31 2       2             -
```

TABLE 12, CONTINUED

```
BELGIU  -        -   3  --                  3 -31          2 2  2          -
LUXEMB  -        -   3  --    5             3 -31      55  252             -
NEW ZE  -        -   3  --       3          3 - 1          3 2  2          -
UN STA  -        - 3 3  --         33    3 33 -31          3321 2          -
UN KIN  -        - 3 3  --       3       3533 -32 1        3 2  3          -
AUSTRA  -        -   3  --       3 33      33 -            3 2112          -
SO AFR  - 23     - 333  --33333 5        3 33 -    2       3 2  2          -
PORTUG  - 53     - 333  --33335 5 1   53 33 -222233155 53 255              -
MALAWI  -    2   - 333  -- 5     3 3   5  5 2-                             -
BOTSWA  - 55555 -5553   --5 555 355555555552-            2  222            -
LESOTH  -    5  -5553   --55151 323 22152552-     55    5555 5555          -
SWAZIL  -    5  -2553   --55155  555522 2555- 2  5                555      -
CANADA  -        -   3  --1 3   331111        -32 2     5    1555          -
FRANCE  - 22 2   -   3  --        311112 2  2-2 31222     3 2112           -
FINLAN  -        -   3  --1       211222      -2 31      2 2  22           -
SWEDEN  -        -   3  --11    3  122        -2 31    22 22         5     -
DENMAR  -        -   3  --1     3   22        -2 31       2          5     -
NORWAY  -        -   3  --1         22        -3231       2   2            -
ICELAN  -        -   3  --1         22        -31 2       2   2            -
AUSTRI  -       -2   3  --1       2 1222      -2222       2   1            -
        -       ---------------------------------         5                -
        -         ------                                                   -
EQ GUI  -     52  111133  5                  50   2 232 55 0    55555      -
GUYANA  -        -2      - 5     32   22    2   22 2321    5      5   522  -
TRINID  -        -2      -     2 22   222   5 2 32 2 21                    -
JAMACA  -        -2      -       32   222 22 2 31 2 21                     -
BARBAD  -        -2      - 5   3 32   332   5 2 212        5          222  -
SENEGA  -        -222    -       22        2   22 2222     2    1    5 22  -
NIGER   -        -555    -   55  222222    5       322          112        -
DAHOME  -        -       -        1222       2     222     5    11    2    -
UP VOL  -     5  -       -       222222            322 2 2   233555        -
JORDAN  - 5      -       -                  5      33222 2   21 2     222  -
CAMERO  -        -       -       2          5      32222      5 5    3     -
MOROCC  -        -2222   -       2  2           22213    -2   55     2     -
CHAD    -        -2  55  -  5    2    2             2255      55   5555    -
GABON   -    5   -555    - 5 55  215333 55     22     2    2                -
LIBERI  -        -2      - 5     313222  2   3 22                          -
CONGOD  -        -       -        12222     5       3 5    2          22   -
RWANDA  -        -2      -        1  2              3           1 2        -
PHILIP  -        -3   1  -        1  2 2                        2     22   -
MADAGA  -        -2  555-        21  222    5     2            211   5555  -
LIBYA   -     5 -555     -   55            55    22 23   22 2  555    2    -
TOGO    -         -  2   -   5   22                2  2                    -
THAILA  -        -222    -       222    2 22 2                             -
INDONE  -        -2   2  -       2               5555      2 2             -
GAMBIA  - 555 5 -555555- 55555 555555555555 2    555      555112   5555   -
SAUDIA  -        -222  5-                2       2  3   55 2        2      -
        -         -       -                     ----------                 -
AFGANI  -        -222 1  -       2     2  2   -  3122122-2             2  -
NIGERI  - 5      -255    -    552 22    225   -   2 3    -2                -
TUNISI  -        -2      -        1  22      -31 2 13    -    11           -
GHANA   -    2   -5      -       2 2 22       -  2 13    -    11     2     -
KUWAIT  - 55     -555    -  5555 5 2   555   5-  25 355-2   55      2 2   -
CYPRUS  -        -    212-        2           -31121 211-          2       -
ETHIOP  -        -2      -       22        2  -   1 211-     11      25    -
PAKIST  -        -2      - 5                  -   1    -2    11     2  2   -
INDIA   -        -       -       2         2 5-   1    -2    11            -
```

TABLE 12, CONTINUED

```
SIERRA    -       -2 2   -       32      255 -21131    -5  11          -
ZAMBIA    -       -555   -   55          5   -    355  -5              -
          -       -      -                   -----------               -
UAR       -       -      -                   -  3133123-2  2132        -
SUDAN     -       -      -                   -     2   -1  213    2    -
SYRIA     -       -      -                   -         -1  2132 2  222 -
YUGOSL    -       -      -                   -    2    -1  11          -
SO YEM    -       -    5 -                   -       1 -5   33       5 -
UGANDA    -       -      -       2        5  -        2-2     2        -
GUINEA    -    2  -      -         5         -     2 3 -2  213222 5    -
YEMEN     -       -      - 555            5  -       3 -1    32        -
ALBANI    - 55555 -555   - 55555 555555555555-    5 53 -155555555 2555 -
TANZAN    -    2  -      -                   -       55-2   1      222 -
KENYA     -       -      -       2     2     -       55-2   1          -
SOMALI    -       -      - 5              5  -55    255-2   33 2       -
IRAQ      -     5 -      -                   -     22  -1  5112    222 -
BURUND    -       -      - 5              5  -     22  -2            1 -
CAMBOD    -    52 -      -                55 -    525  -1 52  555  222 -
CUBA      -  5 22 -555   - 25525       55  55-    5 23 -1552  222 3555 -
CONGOB    - 5     -555   -  555         555 5-    5223 -1  2  2        -
MALI      -       -      -                5  -      23 -2   1       22 -
ALGERI    -       -      -                   -     223 -2   332        -
MAURIT    -    2  -      -                   -     223 -2   1 222 2    -
          -       -      -                   -         --------------- -
HUNGAR    -     2 -   222-                   -       3 -123311323 311---
POLAND    -       -   222-                   -2      3 -             ---
CZECHO    -       -   222-                   -2      3 -             ---
USSR      -     2 -   222-                   -2      3 -             ---
UKRANI    -     2 -   222-                   -2        -             ---
BSSR      -     2 -   222-                   -2      3 -             ---
BULGAR    -       -   222-                   -       3 - 1           ---
MONGOL    -       -   222- 5                 -       3 -     2  1    ---
ROMANI    -       -   222-                   -      2  -  1   1 1    ---
          --------------------------------------------------------------
```

DATA SET DESCRIPTION

DATA - UNITED NATIONS, VOTES IN 1968-1969 (J. HARTIGAN)

CASES AND VARIABLES

126 CASES (NATIONS)
52 VARIABLES - 50 MEASUREMENT (VOTES), 1 CATEGORICAL (NATION IDENTIFICATION), 1 ALPHABETICAL (NATION)

BRIEF DESCRIPTION

THESE DATA REPORT THE VOTES (YES, NO, ABSTAIN, ABSENT) OF THE 126 NATIONS IN THE UNITED NATIONS ON THE 50 QUESTIONS IN THE GENERAL ASSEMBLY IN 1968-1969.

PURPOSE OF STUDY

THE PURPOSE OF THE STUDY IS TO DISCOVER BLOCS OF SIMILAR VOTING NATIONS, TO DISCOVER BLOCS OF SIMILAR ISSUES (SUCH AS THE SEVERAL QUESTIONS RAISED IN PARLIAMENTARY MANEUVERING ON THE CHINA ADMISSION ISSUE), AND TO DISCOVER RELATIONS BETWEEN THE VOTES ON THE VARIOUS ISSUES (FOR EXAMPLE, A ''YES'' ON ONE QUESTION MAY CORRESPOND TO A ''NO'' OR ''ABSTAIN'' ON ANOTHER).

REFERENCE

YEARBOOK OF THE UNITED NATIONS, 1968-1969.

DETAILED DESCRIPTION

VAR	COL	FORMAT	DESCRIPTION	
***	***	******	***********	
1	1-3	XXX.	NATION	
			002 = UNITED STATES	440 = EQUATORIAL GUINEA
			020 = CANADA	450 = LIBERIA
			040 = CUBA	451 = SIERRA LEON
			041 = HAITI	452 = GHANA
			042 = DOMINICAN REPUBLIC	461 = TOGO
			051 = JAMAICA	471 = CAMEROON
			052 = TRINIDAD	475 = NIGERIA
			053 = BARBADOS	481 = GABON
			070 = MEXICO	482 = CENTRAL AFRICAN REPUBLIC
			090 = GUATAMALA	483 = CHAD
			091 = HONDURAS	484 = CONGO (BRAZZAVILLE)
			092 = EL SALVADOR	490 = CONGO (KINSHASA)
			093 = NICARAGUA	500 = UGANDA
			094 = COSTA RICA	501 = KENYA
			095 = PANAMA	510 = TANZANIA
			100 = COLOMBIA	516 = BURUNDI

101 = VENEZUELA	517 = RWANDA
110 = GUYANA	520 = SOMALIA
130 = ECUADOR	530 = ETHIOPIA
135 = PERU	551 = ZAMBIA
140 = BRAZIL	553 = MALAWI
145 = BOLIVIA	560 = SOUTH AFRICA
150 = PARAGUAY	570 = LESOTHO
155 = CHILI	571 = BOTSWANA
160 = ARGENTINA	572 = SWAZILAND
165 = URUGUAY	580 = MADAGASCAR
200 = UNITED KINGDOM	590 = MAURITANIA
205 = IRELAND	600 = MOROCCO
210 = NETHERLANDS	615 = ALGERIA
211 = BELGIUM	616 = TUNISIA
212 = LUXEMBOURG	620 = LIBYA
220 = FRANCE	625 = SUDAN
230 = SPAIN	630 = IRAN
235 = PORTUGAL	640 = TURKEY
290 = POLAND	645 = IRAQ
305 = AUSTRIA	651 = UNITED ARAB REPUBLIC
310 = HUNGARY	652 = SYRIA
315 = CZECHOSLOVAKIA	660 = LEBANON
325 = ITALY	663 = JORDAN
338 = MALTA	666 = ISRAEL
339 = ALBANIA	670 = SAUDI ARABIA
345 = YUGOSLAVIA	678 = YEMEN
350 = GREECE	680 = SOUTHERN YEMEN
352 = CYPRUS	690 = KUWAIT
355 = BULGARIA	700 = AFGANISTAN
360 = ROMANIA	712 = MONGOLIA
365 = USSR	713 = CHINA
369 = UKRAINE	740 = JAPAN
370 = BELORUSSIA	750 = INDIA
375 = FINLAND	770 = PAKISTAN
380 = SWEDEN	775 = BURMA
385 = NORWAY	780 = CEYLON
390 = DENMARK	781 = MALDIVE ISLANDS
395 = ICELAND	790 = NEPAL
420 = GAMBIA	800 = THAILAND
432 = MALI	811 = CAMBODIA
433 = SENEGAL	812 = LAOS
434 = DAHOMEY	820 = MALAYASIA
435 = MAURITIUS	830 = SINGAPORE
436 = NIGER	840 = PHILIPPINES
437 = IVORY COAST	850 = INDONESIA
438 = GUINEA	900 = AUSTRALIA
439 = UPPER VOLTA	920 = NEW ZEALAND

FOR VARIABLES 2 TO 51 THE CODE IS
0 = UNKNOWN
1 = YES
2 = ABSTAIN
3 = NO
5 = ABSENT

2	4	X.	KOR1 - ADOPT USSR PROPOSAL TO DELETE ITEM ON KOREA UNIFICATION
3	5	X.	UK1 - CALL UPON 15 TO USE FORCE AGAINST RHODESIA
4	6	X.	CHI1 - DECLARE THE CHINA ADMISSION QUESTION AN

				IMPORTANT QUESTION
5	7	X.	CHI2	- RECOGNIZE MAINLAND CHINA AND EXPEL FORMOSA
6	8	X.	CHI3	- MAKE STUDY COMMISSION ON CHINA ADMISSION IMPORTANT
7	9	X.	CHI4	- FORM A STUDY COMMISSION ON CHINA ADMISSION
8	10	X.	HUM1	- CONVENTION ON NO STATUTORY LIMITS ON WAR CRIMES AGAINST HUMANITY
9	11	X.	POR1	- CONDEMN PORTUGUESE COLONIALISM AND URGE NATO TO WITHHOLD MILITARY AID
10	12	X.	SOA1	- DEFER CONSIDERATION OF SOUTH AFRICAN EXPULSION FROM UNCTAD
11	13	X.	SOA2	- SOUTH AFRICAN EXPULSION FROM UNCTAD AS IMPORTANT QUESTION
12	14	X.	SOA3	- ADD HUNGARIAN PREAMBLE TO SOUTH AFRICAN EXPULSION
13	15	X.	SOA4	- REPLACE LAST PARAGRAPHS OF PREAMBLE BY HUNGARIAN AMENDMENT
14	16	X.	SOA5	- HUNGARIAN AMENDMENT OF PARAGRAPHS 1 AND 2 OF SOUTH AFRICAN EXPULSION
15	17	X.	SOA6	- MINOR AMENDMENT TO UNCTAD RESOLUTION
16	18	X.	SOA7	- ADOPT SOUTH AFRICAN EXPULSION
17	19	X.	SOA8	- CONDEMN SOUTH AFRICA FOR OCCUPATION OF NAMBIA
18	20	X.	UK2	- CALL ON UNITED KINGDOM TO IMPLEMENT RESOLUTION ON OMAN AND CALL FOR REPORT TO GENERAL ASSEMBLY
19	21	X.	UK3	- SINGLE OUT UNITED KINGDOM AND PORTUGAL FOR THEIR FAILURE TO REPORT TO UN ON COLONIES
20	22	X.	COL1	- ADOPT RESOLUTION ON FOREIGN INTERESTS IMPEDING COLONIAL INDEPENDENCE
21	23	X.	COL2	- REQUEST INTERNATIONAL BANK FOR RECONSTRUCTION AND DEVELOPMENT (IBRD) AND INTERNATIONAL MONETARY FUND (IMF) TO IMPLEMENT RESOLUTION ON COLONIAL INDEPENDENCE
22	24	X.	POR2	- APPEAL TO IBRD AND IMF TO WITHHOLD ALL AID TO PORTUGAL AND SOUTH AFRICA
23	25	X.	POR3	- APPEAL TO IBRD TO WITHDRAW LOANS TO PORTUGAL AND SOUTH AFRICA
24	26	X.	POR4	- ADOPT GENERAL RESOLUTION ON ABOVE ITEMS REGARDING IBRD AND IMF
25	27	X.	PAP1	- RIGHT OF PAPUA AND NEW GUINEA TO INDEPENDENCE (IN PRINCIPLE)
26	28	X.	PAP2	- CALL FOR POWERS TO TURN GOVERNMENT OVER TO PAPUA AND NEW GUINEA
27	29	X.	MOR1	- TRANSFER OF IFNI TO MOROCCO AND SPANISH SAHARA STUDY
28	30	X.	INT1	- INTEGRITY OF 24 TERRITORIES
29	31	X.	INT2	- INTEGRITY OF 24 TERRITORIES
30	32	X.	SOA9	- APPEAL FOR CASH FOR SOUTH AFRICAN EDUCATION PROGRAM
31	33	X.	GIB1	- FREE GIBRALTAR
32	34	X.	NUC1	- CONFERENCE OF NON-NUCLEAR STATES
33	35	X.	NUC2	- CONFERENCE ON NON-NUCLEAR STATES AND BAN ON WEAPONS IN LATIN AMERICA
34	36	X.	NUC3	- INTERNATIONAL ATOMIC ENERGY PROJECT ON PEACEFUL USES OF NUCLEAR EXPLOSIVES
35	37	X.	DIS1	- DISARMAMENT CALL
36	38	X.	KOR2	- CALL FOR EASED TENSIONS IN KOREA
37	39	X.	KOR3	- REAFFIRM THE UN MISSION IN KOREA

```
38      40     X.    COL3 - MISCELLANEOUS ON COMMITTEE FOR ANNIVERSARY
                            OF COLONIAL INDEPENDENCE
39      41     X.    COL4 - MISCELLANEOUS ON COMMITTEE FOR ANNIVERSARY
                            OF COLONIAL INDEPENDENCE
40      42     X.    SEA1 - PEACEFUL USES OF SEABED
41      43     X.    SEA2 - PEACEFUL USES OF SEABED
42      44     X.    MON1 - CALL FOR ORGANIZATIONS TO USE THE SAME
                            METHOD OF ASSESSMENT AS THE UN
43      45     X.    MON2 - PATTERN OF CONFERENCES
44      46     X.    MON3 - UN BOND AMORTIZATION STUDY AS AN IMPORTANT
                            QUESTION
45      47     X.    MON4 - ADOPT PROPOSAL FOR STUDY OF BOND
                            AMORTIZATION
46      48     X.    RUS1 - MAKE RESOLUTION ON RUSSIAN AS WORKING
                            LANGUAGE IN GENERAL ASSEMBLY AN IMPORTANT
                            QUESTION
47      49     X.    RUS2 - ADOPT RESOLUTION ON RUSSIAN LANGUAGE
48      50     X.    RUS3 - ADOPT PARAGRAPH ON RUSSIAN AS WORKING
                            LANGUAGE IN SECURITY COUNCIL
49      51     X.    RUS4 - ADOPT PARAGRAPH ON RUSSIAN AS WORKING
                            LANGUAGE IN SECURITY COUNCIL
50      52     X.    RUS5 - ADOPT WHOLE RESOLUTION ON RUSSIAN LANGUAGE
51      53     X.    RUS6 - PERSONAL LINGUISTIC REQUIREMENT IN UN
                            HIRING
52   54-59 XXXXXX  NATION (ALPHABETIC)

FORMAT IS (F3.0,50F1.0,A4,A2)      N = 126
```

LOCATION OF DATA

```
CARD IMAGE - FS.C073.CUNIT1
```

LISTING OF DATA (THE FIRST TWO LINES ARE COLUMN GUIDES)

```
0          1          2          3          4          5
123 4567890123 4567890123 4567890123 4567890123 4567890123 456789

    KUCCCCHPSS SSSSSSUUCC PPPPPMIISG NNNDKKCCSS MMMMRRRRRR
    OKHHHHUOOO OOOOOOKKOO OOOAAONNOI UUUIOOOOEE OOOOUUUUUU
    RIIIIIMRAA AAAAAA23LL RRRPPRTTAB CCCSRRLLAA NNNNSSSSSS
    1 12341112 345678   12 2341211291 1231233412 1234123456

002 3313313211 3331323222 3333113212 1121112112 3313133331 UN STA
020 3212322111 3331323322 2323113213 1121111112 1313311111 CANADA
040 1531131233 5553151525 5555515551 2222332222 5532311111 CUBA
041 5113312121 5551115551 1212111111 1111115511 1122111211 HAITI
042 1113135515 2221211112 2311111111 1111112211 1113155551 DOMINI
051 3112312133 2221112111 1212112113 1111112211 1121211221 JAMACA
052 3112325133 3331111111 1212112212 1111112211 1121211221 TRINID
053 5123215533 2235111111 1212112313 1111112211 1113211331 BARBAD
070 3113311221 2221112112 2212122112 1111112211 1131311111 MEXICO
090 3113312111 2111211112 3211211511 1111112211 1131311111 GUATAM
091 3213233511 3333311512 3332211111 1111112211 1153211311 HONDUR
092 3213133111 3333311111 1211111111 1111111211 1213311111 ELSALV
093 3213322511 3333315555 5555515551 1111115555 1553111315 NICARA
094 3113322511 2225552512 2235515551 5555112211 1555211315 COSTA
095 3113122511 3331311112 2212115511 1111112211 1113111111 PANAMA
100 5213312111 2221212112 2211111111 5511112211 1131211111 COLOMB
```

```
101 3113312122 2221111112 2311212111 1111112211 1222211211 VENEZU
110 5112222533 2251111111 1211311113 1511112211 1121211221 GUYANA
130 3522312255 5555511111 1511111111 5555552255 1122311111 ECUADO
135 3113122211 2222211112 2211115511 1111112211 1122311111 PERU
140 3213312311 2222312323 3331112111 1111115511 1131311111 BRAZIL
145 5113322111 3335312112 2212112111 1111112211 1112111321 BOLIVI
150 3213135111 2225312112 2221212111 1111115511 1123111321 PARAGU
155 3113311121 2221115152 2311111111 1155112111 1122311111 CHILI
160 3113322111 2221211112 2321111111 1111112211 1131311211 ARGENT
165 3513312111 3335315555 5551111111 5555112211 1155211111 URUGUA
200 3311323211 3331325222 3333113213 1131112212 1313122331 UN KIN
205 3213312111 3331312122 1322115112 1111112212 1313133331 IRELAN
210 3312312211 3331323222 2323113212 1111112211 1213122331 NETHER
211 3313312211 3331323222 2323113212 1121112212 1213122331 BELGIU
212 3313312211 3331323252 2323113212 1111552212 5213122331 LUXEMB
220 3231232211 3331322222 2322222222 1222112212 1222311111 FRANCE
230 3213112221 5555511112 2221211111 1111212211 1121311111 SPAIN
235 5322223311 3331333333 3333355535 1111555512 1331112331 PORTUG
290 1131231122 1113211111 1111311111 3231331113 3231311111 POLAND
305 3222222111 3331323222 2223112212 1111112211 1513221221 AUSTRI
310 1131131122 1113211111 1111311111 3231331123 3231311111 HUNGAR
315 1131231122 1113211111 1111311111 3231331113 3231311111 CZECHO
325 3213312211 3331323222 3323113212 1111112112 1313122331 ITALY
338 5513325511 2225255555 5555555555 1111115511 1113153331 MALTA
339 1531135533 5552155555 5555555555 5555335555 5535555555 ALBANI
345 1131131133 3331111111 1111211111 1111231111 1131311111 YUGOSL
350 3113322111 3221311112 2322113211 1121112211 1113211221 GREECE
352 3112311121 3332211111 1111111111 1111112211 1122211111 CYPRUS
355 1131131122 1113211111 1111311111 3231331113 3131311111 BULGAR
360 1131131122 1113211111 1111311111 1211231113 1232311111 ROMANI
365 1131231122 1113211111 1111311111 3231331123 3231311111 USSR
369 1131231122 1113211111 1111311111 3231231123 3231311111 UKRANI
370 1131231122 1113211111 1111311111 3231331123 3231311111 BSSR
375 2231232111 3335323222 2323112212 1111222211 1213211221 FINLAN
380 3231232111 3335313222 2323113213 1111112211 1213121221 SWEDEN
385 3231322111 3331323222 2323113212 1121112211 1213122221 NORWAY
390 3231232111 3331323222 2323113213 1121112211 1213122221 DENMAR
395 3212312111 3331323222 2323113212 1111111211 1213122221 ICELAN
420 3513235555 5555555555 5555555555 1121111155 5555555555 GAMBIA
432 2131135133 2231111111 1111311111 1111332111 1132311111 MALI
433 2112221133 3331112112 2211211112 1111112111 1122211111 SENEGA
434 5113131133 3332111111 1112211111 1111111111 1122122211 DAHOME
435 2131131133 3332111111 1111311111 2222332111 1122311111 MAURIT
436 3113131133 3331115555 5511311112 1121111111 1122222221 NIGER
437 3113131125 5555511112 1211112111 1121111111 1113222221 IVORYC
438 2131131133 3335111111 1111311111 2222333112 1121351111 GUINEA
439 2113131133 3331111111 1111311112 5555123312 1122222221 UP VOL
440 0012125533 3331111111 1111311111 5555555521 1123311111 EQ GUI
450 3113221533 3331112111 1213111113 1111112211 1113132221 LIBERI
451 5513215133 3331112112 1211111113 1111221111 1121211111 SIERRA
452 3132131133 3332111111 1511211112 1111221111 1113321222 GHANA
461 3113121133 3331111512 1111211112 1111112211 1113211111 TOGO
471 3113135133 3333111111 1111311112 1151222511 1122311111 CAMERO
475 2132131133 3331115555 5221252212 1111222211 1131211111 NIGERI
481 2113221533 3331115555 5551111112 1111112211 1112153335 GABON
482 3113131511 2322211112 2212112112 1222111211 1122122221 CENAFR
483 3113131135 5555551111 1211112111 1111555511 1122211111 CHAD
484 1131135133 3331155555 5555551111 1121332212 1122311111 CONGOB
490 2113135133 2231111111 1111311111 1111112211 1115122221 CONGOD
500 2131135133 3331111111 1111311112 1121222211 1131311111 UGANDA
```

```
501 2131131133 3331111111 1111312112 1111552111 1131311111 KENYA
510 2131131133 2221111111 1111311111 1112552111 1131311111 TANZAN
516 2131135533 1331111111 1111311111 1111232211 1122311111 BURUND
517 3113131133 3331111111 1211311111 1121112111 1113111211 RWANDA
520 2131555533 3331111111 1111311111 1211553311 1132311111 SOMALI
530 3131131133 3521112111 1211111112 1111111111 1122211111 ETHIOP
551 5131131133 3331115555 5511311111 1111222211 1155311111 ZAMBIA
553 3213135211 3331353223 3322115213 2222112212 1113132331 MALAWI
560 3313133311 3331333333 3333223335 1121112212 1313122331 SO AFR
570 3513135511 3331352155 5552111113 5555112211 1113232221 LESOTH
571 5513135511 3331325555 5552155553 5555115255 5555555555 BOTSWA
572 5513125511 3331352155 5225512512 5555111211 1113555521 SWAZIL
580 3113125155 5555511111 1211112112 1111111112 1113111221 MADAGA
590 3112215511 2222212111 2211111113 1111115511 1122121211 MAURIT
600 2121221133 3332211112 2211311111 1111115511 1113211211 MOROCC
615 2131131133 3331111111 1111311111 1121333311 1122311111 ALGERI
616 3132311133 3331111111 1211211111 1111221111 1113111221 TUNISI
620 2112221133 3332115555 5551311111 1111225555 1113311111 LIBYA
625 1131131133 3332111111 1111311111 1111233112 1121311111 SUDAN
630 5112221121 2221212112 1211112111 1111111111 1112211111 IRAN
640 3113312121 2222211112 3321111111 1111112211 1223211111 TURKEY
645 1131131133 2221111111 1111311111 1121231155 1122311111 IRAQ
651 2131131133 3331111111 1111311111 1121233112 1131311111 UAR
652 1131131133 2221111111 1111311111 2121233112 1131311111 SYRIA
660 3112211125 5555511112 2211111111 1111222111 1113311111 LEBANO
663 2113135133 2221111111 1111351111 1121222112 1132311111 JORDAN
666 3113311121 5555513112 2312112112 1121112212 1213111311 ISRAEL
670 2113121153 3332111112 2211312111 1111552211 1113311111 SAUDIA
678 1131135533 3331151511 1111311111 1121333211 1131311111 YEMEN
680 5131131135 5331111111 1111311111 1111133311 1131311111 SO YEM
690 2132131133 2321155555 5555555555 1111555511 1123321111 KUWAIT
700 2131132131 2331111112 2211212111 1111222211 1121211111 AFGANI
712 1131131522 1113211111 1111311111 1231332113 3231311111 MONGOL
713 3113131111 5331212112 2312112111 1121112212 3213123331 CHINA
740 3213312111 3331312112 2322113212 1111112211 1213122332 JAPAN
750 2131131133 3331112111 1115111112 1111221111 1121311111 INDIA
770 2131131533 2332111111 1211111111 1111221111 1121311111 PAKIST
775 2131231115 5555512111 1215511115 1111222211 1222311111 BURMA
780 2131131122 2222211111 2212211112 1111221111 1121211111 CEYLON
781 3112211121 2222212112 2223112213 1111112211 1112221111 MALDIV
790 2131131111 2222212111 1111111112 1111225511 1112311111 NEPAL
800 3113132133 3331112112 2212112112 1111112211 1113221111 THAILA
811 1531135133 2221111111 1111511111 5555232222 5125311111 CAMBOD
812 3112212111 1222212111 2212212112 1111112211 1112211111 LAOS
820 3113221111 1112212111 2212112113 1111112211 1113222221 MALAYS
830 2132231122 2222211111 2212112113 1111222211 1113311111 SINGAP
840 3113131131 2231111111 1311112111 1121112211 1113111211 PHILIP
850 2155551132 3331111111 1211111111 1111122211 1113211111 INDONE
900 3313133211 3331323222 2323113213 1121111112 1313133331 AUSTRA
920 3313112211 3331323222 2323113213 1121112212 1313122331 NEW ZE
```

REFERENCES

Alker, H. R. (1964). Dimensions of conflict in the General Assembly. Amer. Pol. Sci. Rev. 58, 642-657.

Ball, M. M. (1951). Bloc voting in the General Assembly. Internat. Organ. 5, 3-31.

Hartigan, J. A. (1972). Direct clustering of a data matrix. J. Amer. Statist. Assoc. 67, 123-129.

Hovet, T. (1960). Bloc Politics in the United Nations. Cambridge, Harvard University Press.

Lijphart, A. (1963). The analysis of voting in the General Assembly. Amer. Pol. Sci. Rev. 57, 902-917.

Riggs, R. E. (1958). Politics in the United Nations. Champaign, University of Illinois Press.

Russett, B. M. (1966). Discovering voting groups in the United Nations. Amer. Pol. Sci. Rev. 66, 327-339.

DISCUSSION OF THE HARTIGAN CHAPTER

L. Jaeckel

This is a simple-minded technique that can be applied early in the game when not much is yet known about the data to see if there are any gross overall patterns in the data. Cluster analysis should and does get at some of the fine structure; the principal components analysis is not inconsistent with the results of the cluster analysis.

I changed all 5's and 0's to 2's, chose 25 of the 50 questions voted on (every other one, beginning with the second in table 1).

and then ran the data through the standard BMD program. (I also ran the other 25 questions, with nearly the same results.) The first few eigenvalues were: 8.63, 3.77, 2.27, 1.62, 1.02, 0.96, 0.91. (The sum of the eigenvalues is of course 25.)

The first principal component ordered the countries more or less by ideology, with the Communist bloc on one end and the Western powers on the other. No single question or group of questions contributed greatly to the first eigenvector; all of its components were small. The ordering of the countries by the first principal component (omitting the jumble in the middle) is:

Hungary, USSR, Byelorussian SSR, Bulgaria, Poland, Czechoslovakia, Ukranian SSR, Mongolia, Syria, Iraq, Cambodia, Romania, Tanzania, Mauritania, Guinea, Yemen, Mali, Algeria, Yugoslavia, UAR,

Canada, Austria, Malta, Sweden, Denmark, Norway, Japan, Malawi, Ireland, Portugal, Luxembourg, New Zealand, Iceland, Belgium, Netherlands, Australia, Italy, United Kingdom, USA, South Africa.

In the ordering of the countries by the second principal component, the countries at the two extremes of the ordering given above were mixed together at one end. This seems to indicate some curvature in the cloud of points in the 25-dimensional space. Nothing was apparent in the other principal components.

Although principal components is too crude a method for the detailed structure here, it suggests that there may be aspects of the data that cannot be exhibited by cluster analysis. For example, suppose the Communist bloc forms a cluster, and within this cluster there are pro-Russian and pro-Chinese tendencies. It may turn out that because of this the cluster breaks into two subclusters; if so, cluster analysis will exhibit the split. But it may be that the cluster is more like a ball with pro-Russian countries tending to be on one side and pro-Chinese countries on the other side, with no clear division into subgroups. In this case some other method of analysis, applied to this cluster only, might bring out its internal structure. In either case we might ask whether the pro-Russian or the pro-Chinese part of the cluster is "closer" to some other cluster, and by how much.

Standard cluster analysis methods do not answer questions of this kind. (Perhaps some form of multidimensional scaling should be applied to the data.) The point I want to make is that after doing the cluster analysis, we should look again at the data, in the light of the clustering results, to see where to go from there.

R. Moore

Even though the presentation and special techniques were lauded when this paper was presented at the Conference, some concern was expressed by attendees that the subject matter and purpose of the investigation did not come through clearly enough. This condition was rectified in the proceedings by an expanded introduction describing other attempts at determining associations from international voting patterns.

Nonetheless, the original presentation had the flavor of something we all have run into: A researcher appears in the door of the data analyst's office, dumps a mountain of paper on the desk, and says in effect, "Quick, please, a miracle! I need some numbers to fill in the last four lines of a paper I'm scheduled to give in the morning!"

What's a data analyst to do in a situation like this? In my opinion, he is better off doing nothing with large masses of data when he is under duress than he would be if he attempted to meet unrealistically short deadlines. Large masses of data require time and effort to yield their messages. Tools such as the clustering techniques presented in this paper are indispensable in giving the subject matter specialist and the data analyst something to mull over — but the conclusions do not necessarily come quickly just because a computer output is available.

More specifically, I find little reason to criticize Hartigan's treatment. He has used real data, he has applied an analytical tool to examine the data, and he has presented some conclusions. Given the BMDP clustering computer programs, an analyst can follow Hartigan's methodology with almost any set of data.

The fact that the conclusions in the paper are obvious (i.e., that there are political blocs in the U.N. and that several questions are rephrasings of the same question) is not disturbing. Indeed, it is comforting to find recognizable patterns in such data; it keeps alive the hope that surprising conclusions may some day lead to basic breakthroughs in understanding a particular phenomenon.

CHAPTER 3

CRAWFORD HILL RAINFALL DATA

LOUIS A. JAECKEL
Bell Telephone Laboratories, Incorporated, Murray Hill, New Jersey

and

JOHN D. GABBE
Bell Telephone Laboratories, Incorporated, Murray Hill, New Jersey

INTRODUCTION AND SUMMARY OF PREVIOUS WORK

The principal purposes of the rainfall experiment conducted by Bell Telephone Laboratories in the Crawford Hill, New Jersey area during the last several years were to explore the relationship between heavy rainfall and the attenuation of microwave radio transmission in the 10 to 30 GHz frequency range, and to obtain information about the spatial and temporal characteristics of the rain. The feasibility of transmission systems in the 10 to 30 GHz frequency range is of interest to the Bell System because the microwave radio spectrum below 10 GHz is becoming crowded. Heavy rainfall, which has little effect below 10 GHz, can substantially degrade transmission in the higher-frequency range.

The principal mechanism of attenuation at these frequencies is absorption (there is some scattering) (Medhurst, 1965), and the amount of absorption depends on the quantity of water that the signal must penetrate along the path from the transmitter to the receiver. This in turn depends on the rate at which rain is falling through the path.

The plan of the experiment was to

- explore the relationship between rain rate and attenuation by measuring both simultaneously along a path;

- examine the spatial distribution, especially the correlation of rain rates along nearby paths, by simultaneously measuring rain rates in a grid (summer thunderstorms, typical of very heavy rain, form cells approximately 1 km in extent which last about 20 minutes — thus, while transmission on one path is blocked, nearby alternate routes may be open); and to

- relate the path rain rates to point rain rates with the objective of extending the results of the experiment to other parts of the country for which point (single-gauge) rain rate measurements could be obtained.

Apparatus and Data Collection

Ninety-six rain gauges capable of measuring rain rate continuously were built and placed in a grid as shown in figure 1 (Semplak, 1966; Semplak and Keller, 1969). The grid squares are about 1.3 km on a side. The gauges were mounted on poles about 7.5 m above the ground. The area of the gauge is 478 cm (Semplak and Keller, 1969), and the response time constant is somewhat less than one second. This version of the gauge was not designed to record rain rates of less than 10 mm/hr accurately.

During periods of rain the network is scanned once every ten seconds. Five pairs of gauges are read every second in the order given in figure 1. In addition the reading of gauge 33 is recorded every second as is the attenuation of the microwave signals on the path T-R. Time, wind velocity and scan number are also recorded at one- or two-second intervals.

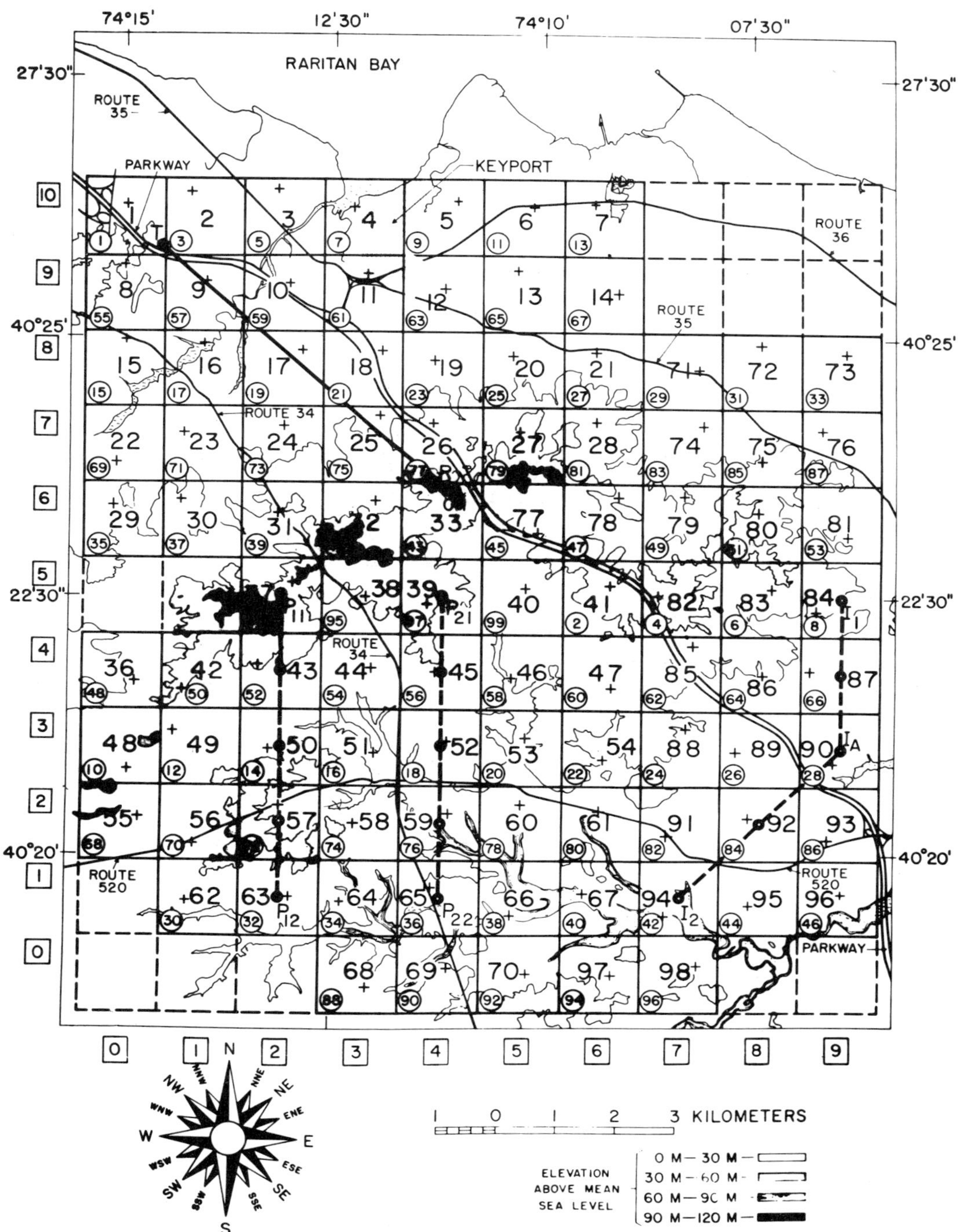

Figure 1. The rain gauge network, showing the station numbers, scanning order (small circled numbers), and row numbers (in squares). The grid spacing is approximately 1.3 km. Station 33 (CH) is at Crawford Hill, New Jersey. The solid line between the transmitter (T) and receiver (R) is an experimental microwave transmission path.

The equipment worked reasonably well. More than 80% of the gauges were operational most of the time. The telemetry was somewhat noisy on occasion, and there is some evidence that the gauge calibrations drifted somewhat.

Data Inspection and Selection

Isometric plots (Freeny and Gabbe, 1969), shown in figure 2, were used to aid in selecting 110 hours of "interesting" rain from 1967 data and 87 hours from 1968 data. Rain is interesting if several gauges show readings above 35 mm/hr over a half-hour period. Isometric plots were also used to help identify malfunctioning gauges, and histograms and probability plots were used to identify outliers.

The selected data contain measurements on 70 rainfalls totaling about 3.4×10^6 observations from 1967 and 2.7×10^6 from 1968. Three subsets of these data are on file at the Health Sciences Computing Facility (HSCF), University of California, Los Angeles, all from the 1967 rainfalls:

- the gauge data, which associate the gauge number with the scanning order, the nominal X, Y coordinates, and the true X, Y coordinates (see figure 1);
- the storm data, which contain 30-second (3-reading) averages of the rain rates of each gauge (and also of the attenuation) for 1967 rainfalls — 7/21, 7/25, 7/28A, 8/9, 9/28, 10/25 and 11/3; and
- the attenuation data, which contain all the rain rates for the four rain gauges on the microwave signal path and 10-second (10-reading) averages of the attenuation readings for 1967 rainfalls — 7/14B, 7/28B, 8/25A, 9/21 and 10/10.

References to these rainfalls are marked "*" (storm data), or "#" (attenuation data) in the tables and figures.

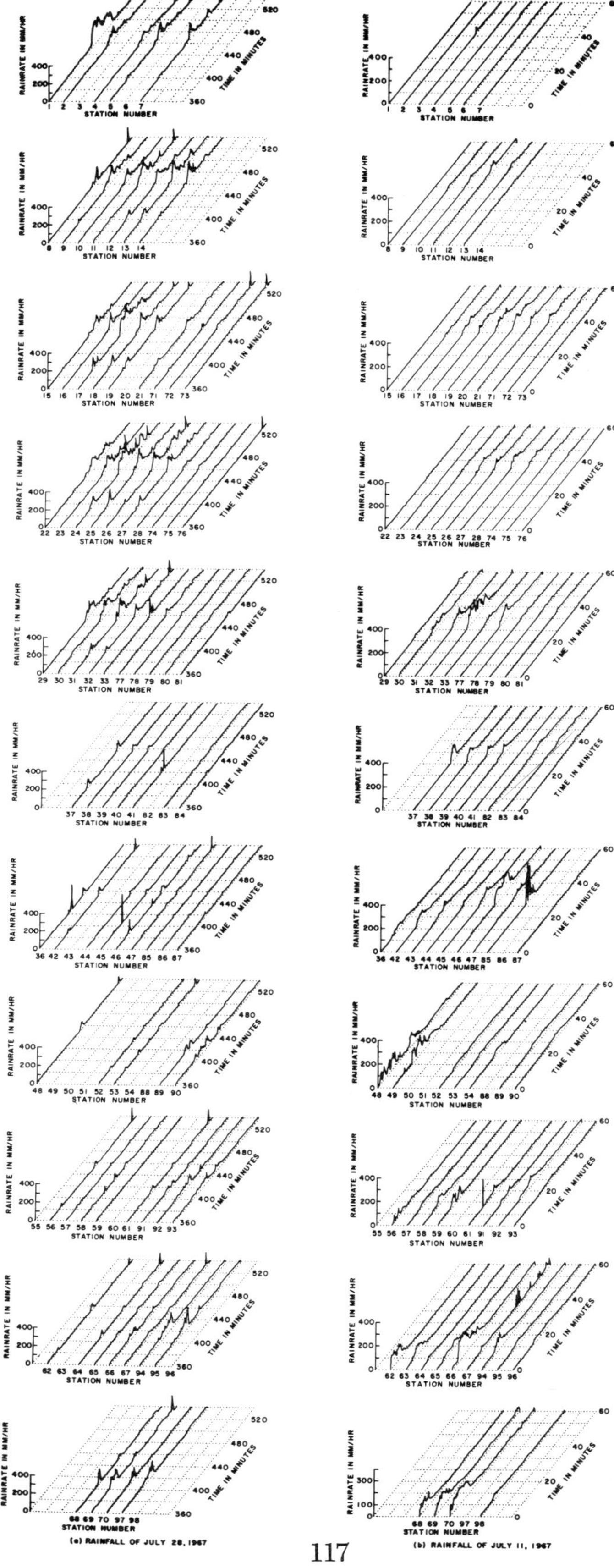

Figure 2. Isometric plots of rainfalls (a) 7/28A, and (b) 7/11. Each trace represents the data from a single station. The positions of the base lines of the traces correspond to the geographic positions of the stations in figure 1. Scales are (a) six-reading averages, and (b) two-reading averages.

A more detailed description of the data accompanies the HSCF file. The Data Set Description at the end of this chapter includes a listing of a portion of this file: data definition, the full set of gauge data, a sample of storm 7/28A rainfall data and attenuation data for storm 9/21.

Previous Analyses and Results

Some aspects of the data were examined fairly extensively by Freeny and Gabbe (1969) and Freeny (1969). The relationship between attenuation and rain rate was first studied by Semplak and Turrin (1969).

Point rain rate distributions for 1967 were examined by Freeny and Gabbe (1969) for each gauge during each rainfall, for all gauges pooled over each rainfall, for each gauge pooled over all rainfalls, and for the grand pool of the data. Heavy rain rates are relatively rare events; they come in irregular bursts and are not amenable to description by simple analytic distributions.

Empirical joint probabilities that rain rates at different times or at different points in space exceed various values were used as measures of the spatial and temporal properties of heavy rain. Similarly, empirical joint probabilities that the average rain rates along pairs of paths exceed various values were used as measures of the benefits of alternate routing of microwave signals.

Dispersion among subsets of the data was used as an informal measure of variation.

Current Work

Current work on the 1967 rainfall data is concentrated on three aspects of the problem: estimation of the drift of a storm across the network; correlation of rainfall intensity with attenuation of a microwave signal transmitted through a part of the network; and a method for giving a compact description of a high-rain-rate event.

DRIFT AND ATTENUATION

Selection of Data, Truncation and Smoothing

We were confronted with a very large quantity of noisy rainfall data; consequently, we had to select subsets of the data which we considered interesting and focus our attention on them. These subsets consist of data covering segments of time ranging from about 15 minutes to two hours, during which a large number of gauges recorded a significant amount of rain.

The selection procedure was essentially subjective. It consisted of examining the isometric plots visually to find periods of substantial rainfall, and then choosing initial and final times for the chosen period so the time segment included a little time before the storm began and a little after it died down. Some 20 or 30 periods of rainfall were chosen for study by this method.

We then eliminated the data from gauges that appeared, from the isometric plots, to be malfunctioning during a storm. Occasionally a gauge which otherwise seemed to function properly would attempt to report a value beyond its range, due either to a sudden fluctuation in the rain or to a momentary malfunction of the equipment. In either case, we felt that the recorded number probably did not accurately represent the rain rate at that moment. In the drift analysis, these events were omitted from the data. In the attenuation analysis, they were omitted only if the numbers were greater than 300 in magnitude; otherwise the number was treated normally. We have no particular reason for this difference in treatment. In any case these events were very infrequent (no outlier events occurred during the storms on file at HSCF).

To limit the effect of outliers, the number 300 was chosen arbitrarily as an upper bound for the rain rates admitted in the analysis: in the drift analysis higher values were changed to 300; in the attenuation analysis, which is more sensitive to the values used, they were omitted. Values above 300 were very infrequent.

Next we considered whether to apply some form of smoothing to the data. Examination of some of the data showed that the values reported by each gauge fluctuated greatly from one reading to the next (each gauge was read once every 10 seconds), but over a time scale of several minutes a longer-term pattern was discernible. In other words, a considerable amount of high-frequency noise was present, along with the more slowly varying pattern of true rainfall, in which we were interested. Some experimentation with 3-reading averages and 6-reading averages convinced us that 3-reading averages would substantially reduce the noise without doing much harm to the underlying rainfall pattern. We felt that six readings were too many to lump together because if the data points are viewed six at a time a trend within the six consecutive points is often apparent.

Therefore, we used 3-reading averages in the drift and attenuation analyses. This choice was not based on any quantitative criterion, but we believe that with these data the exact form of smoothing is not very critical. Note that we did not use true moving averages; we used one 30-second average for each disjoint 30-second period. In cases where a number was omitted, as described in the preceding paragraph, the entire 3-reading average was omitted.

Estimation of Drift

The isometric plots indicate that the pattern of rainfall at adjacent gauges is often similar, but there is a time lag from one gauge to the next. The apparent consistency of these lags suggests that they represent a drift of the rainstorm over the area covered by the gauge network. A procedure for estimating the velocity and direction of the drift was developed and incorporated into a computer program.

We should emphasize that the drift is measured by phenomena observed at ground level. Since we have no direct knowledge of the mechanism in the rain cloud that triggers rainfall, we cannot say whether the observed drift is due to the actual movement of the cloud over the network, or to the propagation through the cloud of physical conditions that cause the rain to fall, or to some combination of these two factors. If we knew the wind speed at the altitude of the cloud, we would know the motion of the cloud. However, the speed and direction of the wind during a rainstorm are very different at different altitudes, so a measurement of wind at ground level would not tell us the motion of the cloud overhead.

To estimate the time lag between two gauges, a simple procedure would be to compute the lagged correlation coefficient of the two series of observations for various amounts of lag and then find the time lag for which the correlation is greatest. This procedure would give a number that (if the second series were shifted forward by that amount) would make the two series look most alike. When more than two gauges are considered, however, the pairwise lags found by this method would generally not be consistent with each other because of random error in the observations. By consistency we mean that the lag from gauge i to gauge j plus the lag from gauge j to gauge k should equal the lag from gauge i to gauge k. Thus we wanted a procedure which would maximize the pairwise lagged correlations, subject to the condition that the pairwise lags be consistent.

Some preliminary examination of the data showed that the farther apart two gauges are, the more tenuous is the relationship between the observed rainfall at the two points. So, rather than consider all possible pairs of gauges, we restricted our attention to pairs of adjacent gauges. Two gauges are adjacent if, on the map, the squares containing them have a common edge or a common vertex; thus each gauge is adjacent to as many as eight others.

We made the hypothesis that the drift is a linear motion with the same direction and velocity at all points in the network, and is constant for the duration of the storm. The drift may then be expressed as a vector (v_x, v_y), whose components are the west-to-east and south-to-north components of the drift. Assuming this, we can calculate the time lag from any point (x_1, y_1) to any other point (x_2, y_2). A little algebra shows this time lag to be

$$\ell(x_1, y_1; x_2, y_2) = \frac{(x_2 - x_1)v_x + (y_2 - y_1)v_y}{v_x^2 + v_y^2} .$$

Now we let

$$\alpha = \frac{v_x}{v_x^2 + v_y^2}$$

and

$$\beta = \frac{v_y}{v_x^2 + v_y^2} ,$$

so the time lag above is

$$\ell(x_1, y_1; x_2, y_2) = (x_2 - x_1)\alpha + (y_2 - y_1)\beta .$$

This set of lags is clearly consistent, since

$$\ell(x_1, y_1; x_2, y_2) + \ell(x_2, y_2; x_3, y_3) = \ell(x_1, y_1; x_3, y_3).$$

Note that α and β are reciprocals of velocities; they represent the time it takes for the storm to move one unit of distance eastward or northward. Given α and β, we can compute the velocity and direction of the drift. We have

$$v = \sqrt{v_x^2 + v_y^2} = \frac{1}{\sqrt{\alpha^2 + \beta^2}}$$

and

$$\theta = \tan^{-1}\left(\frac{\alpha}{\beta}\right) ,$$

where θ is measured clockwise from north.

The goal of our procedure was to estimate α and β. From these estimates we could then compute estimates of v and θ by the formulas above.

The procedure works as follows: For each adjacent pair, say gauges i and j, we compute $\rho_{ij}(\ell)$, the correlation of the rain rates at gauges i and j with lag ℓ; that is, the correlation of $R_i(t)$ and $R_j(t+\ell)$. If the coordinates of these gauges are (x_i, y_i) and (x_j, y_j), then for given α and β the time lag from gauge i to gauge j is

$$\ell_{ij}(\alpha,\beta) = (x_j - x_i)\alpha + (y_j - y_i)\beta.$$

Since we could not maximize all of the pairwise correlations simultaneously, we maximized some aggregate quantity, so, on the whole, the correlations were as large as possible. To do this we formed the sum

$$S(\alpha,\beta) = \sum_i \sum_j \rho_{ij}[\ell_{ij}(\alpha,\beta)],$$

where the sum is taken over adjacent pairs only. We then estimated α and β by finding the α and β that maximize $S(\alpha, \beta)$. Thus we made the correlations as large as possible, subject to the assumptions on the nature of the drift and to the condition that the set of lags be consistent.

Note that the unit of time here is 30 seconds, and $\rho_{ij}(\ell)$ is defined only for integer values of ℓ. Since the gauges are read sequentially during each 10 second period, corresponding readings of different gauges were taken at slightly different times; a correction term was added to $\ell_{ij}(\alpha,\beta)$ to compensate for this discrepancy.

Our reasons for using correlations, rather than some other function, like the covariance, were as follows. We wanted the information on the drift from each gauge to enter into the computations on an equal basis, even though the intensity of the rainfall during a storm is greater in some parts of the network than in others. In addition, we suspected that some of the gauges were out of calibration and reported rain rates that were consistently too high or too low. Since the correlation is insensitive to location or scale changes in the data from each gauge, its use tended to compensate for the errors in calibration. However, if there was very little activity at a gauge variations in the data might be magnified out of proportion to their true meaning, so we rejected the data from gauges that showed very little activity during the storm. Experimentation with the strictness of the criterion for rejection showed that varying the criterion changed the estimates by about 1%, an amount we consider negligible.

The maximization problem is simplified by the following device. Suppose $\rho_{ij}(\ell)$ can be approximated by a parabola, at least for ℓ in a limited range. If we replace each $\rho_{ij}(\ell)$ by

$$r_{ij}(\ell) = p_{ij}\ell^2 + q_{ij}\ell + s_{ij},$$

which is defined for all ℓ rather than only for discrete values of ℓ, then, letting $u_{ij} = x_j - x_i$ and $v_{ij} = y_j - y_i$, we have

$$r_{ij}[\ell_{ij}(\alpha,\beta)] = r_{ij}(u_{ij}\alpha + v_{ij}\beta)$$

$$= p_{ij}(u_{ij}\alpha + v_{ij}\beta)^2 + q_{ij}(u_{ij}\alpha + v_{ij}\beta) + s_{ij}.$$

Then, omitting the subscripts, $S(\alpha,\beta)$ becomes

$$\alpha^2\Sigma\Sigma pu^2 + 2\alpha\beta\Sigma\Sigma puv + \beta^2\Sigma\Sigma pv^2 + \alpha\Sigma\Sigma qu + \beta\Sigma\Sigma qv + \Sigma\Sigma s$$

$$= A\alpha^2 + 2B\alpha\beta + C\beta^2 + D\alpha + E\beta + F,$$

where these coefficients are defined to be the corresponding sums above. By differentiating, we see that the maximum must occur when

$$2A\alpha + 2B\beta + D = 0$$

and

$$2B\alpha + 2C\beta + E = 0;$$

that is, when

$$\alpha = \frac{BE - CD}{2(AC - B^2)}$$

and

$$\beta = \frac{BD - AE}{2(AC - B^2)} \ .$$

So we can find the maximum of this approximation for the sum of correlations in a simple and direct way.

The maximization procedure is as follows. First, for each pair of gauges used, $\rho(\ell)$ is computed for integer values of ℓ, $|\ell| \leq 12$; that is, for $|\text{lag}| \leq 6$ minutes. Then the ℓ_o which maximizes $\rho(\ell)$ is found, and if $|\ell_o| \leq 10$, a parabola is fitted by least squares to the five values $\rho(\ell_o-2)$, $\rho(\ell_o-1)$, $\rho(\ell_o)$, $\rho(\ell_o+1)$ and $\rho(\ell_o+2)$. Some examples of plots of these correlations are given in figure 3. It can be seen from these plots that fitting such a five-point parabola generally gives a reasonably good fit to $\rho(\ell)$ for a short span of values of ℓ. Now the maximization device described above is applied to these parabolas, temporarily omitting pairs for which a parabola was not computed, giving us a first approximation, say α_1 and β_1, to α and β.

These values give us an estimate of the time lag for each pair: $\ell_{ij}(\alpha_1, \beta_1) = (x_j - x_i)\alpha_1 + (y_j - y_i)\beta_1$. A new five-point parabola is now computed for each pair, centered this time on $\ell_{ij}(\alpha_1, \beta_1)$, and the maximization device is applied again, yielding new estimates α_2 and β_2. The part of the procedure described in this paragraph is iterated several times, after which the estimates of α and β converge.

The velocity and direction of the storm are then computed as explained above and printed out with a variety of other information. Plots are made of the correlation functions for each pair, including the estimated lag for that pair based on the final α and β, and the fitted parabola centered at that lag (see figure 3). For each pair, the loss in the correlation (that is, the difference between the maximum correlation and the correlation at the time lag based on α and β) is computed. Most of these differences are very small; the larger differences tend to be associated with a handful of gauges at which something peculiar must have been happening.

The results of the drift analysis for most of the heavy rainstorms and some of the moderate ones during 1967 are given in table 1.

Drift Estimation for Subsets of Gauges

We attempted to test the accuracy of the drift estimates by running the drift program on various subsets of the gauges functioning during a particular storm and comparing the results.

One attempt was a "pseudojackknife": For each functioning gauge, we ran the drift program using the data from all of the other gauges, but omitting the data from that one gauge. The results of this procedure for storm 7/25 are shown in figure 4. The coordinates are v_x and v_y, the components of the drift vector. For each run, the gauge number of the omitted gauge is plotted on the graph as the drift estimate obtained by omitting that gauge. A zero appears on the graph at the drift estimate using all the functioning gauges. We see from the graph that there is a certain amount of consistency in these estimates. They all fall in an area which is not very large — but not very small, either. Since we do not have independent identically distributed observations, we do not call this a jackknife, and we do not know how to interpret it.

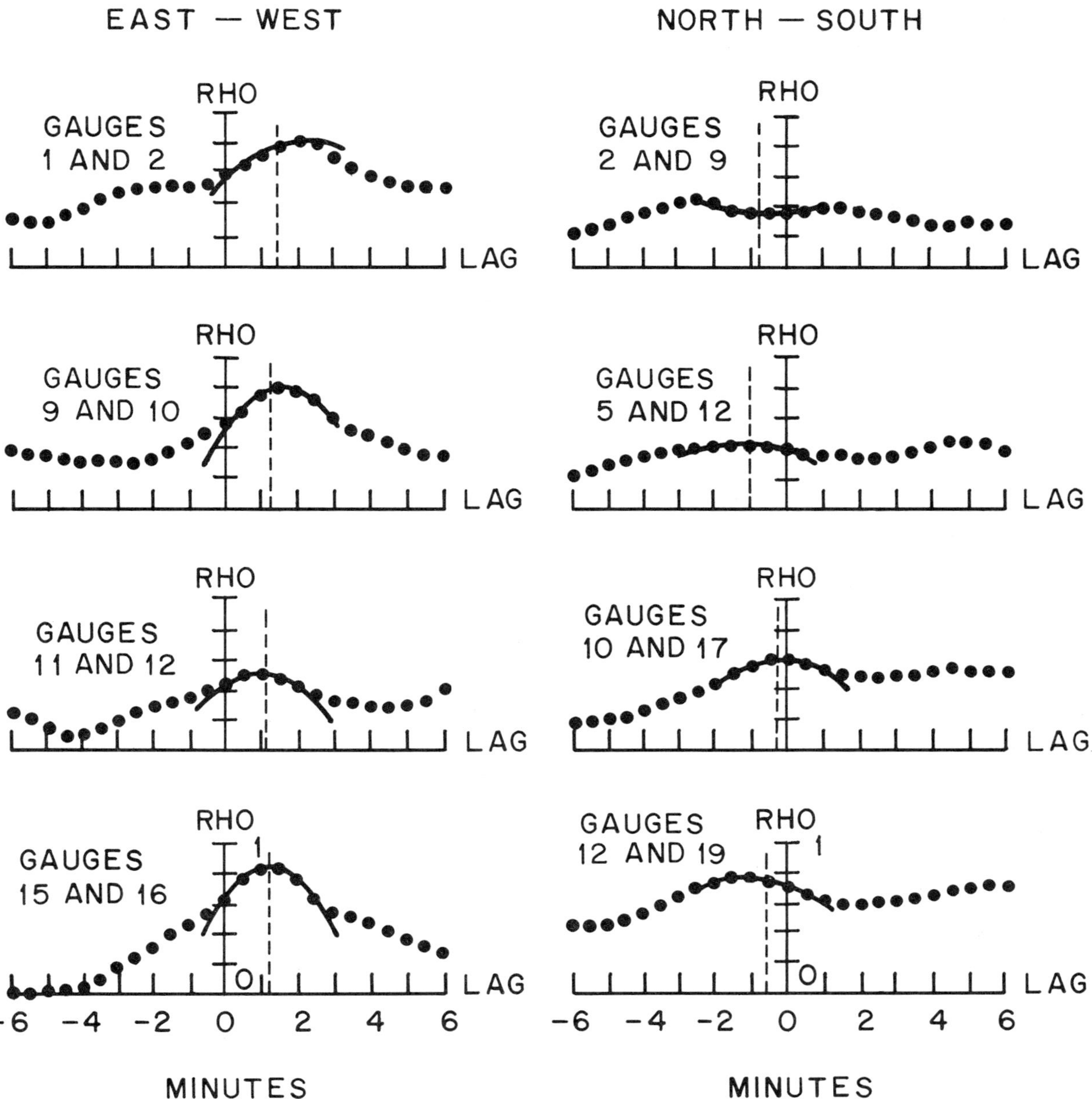

Figure 3. Examples of the pairwise correlations as functions of lag, for storm 7/25*. The column headings refer to the orientation of the gauge pair. The dashed vertical line represents the estimated lag for that pair based on the final estimates of α and β. The parabola is fitted to the values of the correlation function near the dashed line.

*Storm data on file at HSCF.

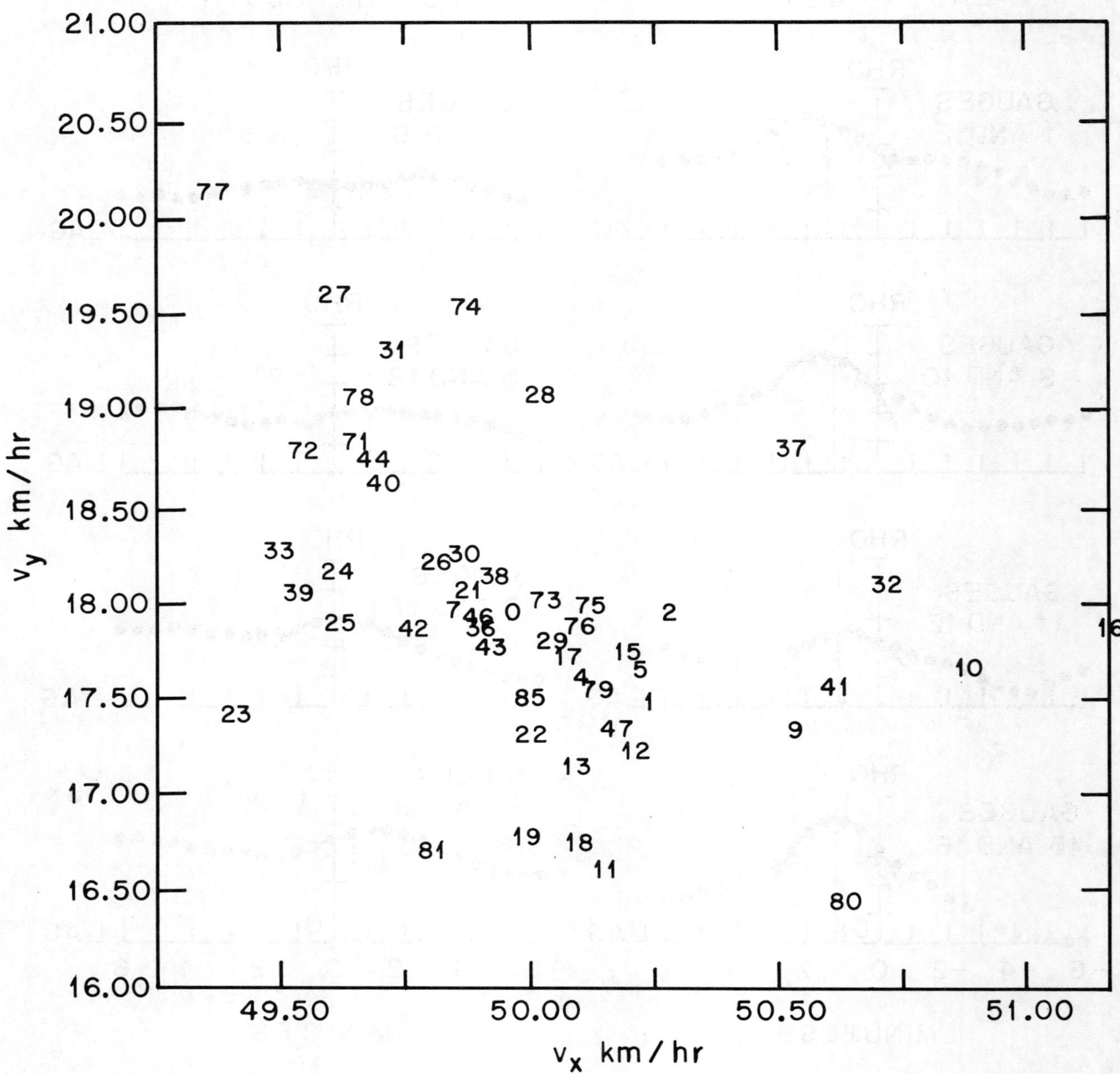

Figure 4. Results of the "pseudojackknife" for rainfall 7/25*. The number plotted identifies the omitted gauge. Only the first seven rows of gauges were used in the calculations.

---

*Storm data on file at HSCF.

Table 1. Estimated drift for rainstorms in 1967. The durations given are the time periods used in the drift analysis.

Date	Duration (minutes)	Velocity (km/hr)	Direction (clockwise from north)
6/29	80	39.7	90.5°
7/3	60	65.1	56.1°
7/14B	80	31.8	62.1°
7/21*	50	37.7	36.9°
7/25*	80	54.7	74.4°
7/28A*	80	58.9	34.1°
7/28B	60	49.9	40.2°
8/9*	70	53.9	81.8°
8/25A	80	36.9	57.7°
9/21	60	82.4	62.4°
9/28*	60	65.2	353.5°
10/10	70	41.2	10.5°
10/18	40	46.0	65.6°
10/25*	60	79.1	20.5°
11/3*	40	48.6	352.3°
11/17	60	55.7	118.0°

*Storm data on file at HSCF.

Various other methods of choosing subsets of the gauges were suggested in the discussions at the Conference. One suggested method, which we tried on the computer at UCLA, was to divide the network into four disjoint quadrants and compute the drift for each quadrant. Some results are given in table 2. The quadrant estimates differ considerably among themselves. It is not clear whether this is due to variability inherent in the procedure or to real differences in the drift velocity over different parts of the network, but we suspect it is the latter.

Table 2. Estimated drifts for the quadrants of the grid for two rainstorms in 1967.

	7/21*		7/28A*	
Quadrant	Velocity (km/hr)	Direction	Velocity (km/hr)	Direction
NW	38.3	22.3°	56.0	26.1°
SW	33.3	36.1°	61.6	47.0°
NE	31.2	48.8°	66.1	35.3°
SE	38.1	51.4°	55.4	47.4°
Entire network	37.5	38.0°	58.9	34.2°

*Storm data on file at HSCF.

We also experimented with the following method of choosing subsets of gauges. A 5 x 5 square made of 25 adjacent grid squares can be extracted from the grid shown in figure 1 in 42 ways. For example, the corners of one such square are at gauges 23, 27, 49 and 53. Although these squares could contain as many as 25 functioning gauges, they generally do not, and sometimes they contain considerably fewer than that. For a given storm we estimated the drift for each of the 42 5 x 5 squares, giving us 42 pairs of numbers. In most cases these numbers changed gradually from one 5 x 5 square to the next because of the large overlap in adjacent 5 x 5 squares. However, the drift estimates for widely separated 5 x 5 squares were often substantially different. Since the four extreme 5 x 5 squares are almost the same as the four regions used in table 2, that table can serve as an example of these differences. Figure 5 gives the 42 drift vectors for one storm (July 28A), arranged in a 7 x 6 array according to the positions of the 5 x 5 squares.

Although we have no quantitative measure of accuracy of the drift estimates, we are confident that the phenomenon is real and that our procedure gives a reasonable estimate of the true drift.

Correlation of Rainfall and Attenuation

A microwave signal at a frequency of 18.5 GHz was transmitted over a 6.4 km path through part of the network. The path, indicated on figure 1, passes near gauges 9, 17, 25 and 33. Measurements of the attenuation in the signal were taken every second and recorded with the rain rate data from the gauges.

It has been known for some time that microwave attenuation is directly related to intensity of rainfall along the transmission path. Some references to the theoretical and empirical work done on this problem are given by Semplak and Turrin (1969). Their paper contains a more detailed description of the equipment used in the experiment and the results of their analysis of the same data that we studied.

Let $A(t)$ be the attenuation at time t, measured in decibels, and let $R_i(t)$, $i = 1, 2, 3, 4$, be the rain rates at gauges 9, 17, 25 and 33, respectively. For each storm we perform a linear regression of the attenuation on these four rain rates. That is, we fit the following equation, containing the five variables b_i, $i = 0, 1, 2, 3, 4$, to the data by least squares:

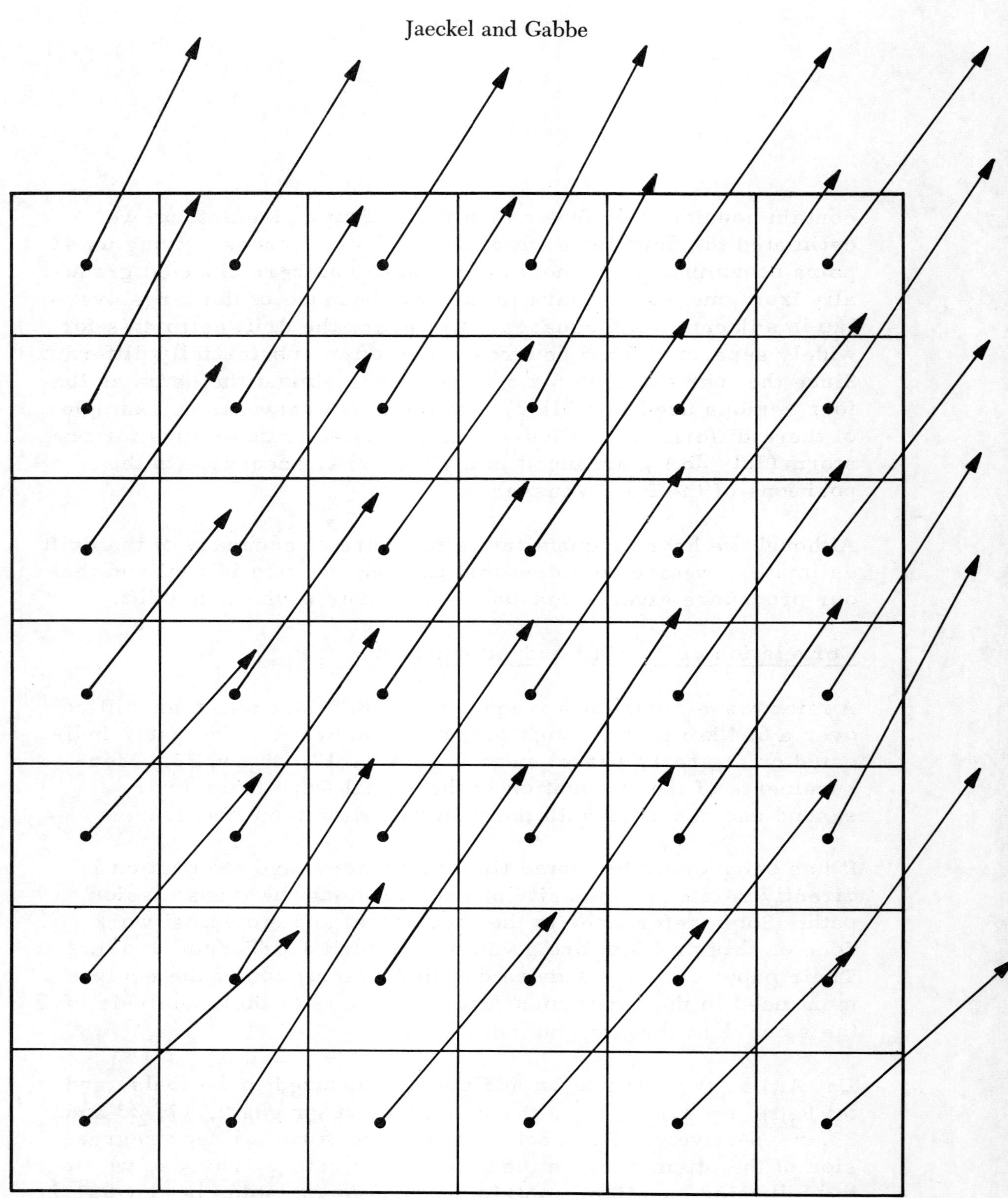

Figure 5. Drift vectors for each of the 42 5 x 5 squares, arranged according to position of the square (storm 7/28A*).

---

*Storm data on file at HSCF.

$$A(t) = b_o + \sum_{i=1}^{4} b_i R_i(t) + \text{error}. \tag{1}$$

We note that theoretically the equation that should be fitted to the data is

$$A(t) = c \sum_{i=1}^{4} w_i [R_i(t)]^{1.09} + \text{error},$$

where the w_i are fixed, depending on the spacing of the gauges, so that c is the only quantity to be estimated. (See Semplak and Turrin, 1969.)

We chose to work with equation (1) involving the estimation of five quantities, for the following reasons: We suspected that some of the gauges are not properly calibrated, or that the error in calibration varies over a period of time. Thus the rain rate reported by a gauge might be significantly greater than or less than the true rain rate. The use of a different regression coefficient for each gauge is intended to compensate somewhat for the lack of calibration.

A few considerations led us to include variable b_o. Due to the nature of the calibration method, each gauge reports a different value when no rain is falling, so zero rain does not mean zero reported rain rate. The curve $y = x^{1.09}$ is nearly linear over a large segment of its range and has a slight hook near the origin; since there is uncertainty about the rain rate measurements when the rain is near zero anyway, we decided to dispense with the exponent and fit a linear function that is not constrained to go through the origin. At any rate, the values of b_o were generally small.

It soon became clear that for most storms gauge 33 did not function properly, so the regression was done both with and without it. During most storms, little rain was reported by this gauge and b_4 was almost always very small, and in a few cases was negative.

Also, the goodness of fit generally was not much improved by including gauge 33, so subsequent analysis was done omitting this gauge.

Computations were performed on 30-second averages, which means 3-reading averages for the rain gauges and 30-reading averages for the attenuation measurements. Rain rates above 300 were omitted. We also omitted from the fitting procedure those time points for which A(t) was less than 2, which we considered insignificant, and those for which A(t) was greater than 45, because above this point the measuring equipment seemed to be overloading. An example of this apparent overloading may be seen in the graph in figure 6 for storm A on July 28. Some experimentation with the upper cutoff point for A(t) showed that the regression coefficients were not very sensitive to its value.

Three graphs were made for each storm: a plot of attenuation and predicted attenuation (that is, $b_o + \Sigma b_i R_i(t)$) as functions of time; a scatter plot of attenuation against predicted attenuation; and a plot of residuals against time. All points in the chosen time segments are plotted on the graphs whether or not they were used in fitting the equation. These graphs for storm A on July 28 are presented in figures 6, 7 and 8.

Examination of the residuals shows a definite pattern of serial correlation. This is probably due largely to the rainfall between the gauges and on the southeastern part of the transmission path, where we do not have measurements. Such unmonitored rainfall can be expected to introduce some bias for as long as a particular pattern of rain persists, typically for a period of a few minutes.

Table 3 gives the principal results from the fitting procedure. The means are averages over the time points actually used in the fits. R^2 is the multiple correlation coefficient, or the proportion of explained variance. The last column is the ratio of attenuation to average rainfall computed as explained below. Three storms are listed separately because of their short duration, low attenuation or low R^2. The totals given are for the eleven storms above them.

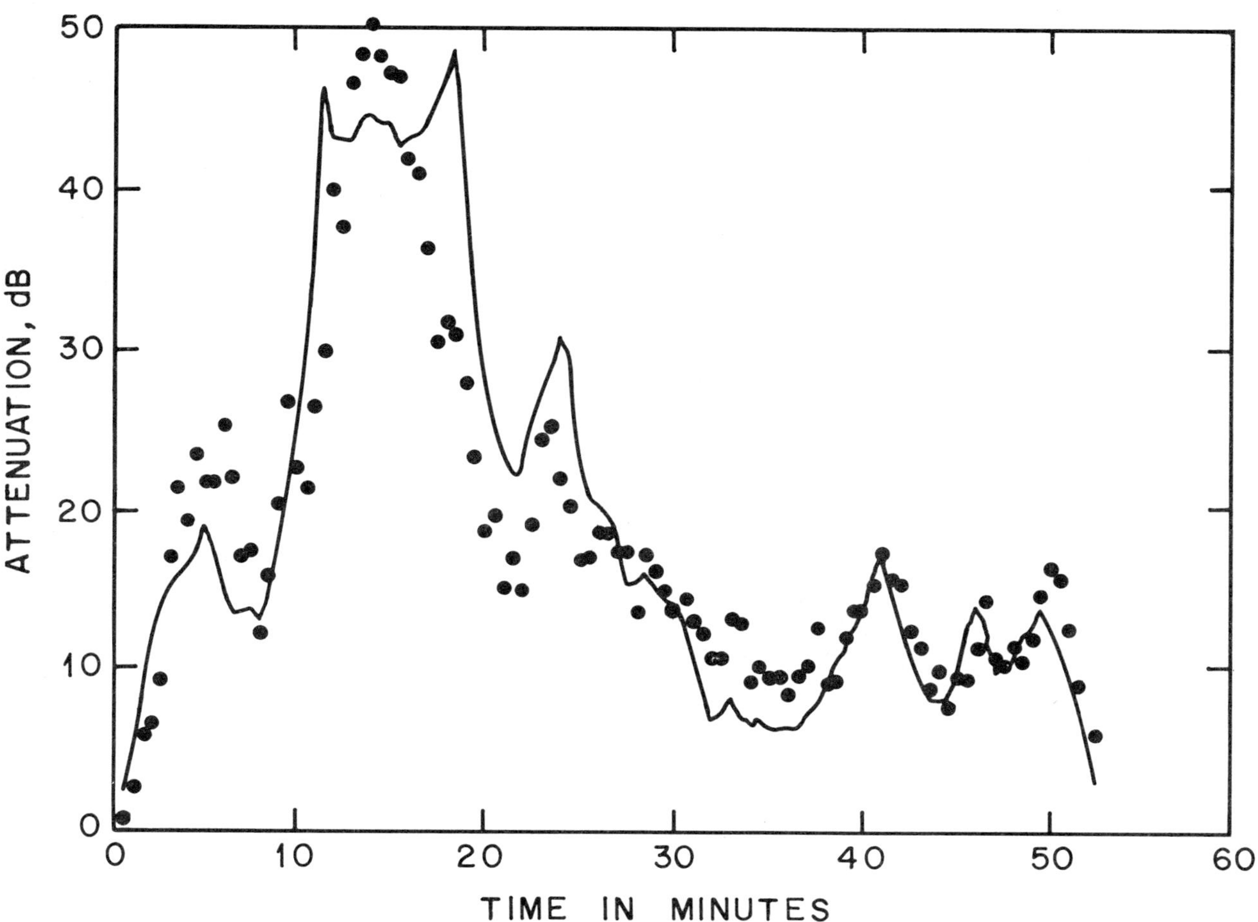

Figure 6. Points indicate predicted values of attenuation from fitting the rain rates of gauges 9, 17 and 25 for rainfall 7/28A*. Curve is the observed attenuation.

* Storm data on file at HSCF.

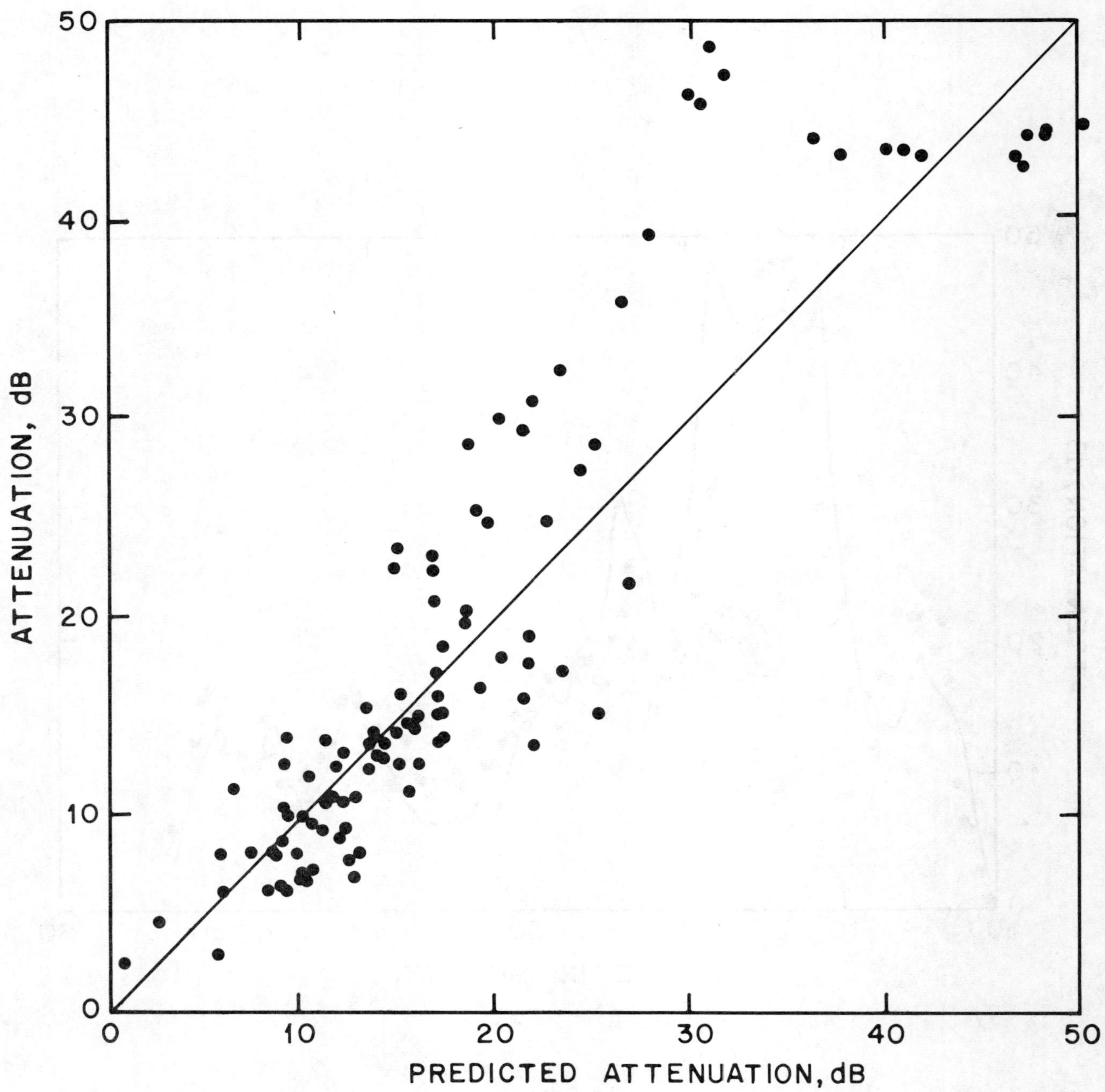

Figure 7. Predicted versus observed attenuation from fit of rain rates of gauges 9, 17 and 25 to the attenuation for rainfall 7/28A*. Diagonal line indicates predicted equal observed.

* Storm data on file at HSCF.

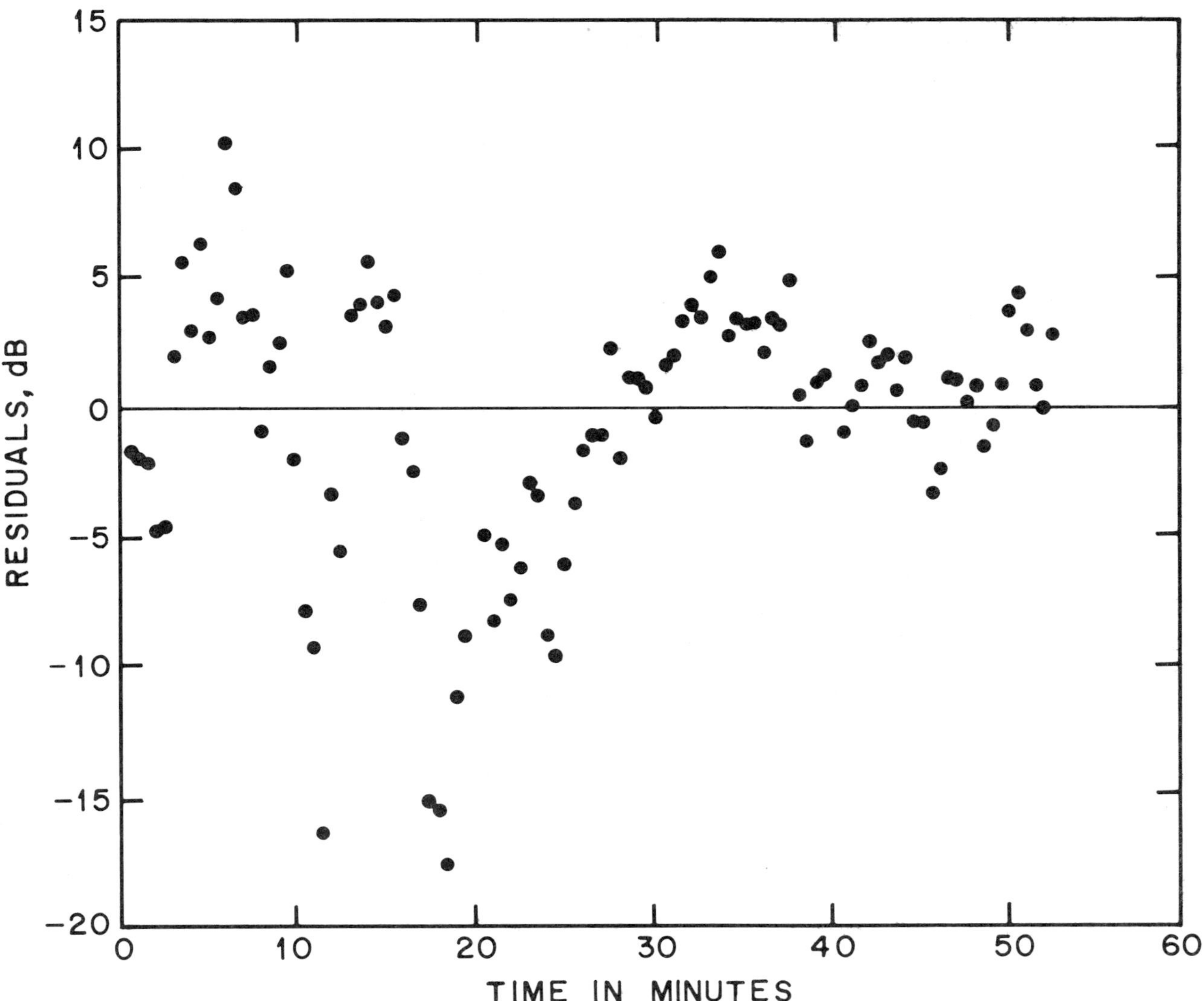

Figure 8. Residuals of the fit of rain rates of gauges 9, 17 and 25 to the attenuation for rainfall 7/28A*.

*Storm data on file at HSCF.

It appears from the regression that b_2 tends to be smaller than b_1 and b_3 for the same storm. It also appears that the overall level of these coefficients differs from storm to storm. These observations were tested by performing a two-way analysis of variance on the logarithms of the coefficients. The analysis showed that b_1, b_2 and b_3 are significantly different, but was inconclusive about the difference between storms.

Rather than give the details of the analysis of variance, we present a simpler analysis which will make the same points. The totals given in table 3 report rain rate at gauge 17 as almost exactly double the rate for either gauge 9 or gauge 25. And b_2 is on the average almost exactly half b_1 or b_3. In fact, for almost every storm, gauge 17 has the most reported rain and the smallest regression coefficient. Since we assume that the rain rate reported by each gauge is approximately representative of the rainfall in its vicinity, the relation of the readings from gauge 17 to the attenuation should not be much different from that for the other two gauges. That is, if gauge 17 were accurate, b_2 would be about the same size as b_1 and b_3, even if more rain had fallen on it than on the others. Thus we conclude that, relative to the other gauges, gauge 17 reported values approximately twice as great as it should have.

It follows that the sum of adjusted gauge means

$$m_9 + \tfrac{1}{2} m_{17} + m_{25}$$

will be a measure of the overall amount of rainfall for each storm. To obtain a simple measure of the ratio of attenuation to rainfall for each storm, we divided the average attenuation by this sum. The results are given in the last column in table 3.

Ten of the eleven ratios above the row of totals are clustered together, while the ratio for October 25 is almost twice as great as the next largest one. We concluded that something was different about this storm. Semplak and Turrin (1969) proposed that this difference may be caused by an updraft; we believe, however, that the difference is too great to be accounted for in this way.

Table 3. Results of fitting attenuation to rain rate.

Date 1967	Length (min)	Mean Rainfall m_9	m_{17}	m_{25}	Mean Atten-uation	Regression Coefficients b_1	b_2	b_3	b_0	R^2	Ratio
7/14A	41	41.3	99.0	48.7	12.8	.0573	.0455	.0882	1.67	.71	.0917
7/14B	66.5	101.2	135.6	55.9	23.2#	.0904	.0749	.1245	-3.08	.86	.1031
7/25*	67	24.0	77.6	75.1	21.6	.2086	.0674	.1032	3.65	.67	.1566
7/28A*	52.5	76.0	125.4	69.3	17.9	.0755	.0666	.1146	-3.01	.86	.0904
7/28B	46	20.3	63.7	31.4	11.8#	.2315	.0647	.1051	-0.27	.75	.1411
8/9*	48	50.3	80.0	23.3	12.5	.1776	.0691	.4144	-11.59	.90	.1100
8/25A	23	22.5	60.3	46.2	15.3#	.2261	.0874	.1240	-0.76	.77	.1547
9/21	31.5	21.7	72.1	28.6	8.3#	.4846	.0435	.0780	-7.55	.66	.0960
9/28*	39	9.6	44.3	23.2	6.4	.1802	.1284	.0611	-2.41	.72	.1163
10/10	46	20.4	62.7	29.6	8.9#	.0794	.0385	.2476	-2.47	.89	.1093
10/25*	33	27.2	29.4	31.8	20.0	.2806	.2679	.3764	-7.45	.92	.2713
Totals for 11 storms		414.5	850.1	453.1		2.0918	0.9539	1.8371			
7/21*	10.5	16.3	33.5	62.7	24.4	-.0126	.2694	.2106	2.37	.89	.2546
8/25B	51	23.3	29.5	24.0	3.6	.0427	.0654	.0984	-1.66	.52	.0579
11/3*	12.5	11.2	24.8	19.0	5.1	.0936	.0455	.0811	1.36	.38	.1197

*Storm data on file at HSCF.

Attenuation data on file at HSCF.

One possibility is that the distribution of raindrop sizes was different for this storm than for the others. Another is that the equipment for measuring attenuation was out of calibration that day. A third possibility is that the rainfall on the southeastern part of the path was very heavy. In support of the last possibility, we see that this is the only storm among the eleven for which $m_{25} > m_{17}$. Moreover, the short storm on July 21 also has both a high m_{25} and a high ratio of attenuation to rainfall. But without additional information, we are unable to say why the October 25 storm is different from the others.

COMPACT DESCRIPTION

We define a compact description as an interpolation function, $F(x_i; \theta_j)$, where the $x_i (i = 1, \ldots, m)$ are independent variables, the $\theta_j (j = 1, \ldots, n)$ are parameters and n is very much smaller (more compact) than the number of data points. In the present problem, the interpolation function takes the form $R(x, y, t; \theta_j)$ which gives rain rate, R, as a function of position, x and y, time, t, and the parameters θ_j. We refer to $R(x, y, t; \theta_j)$ as the model, and determine the θ_j by least squares fitting of the model to the data. Our model is quite different from that of Amorocho and Brandstetter (1967).

In the absence of a true model, a descriptive model offers the following:

- an algebraic expression that greatly simplifies interpolation, calculations (such as average rain rates along various paths and total quantity of water dumped by the event), and the communication of the essentials of the data base;

- a characterization of the event, implied by the estimates of the θ_j and their covariances, and an insight into the overall physical pattern of the event which may be contained in some functions of the θ_j.

In practice, a compact description is usually valid over a quite restricted range of the independent variables, and ignores much of the fine structure that may be present in the data. Also, the physical significance of the parameters of an interpolation function should be regarded with due caution.

Few rain rate events are important, complete and simple enough to be candidates for the modeling effort. An event is important enough if the rain rate at the peak is high (greater than 150 mm/hr) and the event covers several gauges in the network; it is complete enough if the peak remains on the network for several minutes; and it is simple enough if it appears that no more than approximately 25 parameters are needed to provide a reasonably good description of the entire event. The choice of 25 as an upper limit to the number of parameters is somewhat arbitrary; its implications in terms of the present models are discussed later.

We have chosen a segment of the July 28A storm as an example of a high rain rate event. The data comprise available readings for 57 consecutive scans from a 7 x 7 square consisting of the top seven rows and leftmost seven columns of the grid shown in figure 1. The 49 grid squares contain 47 rain gauges (30 cm in diameter) of which 42 were operating during this 9½-minute period. The 2394 data points arranged by scan and grid position are presented in table 4; these numbers rather than those in HSCF (under 28A) are analyzed below.

These data were time smoothed using a 9-scan running triangular filter with weights proportional to 1, 2, 3, 4, 5, 4, 3, 2, 1 to remove some of the rapid small-scale fluctuation characteristics of heavy rain rates. The upper third of table 5 gives the 9-scan time smoothed rain rates at one minute intervals (every sixth scan). We sketched approximate 150 mm/hr, 100 mm/hr and 50 mm/hr rain rate contours on this map of the grid.

The contours on the nine time smoothed scans of table 5 are consistent with the impression that a mountain of rain with its center inside the inner contour is covering the grid. As time goes on

Table 4. Gauge numbers, nominal grid coordinates and rain rates (given in scans 98 to 154 in mm/hr, ten seconds between scans).

Gauge numbers (see Fig. 1)

	2		4	5		7
8	9	10	11	12	13	14
15	16	17	18	19		21
22	23	24	25	26	27	28
29	30	31	32	33	77	78
		37	38	39	40	41
36	42		44	45	46	47

Nominal coordinates (x, y)

	1,10		3,10	4,10		6,10
0,9	1,9	2,9	3,9	4,9	5,9	6,9
0,8	1,8	2,8	3,8	4,8		6,8
0,7	1,7	2,7	3,7	4,7	5,7	6,7
0,6	1,6	2,6	3,6	4,6	5,6	6,6
		2,5	3,5	4,5	5,5	6,5
0,4	1,4		3,4	4,4	5,4	6,4

1x=1y=1.3km

98

	86		41	22		23
36	60	20	63	32	29	25
41	21	174	30	15		16
50	73	137	31	8	9	0
196	191	214	10	165	9	2
		77	69	34	3	0
34	50		11	51	76	32

99

	118		32	50		20
33	88	20	95	29	31	23
36	14	167	35	12		16
79	32	169	35	7	9	0
176	187	241	3	103	9	2
		77	63	108	6	0
43	24		20	56	60	35

100

	124		39	40		26
52	71	17	108	31	34	30
39	15	117	31	13		16
71	68	199	84	15	9	0
206	174	152	4	85	9	2
		72	73	41	27	0
43	19		25	36	83	38

101

	88		39	39		20
46	74	17	97	26	29	25
37	23	150	30	10		16
27	21	188	79	16	9	0
200	212	192	18	144	9	2
		68	84	41	23	0
29	21		30	35	88	45

102

	98		51	44		23
64	71	24	103	23	34	16
38	19	129	31	10		27
41	49	150	77	9	9	0
177	218	212	47	203	9	2
		75	55	41	22	0
36	27		37	39	91	49

103

	72		47	35		26
43	92	18	103	20	36	20
58	19	159	32	15		27
72	71	125	85	22	17	0
170	219	199	46	109	15	2
		137	60	43	37	1
39	25		24	36	87	22

104

	120		57	36		21
53	85	20	58	20	28	16
34	26	153	29	11		25
73	104	109	99	36	17	0
109	174	215	70	15	28	2
		136	62	32	23	6
24	20		27	37	68	14

105

	133		53	30		18
44	103	10	95	15	33	12
36	17	183	26	14		26
118	226	108	125	38	18	0
61	182	180	84	17	41	2
		113	64	40	23	9
8	15		20	30	77	23

Table 4, continued

106

	109		35	71		24
18	91	30	89	23	38	17
45	26	135	19	11		33
74	193	112	76	52	18	0
36	180	189	79	41	38	2
		103	63	40	28	17
12	22		20	15	81	28

107

	129		36	36		19
51	84	34	130	16	30	9
31	36	162	24	10		17
76	196	85	65	71	9	0
42	176	208	100	21	32	2
		97	55	37	33	13
22	15		24	31	74	13

108

	101		40	41		18
4	69	23	107	20	37	16
36	27	174	27	11		28
109	189	194	97	89	29	0
66	181	190	111	29	53	2
		81	62	39	34	5
22	12		26	23	112	32

109

	109		37	49		22
3	88	35	110	17	33	17
62	30	131	32	14		31
126	171	235	48	93	23	0
73	199	187	103	19	70	2
		101	36	54	28	5
10	17		18	29	40	35

110

	63		35	48		16
33	61	48	96	23	37	10
77	31	155	30	14		31
100	184	231	47	72	100	0
125	172	247	100	17	102	2
		87	54	39	23	3
8	13		26	39	80	41

111

	88		36	52		25
23	72	34	103	21	35	14
55	35	212	25	16		38
140	153	211	110	85	83	0
99	153	234	119	14	91	2
		67	42	24	25	11
16	17		19	29	95	34

112

	75		29	50		16
8	105	40	112	20	35	8
58	34	173	27	16		44
130	161	176	85	98	103	0
118	98	193	133	33	94	35
		66	64	46	26	11
17	16		27	18	77	26

113

	73		32	51		23
20	88	22	128	32	32	9
38	30	112	33	16		27
141	176	238	97	84	108	0
114	129	246	140	18	78	34
		26	37	41	25	8
14	20		26	16	58	25

114

	64		33	40		16
9	112	39	117	23	31	8
53	30	120	30	27		79
139	157	212	152	70	64	2
124	125	180	156	23	67	33
		24	46	47	33	3
6	18		23	30	63	28

115

	75		37	26		20
11	135	54	128	17	22	8
55	38	186	30	33		34
139	162	239	99	84	163	4
149	93	178	114	35	52	34
		41	38	40	16	5
5	18		23	22	75	34

Table 4, continued

116						
	79		39	33		21
3	87	45	98	16	20	8
60	38	182	55	29		45
153	152	239	129	141	157	50
142	91	182	132	28	74	15
		32	35	45	23	3
4	15		20	23	49	44

117						
	66		33	40		20
9	91	33	77	16	23	10
82	32	99	60	27		29
136	171	242	156	137	146	39
154	139	167	134	34	91	27
		30	40	54	19	4
6	13		20	23	44	28

118						
	77		34	47		29
3	119	19	116	25	23	6
67	37	212	55	19		70
158	156	183	180	136	152	49
138	84	174	135	44	52	33
		27	30	75	19	0
3	16		20	28	58	23

119						
	74		37	49		39
31	92	51	115	25	22	11
88	58	275	59	22		46
138	160	173	163	108	146	37
116	119	157	167	27	93	26
		21	35	39	25	8
3	12		20	35	78	19

120						
	75		68	32		29
28	79	59	99	38	26	22
81	45	251	47	27		51
117	177	147	164	136	181	51
125	95	102	153	28	82	34
		25	31	39	24	27
2	11		20	27	79	20

121						
	48		51	35		21
3	93	49	112	47	25	13
97	62	270	97	30		72
126	152	230	167	141	169	32
118	152	118	102	17	82	42
		26	26	40	17	12
2	13		18	16	89	32

122						
	77		74	40		22
27	81	50	145	47	20	8
88	68	192	102	58		88
129	164	202	160	184	134	33
143	204	126	88	23	97	39
		18	30	13	33	4
3	5		23	12	75	38

123						
	77		81	51		24
41	92	66	148	78	21	7
83	70	206	101	52		136
154	158	206	174	144	140	31
96	147	101	100	24	83	43
		29	28	39	21	6
2	6		18	18	47	24

124						
	87		82	37		21
22	104	95	166	67	21	5
70	72	234	63	78		97
144	153	159	218	165	155	32
96	140	107	92	25	87	45
		18	29	22	25	3
5	5		16	25	78	31

125						
	86		82	37		23
31	123	127	147	88	19	7
47	85	235	103	89		75
132	147	164	217	191	83	30
99	125	57	68	53	68	40
		16	33	28	28	0
3	8		18	13	62	20

Table 4, continued

126						
	96		81	49		25
46	121	118	157	94	22	19
80	71	220	120	85		69
134	124	214	236	176	107	29
65	165	80	83	27	67	30
		15	28	24	22	4
2	3		14	10	38	40

127						
	94		78	49		25
19	121	72	169	123	27	25
81	74	233	203	86		57
137	156	232	241	191	115	39
116	152	136	79	27	75	32
		20	30	14	29	3
2	3		22	8	56	30

128						
	85		79	54		27
117	121	113	160	122	36	25
101	122	240	183	73		126
132	140	197	242	190	85	26
95	155	43	72	52	92	40
		11	28	46	21	2
4	3		20	13	105	40

129						
	48		86	50		25
42	110	109	160	128	51	28
107	137	240	156	104		107
124	159	181	218	191	58	28
86	179	138	89	41	71	32
		17	27	27	19	1
2	10		26	12	68	41

130						
	75		76	44		29
19	124	69	148	124	78	25
89	120	256	198	126		109
113	148	198	198	197	121	28
126	164	49	54	28	81	28
		14	28	16	24	2
3	5		25	11	35	34

131						
	100		82	58		24
110	145	80	166	136	98	33
97	141	283	188	116		110
125	140	196	227	186	110	28
100	148	116	70	28	57	21
		21	25	15	25	17
3	11		20	13	94	12

132						
	143		74	55		21
65	170	87	166	119	109	90
113	208	249	199	128		113
127	143	173	223	186	126	26
72	154	66	62	30	53	32
		18	28	21	22	34
2	4		21	14	28	36

133						
	92		81	67		21
108	162	146	143	123	121	91
106	220	228	144	118		111
130	126	169	198	174	102	32
79	135	64	55	12	82	25
		15	30	13	25	19
6	5		18	12	36	36

134						
	94		82	52		22
86	166	74	171	124	114	75
115	205	200	173	122		96
120	137	189	225	200	137	39
37	153	93	54	22	54	27
		17	29	14	27	12
2	4		12	6	20	52

135						
	189		79	63		25
81	169	86	166	138	123	59
122	156	219	171	139		76
108	130	173	218	189	157	32
68	163	156	49	13	64	26
		12	27	13	22	10
9	11		12	6	21	38

Table 4, continued

136

	148		81	54		24
94	167	152	166	131	94	49
115	167	203	137	129		88
123	142	192	227	150	149	31
101	160	92	49	10	64	31
		16	25	10	27	7
11	7		11	5	21	47

137

	157		93	84		26
119	161	75	180	155	127	70
121	193	152	168	128		97
123	146	154	218	181	138	21
124	161	158	30	23	43	24
		14	27	14	27	7
3	6		11	8	21	26

138

	233		95	86		24
143	206	115	190	142	114	100
119	141	173	130	133		95
116	144	156	213	179	152	33
127	150	75	41	9	43	26
		19	19	12	20	9
2	12		14	3	34	23

139

	187		84	87		23
145	149	131	199	134	118	96
141	84	172	130	138		85
108	98	164	216	163	155	38
102	171	177	36	20	41	30
		17	17	7	19	8
2	3		13	4	10	24

140

	175		74	80		18
150	243	104	221	162	118	91
142	89	159	117	134		86
115	114	159	228	137	104	38
142	176	121	38	17	47	28
		25	15	9	27	2
5	8		7	2	17	40

141

	183		72	78		45
90	204	125	256	110	137	102
114	91	167	123	131		59
105	69	164	184	155	153	30
137	170	186	33	93	36	29
		19	17	7	25	4
5	6		5	2	20	33

142

	186		70	79		49
137	152	170	242	141	131	107
145	111	176	117	139		75
100	106	160	190	102	135	25
121	184	138	23	168	36	15
		20	11	9	22	6
6	5		7	2	18	22

143

	137		80	98		73
183	153	150	225	120	138	135
147	84	182	121	154		66
102	105	182	159	78	124	30
144	157	167	18	91	37	29
		22	16	9	15	5
2	5		8	1	9	24

144

	158		83	145		76
159	219	110	208	143	134	112
139	104	209	129	114		64
100	60	160	152	100	97	31
122	156	73	18	9	33	21
		18	9	3	15	6
1	5		12	1	18	15

145

	148		114	90		87
168	197	146	211	148	227	140
130	93	192	123	121		74
120	112	159	145	95	78	30
117	152	96	11	9	34	26
		15	8	3	12	4
6	3		7	2	7	17

Table 4, continued

146						
	168		90	145		85
160	185	166	199	134	204	119
118	75	200	125	108		56
93	104	157	144	63	114	36
69	117	98	41	11	35	26
		16	7	6	17	1
9	3		5	1	3	19

147						
	122		114	156		94
138	174	139	193	113	257	137
111	90	203	111	108		61
109	123	284	151	113	47	41
78	169	170	30	23	34	23
		17	4	1	14	1
5	7		6	1	11	18

148						
	136		99	136		93
163	167	121	202	138	264	224
134	82	189	116	115		85
114	100	355	142	101	114	40
88	163	96	41	9	29	17
		17	3	2	14	2
1	3		8	1	12	13

149						
	86		111	144		100
196	137	102	206	131	260	162
96	79	202	95	113		80
92	65	395	134	102	140	30
105	149	60	34	9	25	12
		19	3	0	21	1
1	7		4	1	12	12

150						
	110		117	176		88
172	166	78	211	175	202	228
121	76	192	85	112		95
92	87	415	104	95	103	29
99	147	145	52	16	25	13
		14	12	1	16	1
1	11		3	2	4	10

151						
	119		111	136		178
242	132	104	221	148	211	249
119	44	184	95	109		78
100	89	425	81	125	106	30
87	157	40	44	13	49	21
		14	5	4	10	1
1	4		3	21	8	10

152						
	106		81	181		108
119	109	154	231	148	237	224
108	57	171	103	97		53
110	115	300	120	71	47	28
117	151	137	42	13	31	9
		13	4	3	10	3
0	1		3	1	3	10

153						
	93		106	160		102
140	111	71	216	103	248	175
104	45	179	111	87		71
50	95	176	136	53	68	24
123	123	124	73	13	16	10
		12	4	2	12	0
2	5		3	1	2	8

154						
	124		112	200		126
137	102	98	185	100	259	245
82	53	181	96	80		77
80	49	193	127	120	26	23
131	136	58	55	18	14	14
		13	7	4	13	1
1	1		3	6	3	8

Table 5. Selected time-smoothed data, fitted values and residuals. Contours of 150 mm/hr, 100 mm/hr and 50 mm/hr rain rates have been sketched.

SCAN				102							108							114			
DATA mm/hr		111		43	44		21		78		34	46		20		72		42	40		26
	33	86	25	99	19	34	14	15	91	37	112	22	33	10	13	99	41	107	26	23	10
	42	26	156	26	12		27	55	33	159	30	18		41	75	44	203	58	27		53
	89	170	134	85	55	21	0	131	166	215	95	87	93	3	140	162	203	156	128	152	38
	76	185	198	85	39	40	2	114	135	211	126	23	81	21	135	113	157	136	31	76	30
			106	58	40	28	9			57	47	41	26	7			27	34	49	22	7
	18	18		23	28	79	25	12	17		24	25	73	31	4	14		20	26	65	28
FIT mm/hr		3	3	2	1	0	0		10	11	10	6	3	1		22	26	24	19	12	6
	14	22	22	15	7	2	0	30	43	47	39	24	11	4	49	68	77	71	53	32	16
	48	74	76	52	23	7	1	75	108	117	96	59	27	10	98	134	148	134	98	58	28
	92	143	147	101	46	14	3	114	163	175	142	86	40	14	126	168	183	161	115	67	32
	98	153	159	110	51	15	3	105	149	158	126	76	35	12	104	136	145	125	87	49	23
	58	92	96	67	31	10	2	59	82	86	68	41	18	6	56	71	74	62	42	23	11
	19	30	32	23	11	3	1	20	27	28	22	13	6	2	19	24	24	20	13	7	3
RESIDUALS mm/hr		108		41	43		21		68		24	40		19		49		17	21		20
	19	65	4	84	13	32	14	-14	48	-9	73	-1	21	6	-34	31	-35	36	-26	-6	-5
	-5	-47	80	-24	-10		25	-18	-74	42	-65	-40		31	-21	-89	55	-75	-70		25
	-2	27	-13	-15	9	7		17	3	40	-46	1	53	-10	14	-6	20	-4	12	85	7
	-21	32	38	-24	-11	25	0	8	-13	53	0	-52	46	9	31	-22	12	12	-55	27	7
			9	-8	9	19	7			-28	-20	0	7	1			-45	-27	6	-1	-2
	0	-12		1	18	76	24	-7	-9		2	12	67	29	-14	-9		0	12	58	25

Table 5. Con't (2)

SCAN				120							126							132			
DATA mm/hr		82		77	42		23		88		80	52		25		159		84	69		24
	32	105	87	150	77	22	11	65	135	93	159	124	74	42	107	172	109	175	137	114	74
	76	74	227	105	69		89	99	143	247	180	109		105	120	167	193	153	130		92
	137	151	191	202	170	129	32	125	144	193	220	189	104	29	119	134	172	219	175	142	31
	103	150	97	89	30	81	40	95	158	86	69	31	72	30	94	158	115	45	16	55	27
			20	29	27	25	5			16	28	22	23	10			16	25	12	24	10
	3	6		18	16	66	30	3	6		21	12	59	33	6	7		12	6	23	36
FIT mm/hr		38	46	46	39	29	18		57	68	71	65	53	38		76	90	96	92	79	62
	69	93	106	104	87	61	37	87	114	130	132	117	91	63	102	130	148	152	140	116	86
	115	150	168	159	128	88	52	126	159	176	171	147	110	73	132	161	176	174	154	123	88
	130	165	178	164	128	86	49	128	156	167	157	129	94	60	124	145	152	144	123	94	65
	99	122	128	114	87	56	31	92	108	111	101	80	56	35	84	94	96	87	71	52	35
	51	61	62	54	40	25	13	46	52	52	45	35	24	14	41	45	43	38	30	21	13
	18	21	20	17	12	7	4	16	18	17	14	11	7	4	15	15	14	12	9	6	4
RESIDUALS mm/hr		43		31	3		5		31		9	-12		-12		84		-11	-23		-37
	-35	12	-19	46	-9	-38	-25	-21	21	-36	27	7	-16	-20	5	42	-38	23	-2	-1	-12
	-38	-76	59	-53	-59		38	-26	-15	71	9	-37		32	-11	6	17	-20	-23		4
	7	-13	13	38	41	43	-15	-2	-11	26	63	60	10	-30	-4	-9	20	74	52	48	-32
	4	28	-29	-24	-56	25	9	3	50	-24	-31	-48	15	-4	10	64	20	-41	-54	3	-6
			-41	-23	-11	0	-7			-34	-17	-12	0	-3			-26	-12	-16	3	-2
	-14	-14		1	4	58	26	-12	-11		7	1	52	29	-8	-7		1	-2	17	33

Table 5. Con't (3)

SCAN 138							
DATA mm/hr		171		80	94		53
	145	185	137	225	135	141	111
	136	96	179	122	134		72
	105	97	164	182	116	125	31
	126	166	136	27	69	38	25
			20	13	7	20	5
	4	5		8	2	15	25
FIT mm/hr		93	110	118	116	104	86
	114	141	159	164	154	133	105
	134	159	171	169	153	126	95
	117	133	137	130	112	89	64
	76	82	82	74	61	46	32
	37	33	36	31	25	18	12
	13	13	12	10	8	5	3
RESIDUALS mm/hr		78		-37	-21		-31
	31	45	-21	62	-18	8	7
	1	-61	8	-46	-18		-22
	-11	-35	27	52	4	36	-32
	50	83	54	-46	8	-8	-6
			-16	-17	-17	2	-6
	-8	-7		-1	-5	10	22

SCAN 144							
DATA mm/hr		126		105	145		99
	170	165	122	206	139	234	179
	118	79	195	108	111		75
	103	96	89	133	98	99	34
	93	152	104	36	13	32	19
			16	6	2	15	2
	3	5		5	3	9	14
FIT mm/hr		108	126	136	135	125	106
	123	148	164	169	161	142	116
	133	153	163	160	146	123	96
	110	120	122	115	100	80	60
	68	72	70	63	52	40	28
	32	33	30	26	21	15	10
	12	11	10	8	6	4	3
RESIDUALS mm/hr		18		-30	9		-6
	47	17	-41	36	-21	92	63
	-14	-73	32	-51	-33		-20
	-6	-23	-32	18	-1	19	-24
	24	81	35	-25	-38	-7	-9
			-13	-19	-17	0	-8
	-8	-5		-2	-2	5	11

SCAN 150							
DATA mm/hr		102		100	169		129
	129	110	99	199	110	227	215
	92	51	177	101	74		75
	91	66	129	125	92	55	26
	122	147	103	52	22	21	12
			14	5	3	11	1
	2	4		4	4	7	11
FIT mm/hr		121	139	149	149	140	122
	129	151	166	170	162	145	120
	130	146	152	149	135	115	92
	102	109	108	101	87	71	54
	61	62	59	53	44	34	24
	28	28	25	21	17	12	9
	10	9	8	7	5	3	2
RESIDUALS mm/hr		-18		-48	-20		7
	0	-40	-66	-30	-51	82	95
	-37	-94	25	-47	-60		-16
	-9	-41	21	25	4	-15	-27
	60	85	43	0	-20	-11	-12
			-10	-15	-13	0	-7
	-7	-4		-2	0	3	9

the mountain moves northward and eastward and becomes more spread out.

A function that has the characteristics of a mountain is the bivariate normal

$$R(x, y) = \frac{A}{2\pi\,\sigma_x \sigma_y (1-\rho^2)^{\frac{1}{2}}} \exp\left\{ \frac{-1}{2(1-\rho^2)} \left[\left(\frac{x-\mu_x}{\sigma_x} \right)^2 - 2\rho \left(\frac{x-\mu_x}{\sigma_x} \right) \left(\frac{y-\mu_y}{\sigma_y} \right) + \left(\frac{y-\mu_y}{\sigma_y} \right)^2 \right] \right\} \quad (2)$$

which has been multiplied by A, the "volume" of the mountain. Additional attractions of this function are that it is mathematically tractable, well behaved, positive everywhere, and tabulated. Thus it is easy to fit and to use in calculations. Furthermore, all the parameters, μ_x, μ_y, σ_x, σ_y, A and ρ, have well known, easily understood interpretations and can be readily converted to a set of descriptors that have direct physical meaning. As descriptors, we chose the functions

the position of the peak,

μ_x

μ_y

the peak value of rain rate,

$$P = \frac{A}{2\pi\,\sigma_x \sigma_y (1-\rho^2)^{\frac{1}{2}}}$$

the direction of the <u>major</u> axis of the mountain in the x, y plane,

$$\alpha = \tfrac{1}{2} \operatorname{arc\,tan} \left(\frac{2\rho\sigma_x\sigma_y}{\sigma_x^2 - \sigma_y^2} \right)$$

and the standard deviations along the major and minor axes of the mountain respectively,

$$\Sigma_J = \tfrac{1}{2} (Q_+^{\frac{1}{2}} + Q_-^{\frac{1}{2}})$$

$$\Sigma_I = \tfrac{1}{2} (Q_+^{\frac{1}{2}} - Q_-^{\frac{1}{2}})$$

where

$$Q_\pm = \sigma_x^2 + \sigma_y^2 \pm 2\sigma_x\sigma_y(1-\rho^2)^{\frac{1}{2}}.$$

Equation (2) was fitted to each of the nine scans of table 5 separately, using a nonlinear least squares fitting procedure to determine the parameters (Marquardt, 1963). Table 6 contains the parameters and descriptors for the nine fits, and table 7 the (typical) set of standard errors and correlations for scan 126.

The residual standard errors of the fits are acceptable. (For comparison, the residual standard error <u>in calibration</u> (Freeny, 1969) is 7 mm/hr and the gauges had been out in the field for some months before the event.) Values of multiple R^2 are quite reasonable in view of the simplicity of the model and the evident fine structure of table 5. The standard errors of the coefficients are comfortably small (except for ρ), and all the parameters tend to increase with time. By scan 150 the peak of the event has moved to the edge of the network and the model cannot follow the event any further.

The time trends evident in table 6 led us to add time as a variable and make the parameters linear with time. Thus,

Table 6. Parameters, descriptors and other statistics from the individual fits to equally spaced scans.

Scan	μ_x^+	μ_y^+	σ_x^+	σ_y^+	ρ	A	$P^\#$	α^{++}	Σ_J^+	Σ_I^+	R^{2*}	R.S.E.**
102	5.57	1.64	1.87	1.45	.011	2790	163	89	1.87	1.45	.82	36
108	5.73	1.72	1.55	1.70	.076	2500	152	19	1.72	1.52	.80	41
114	5.95	1.95	1.33	2.43	.067	3380	167	3	2.44	1.32	.85	38
120	6.38	2.23	1.54	2.42	.114	4180	180	7	2.43	1.52	.88	36
126	6.67	2.36	1.51	2.40	.143	4710	209	8	2.41	1.49	.93	30
132	6.97	2.16	1.73	2.85	.066	5960	192	4	2.85	1.73	.93	32
138	7.36	1.95	2.09	3.40	.170	7720	175	9	3.43	2.05	.91	35
144	7.81	3.07	2.20	5.83	.537	10880	160	13	5.96	1.82	.91	34
150	9.54	6.91	3.79	7.23	.854	15440	172	26	7.96	1.79	.89	36

Key: + 1 unit = 1 grid square = 1.3 km; see bottom of table 4 for coordinates

In mm/hr

++ In degrees measured counterclockwise from x axis

* Multiple R^2

** Residual standard error

Table 7. Standard errors and correlations for the coefficients of scan 126.

	μ_x	μ_y	σ_x	σ_y	ρ	A
Standard Errors	.10	.13	.12	.24	.11	694
Correlations						
μ_y	-.18					
σ_x	-.06	-.06				
σ_y	.04	-.25	.04			
ρ	-.13	.01	-.18	-.17		
A	-.07	.20	.24	-.91	.16	

$$\mu_x(t) = \mu_{x0} + \mu_{x1} t$$

$$\mu_y(t) = \mu_{y0} + \mu_{y1} t$$

$$\sigma_x(t) = \sigma_{x0} + \sigma_{x1} t$$

$$\sigma_y(t) = \sigma_{y0} + \sigma_{y1} t$$

$$\rho(t) = \rho_0 + \rho_1 t$$

$$A(t) = A_0 + A_1 t$$

where the 12 coefficients μ_{x0} to A_1 are the new parameters of the model. A linear time dependence is chosen because the results in table 6 do not suggest (or support) a more complicated relationship. The modification is applied to the parameters rather than to descriptors partly because the algebra and programming are simpler, and partly because a linear time dependence in the parameters transforms into a more complicated time dependence in some of the descriptors; the latter is especially important in the case of P, which appears to go through a maximum close to scan 126. This method of introducing time also has some disagreeable consequences. In particular, the values of most of the descriptors, e.g., $\Sigma_J(t)$, $\Sigma_I(t)$ and $\alpha(t)$, are not invariant under rotation of the x and y axes; and the physical consequences of the constraints that linearity with time in the parameters imposes on the descriptors are hardly transparent. These deficiencies are secondary at this stage of development in the model and we do not deal with them here.

New descriptors of interest include:

the position of the peak at scan 102,

μ_{x0}

μ_{y0}

the velocity and direction of motion of the peak, and

$$v = \sqrt{\mu_{x1}^2 + \mu_{y1}^2}$$

$$\beta = \text{arc tan} \frac{\mu_{x1}}{\mu_{y1}}$$

the standard deviation along the major and minor axes of the mountain respectively,

$$\Sigma_J(t) = \tfrac{1}{2} \left([Q_+(t)]^{\frac{1}{2}} + [Q_-(t)]^{\frac{1}{2}}\right)$$

$$\Sigma_I(t) = \tfrac{1}{2} \left([Q_+(t)]^{\frac{1}{2}} - [Q_-(t)]^{\frac{1}{2}}\right)$$

where

$$Q_\pm(t) = \sigma_x^2(t) + \sigma_y^2(t) \pm 2\sigma_x(t)\sigma_y(t)\,(1-[\rho(t)]^2)^{\frac{1}{2}}$$

the peak value of rain rate,

$$P(t) = \frac{A(t)}{2\pi\,\sigma_x(t)\sigma_y(t)\,(1-[\rho(t)]^2)^{\frac{1}{2}}}$$

the direction of the major axis of the mountain in the x, y plane,

$$\alpha(t) = \tfrac{1}{2}\,\text{arc tan} \frac{2\rho(t)\sigma_x(t)\sigma_y(t)}{[\sigma_x(t)]^2 - [\sigma_y(t)]^2}$$

add the instantaneous rate of revolution of the major axis (angular velocity of the event),

$$\omega(t) = \frac{d\alpha(t)}{dt}$$

the rate of change of the peak value,

$$P'(t) = \frac{dP(t)}{dt}$$

the rate of spread along the major axis,

$$\Sigma'_J(t) = \frac{d\Sigma_J(t)}{dt}$$

the rate of spread along the minor axis,

$$\Sigma'_I(t) = \frac{d\Sigma_I(t)}{dt}$$

the area covered by 1 sigma ellipse,

$$C(t) = \pi \Sigma_J(t) \Sigma_I(t)$$

the rate of change of the area of the storm,

$$C'(t) = \frac{dC(t)}{dt}$$

the average rain rate within the 1 sigma ellipse,

$$\overline{V}(t) = 0.4A(t)/C(t)$$

the rate of change of average rain rate,

$$\overline{V}'(t) = \frac{d\overline{V}(t)}{dt}$$

the volume of water precipitated in the event,

$$V = \int_{t_0}^{t_f} \int_{-\infty}^{\infty} \int R(x, y, t; \tilde{\theta}_j) = \int_{t_0}^{t_f} A(t)dt$$

and the volume of water precipitated within the 1 sigma ellipse.

$$V_c = 0.4V$$

In this paper we confine ourselves to calculating v and β and tabulating P(t), $\Sigma_J(t)$, $\Sigma_I(t)$ and $\alpha(t)$.

The twelve parameters were determined by fitting the time-dependent model $R(x, y, t; \theta_1, \theta_2, \ldots, \theta_{12})$ to the 2058 time smoothed values in the 49 scans numbered 102 to 150. Nominal values of x, y and t were used. This is equivalent to assuming that all gauges were at the centers of their grid squares and were scanned simultaneously; we consider effects introduced by this simplification are inconsequential.

The parameters and other statistics of the fit are given in table 8; the fitted value of rain rate for nine scans (150 mm/hr, 100 mm/hr and 50 mm/hr contours are sketched) and the residuals are given in table 5; a normal probability plot of the residuals is shown in figure 9; and some descriptors are tabulated in table 9.

The fit is well behaved. The distribution of the residuals (from the even numbered scans) is seen from figure 5 to be closely normal. The residual standard error (34 mm/hr) is acceptably small and the multiple R^2 (.89) is acceptably high. The parameters are quite precisely defined (except for ρ_0). The correlations are surprisingly small, the only ones of importance (> .5) being across time (e.g., μ_{x0} with μ_{x1}) as underlined in table 8 and the correlation among the components of σ_y and A. The temporal correlations largely reflect the choice of scan 101, rather than the middle scan, 126, as the origin of the time coordinate. This is confirmed by transforming the covariance matrix to correspond to a translation of the time coordinate by 25 scans. The transformed correlations are .52, .34, .72, .43,

Table 8. Parameters and other statistics for the time-dependent fit.

	μ_{x0}	μ_{x1}	μ_{y0}	μ_{y1}	σ_{x0}	σ_{x1}	σ_{y0}	σ_{y1}	ρ_0	ρ_1	A_0	A_1
Value	1.59	.026	6.37	.054	1.56	.053	1.31	.017	-.019	.008	2210	133
Standard error	.07	.003	.04	.002	.08	.005	.04	.002	.044	.002	160	9
Units ℓ = 1.3 km, s = 10 sec	ℓ	ℓ/s	ℓ	ℓ/s	ℓ	ℓ/s	ℓ	ℓ/s		s^{-1}	ℓ^2mm/hr	ℓ^2mm/hr/s
Multiple R^2	.89											
Resid. Stand. Error	34	mm/hr										

Correlation Matrix

	μ_{x0}	μ_{y0}	σ_{x0}	σ_{y0}	ρ_0	A_0	μ_{x1}	μ_{y1}	σ_{x1}	σ_{y1}	ρ_1
μ_{y0}	.18										
σ_{x0}	-.38	-.08									
σ_{y0}	-.10	-.02	.15								
ρ_0	-.08	-.22	.18	.11							
A_0	-.16	-.06	.41	-.74	-.01						
μ_{x1}	-.80	-.16	.28	.08	.03	.14					
μ_{y1}	-.17	-.85	.07	-.08	.16	.13	.25				
σ_{x1}	.22	.06	-.76	-.16	-.14	-.30	-.30	-.08			
σ_{y1}	.07	-.11	-.17	-.78	-.09	.61	-.05	.28	.23		
ρ_1	.04	.17	-.22	-.09	-.84	.01	.02	-.15	.27	.19	
A_1	.12	.15	-.31	.39	.01	-.69	-.22	-.28	.47	-.65	-.08

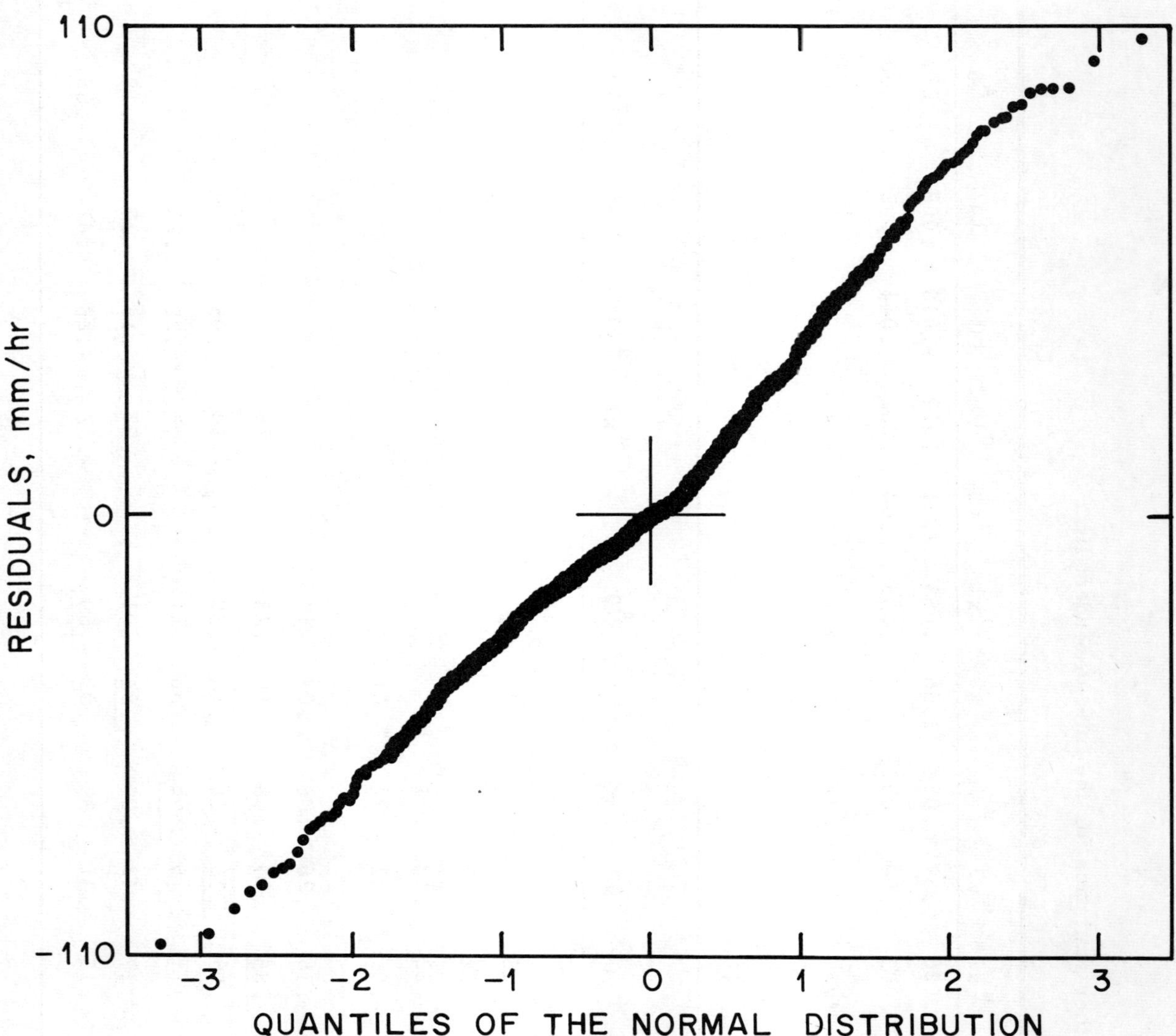

Figure 9. Normal plot of residuals from the fit to 49 time-smoothed scans.

Table 9. Some descriptors of the time-dependent fit.

Scan	Peak Rain Rate#	S. E. Peak#	α^*	Σ_J^+	Σ_I^+
102	171.6	16.8	-3	1.6	1.3
108	180.6	11.4	1	1.9	1.4
114	183.1	8.3	5	2.2	1.5
120	182.4	7.2	7	2.5	1.6
126	180.3	7.4	9	2.9	1.7
132	177.5	8.1	10	3.2	1.7
138	174.6	9.0	11	3.5	1.8
144	171.9	9.7	12	3.8	1.9
150	169.5	10.5	13	4.2	1.9

Note: Velocity of the peak (v) 28 ± 1 km/hr

Direction of motion (β) 25° 16' ± 2° 40' (clockwise from north)

Key: # mm/hr

* Degrees measured counterclockwise from east

\+ Units of 1 grid square (1.3 km)

.08, and -.73 instead of -.80, -.85, -.76, -.78, -.84, and -.69. The correlations between σ_y and A reflect the fact that the mountain is long in the y direction.

The descriptors in table 9 give the velocity of the motion of the peak, the heights of the peak and the change in shape and orientation of the mountain. If the twelve θ_j parameters are transformed to twelve ϕ_i descriptors, $cov(\phi_i) = J\ cov(\theta_i)\ J'$; where J is the Jacobian of the ϕ_i with respect to the θ_j, and J' is the transpose of J. As some of the descriptors are functions of time, J and also $cov(\phi_i)$ are functions of time. This relationship is used to calculate the standard errors of velocity and peak rain rate given in table 9. (Note that because $R(x, y, t; \theta_j)$ is nonlinear in the θ_j, cov (θ_j) is an approximation valid only in the immediate neighborhood of the parameter values corresponding to the local minimum in the surface of sums of squares to which the fit has converged.)

In the present parameterization, the velocity of the peak is constrained to be constant. Both the magnitude and direction appear well determined. To produce much uncertainty in β, the uncertainty in tan $\beta = \mu_{x1}/\mu_{y1}$ would have to be very large. The peak rain rate rises to reach a maximum near scan 114 and then declines less rapidly than it rose. The standard error of the peak rate is about 10% at scan 102, declines to a minimum of 4% near scan 122 and is back to 6% by scan 150. The major axis of the mountain rotates counterclockwise from slightly south of east to somewhat north of east. The spread along the major axis almost triples and that along the minor axis increases by about one-half. As the peak is changing comparatively little in value, the total quantity of rain falling is increasing substantially. All the descriptors are changing more slowly at the end of eight minutes than at the beginning.

The velocity of the peak given in table 9 is less than half the velocity of 58 km/hr given in table 1. There is no reason to expect the two velocities to agree. The velocity in table 1 measures the motion of patterns of rainfall across the grid, while the velocity in table 9 is the velocity with which the apparent center of the bulk of the rainfall moves across the grid.

A weakness of the model is evident from the pattern of the residuals in table 5. Some grid positions always tend to have low residuals, others high. The pooled residuals are close to being normally distributed, and additional probability plots show that this is also true for the residuals from individual scans. The variance changes little over the 49 scans. However, the residuals from individual gauges over 49 scans are not normally distributed, but are strongly bimodal, and show substantial serial correlations. Some of the bimodal distributions have the higher rain rate mode and some the lower rain rate mode early in the event. Table 10, which gives the mean residual for each gauge over the 49 time smoothed scans, clearly displays the biases of the individual gauges. The biases themselves are close to being normally distributed. The use of an over-simple model may contribute to these biases and bimodalities both because the model fails to account for the fine structure of the rainfall event, and because of miscalibrated or malfunctioning rain gauges. Some improvement in certain aspects of the description may be achieved by reparameterizing the descriptors to introduce rotational invariance and more straightforward temporal behavior, and by adding (or subtracting) a second time-dependent bivariate normal to the present function. Beyond this, further complexity is unlikely to improve the description very much, and may make the interpretation of the parameters much more difficult. These considerations, which led us to conclude that events that seem to require more than about 24 parameters (i.e., two time-dependent mountains) to describe, are too complex in character to be candidates for the modeling effort.

In summary, the model $R(x, y, t; \theta_j)$ offers a good compact description of the overall behavior of a high rain rate event by means of twelve parameters (and associated statistics), which may be easily transformed into useful physical descriptors. In doing this the model ignores as fine structure about 11% of the total variation in the (time smoothed) data.

Table 10. Mean residuals over all 49 scans in table 5, in mm/hr, ordered by grid position.

Mean residuals ordered by grid position						
	52		0	7		-2
0	28	-28	46	-10	15	10
-17	-56	42	-41	-38		11
1	-10	18	23	22	35	-21
18	41	21	-19	-39	14	0
		-24	-18	-8	3	-2
-9	-8		1	4	38	24

ACKNOWLEDGEMENTS

Mrs. A. E. Freeny collaborated in the development of the compact description; and Paul A. Tukey assisted with the computer programming for the drift and attenuation analyses.

```
DATA SET DESCRIPTION


DATA - CRAWFORD HILL RAINFALL DATA (L. JAECKEL AND J. GABBE)

CASES AND VARIABLES

   THE USUAL CLASSIFICATION INTO CASES AND VARIABLES DOES NOT APPLY
   TO THESE DATA.

   THE DATA ARE AVAILABLE IN CARD IMAGE FORM ON A STANDARD LABELED
   9-TRACK TAPE.  THE DATA MUST BE REARRANGED IF ONE WISHES TO USE
   THE USUAL PACKAGE PROGRAMS.

   GAUGE DATA
     110 RECORDS (ONE FOR EACH GAUGE LOCATION)
     6 VARIABLES - MEASUREMENT

   STORM DATA
     7/21   - 100 RECORDS
     7/25   - 240 RECORDS
     7/28A  - 160 RECORDS
     8/9    - 140 RECORDS
     9/28   - 120 RECORDS
     10/25  - 120 RECORDS
     11/3   -  80 RECORDS

     4 VARIABLES - IDENTIFICATION
     110 VARIABLES - IDENTIFICATION
     112 VARIABLES - MEASUREMENT (FOR EACH RECORD)

   ATTENUATION DATA
     7/14B  - 780 RECORDS
     7/28B  - 300 RECORDS
     8/25A  - 240 RECORDS
     9/21   - 240 RECORDS
     10/10  - 360 RECORDS

     4 VARIABLES - IDENTIFICATION
     5 VARIABLES - MEASUREMENT (FOR EACH RECORD)


BRIEF DESCRIPTION

   VAR     NAMES OF VARIABLES IN GAUGE DATA
   ***     ***** ** ********* ** ***** ****
    1      NOMINAL X-COORDINATE
    2      NOMINAL Y-COORDINATE
    3      GAUGE NUMBER
    4      SCANNING ORDER
    5      TRUE X-COORDINATE
    6      TRUE Y-COORDINATE

   VAR     NAMES OF VARIABLES IN STORM DATA
   ***     ***** ** ********* ** ***** ****
    1      DATE
    2      NUMBER OF RECORDS IN THIS BLOCK
    3      STARTING TIME
```

```
 4      FINAL TIME
***
1-110   GAUGE OPERATIVE
***
1-110   RAIN RATES
111     ATTENUATION
112     TIME

VAR     NAMES OF VARIABLES IN ATTENUATION DATA
***     ***** ** ********* ** *********** ****
 1      DATE
 2      NUMBER OF RECORDS IN THIS BLOCK
 3      STARTING TIME
 4      FINAL TIME
***
 1      TEN-READING AVERAGE OF THE MICROWAVE ATTENUATION
2-5     RAIN RATES RECORDED AT GAUGES 9, 17, 25 AND 33
```

PURPOSE OF STUDY

THE DATA ARE A SUBSET OF THE 1967 DATA FROM THE CRAWFORD HILL RAINFALL EXPERIMENT. THEY CONSIST OF RAIN RATES, ATTENUATION OF MICROWAVE RADIO TRANSMISSION AND ANCILLARY DATA FOR SELECTED STORMS. THE PURPOSE OF THE STUDY WAS TO EXPLORE THE EFFECT OF HEAVY RAINFALL ON ATTENUATION AND TO OBTAIN INFORMATION ON SPATIAL AND TEMPORAL CHARACTERISTICS OF RAIN.

REFERENCES

FREENY, MRS. A. E. AND J. D. GABBE. A STATISTICAL DESCRIPTION OF INTENSE RAINFALL. BELL SYS. TECH. J. 48, 1969, 1789.

SEMPLAK, R. A. AND H. E. KELLER. A DENSE NETWORK FOR RAPID MEASUREMENT OF RAINFALL RATE. BELL SYS. TECH. J. 48, 1969, 1745.

DETAILED DESCRIPTION

IN THE GAUGE AND STORM DATA SETS COLUMNS 73 TO 80 CONTAIN SEQUENCE NUMBERS BY WHICH THE CARDS ARE ORDERED. IN THE ATTENUATION DATA THE SEQUENCE NUMBERS ARE IN COLUMNS 19 TO 26. IN THE STORM AND ATTENUATION DATA THE FIRST FOUR COLUMNS OF THE SEQUENCE NUMBER CONTAIN THE STORM IDENTIFICATION, THAT IS, THE MONTH AND DAY OF THE STORM. THE SEQUENCE NUMBERS ARE NOT INCLUDED AS VARIABLES IN THE DESCRIPTIONS OF THE DATA SETS.

```
VAR    COL    FORMAT    DESCRIPTION OF GAUGE DATA
***    ***    ******    *********** ** ***** ****
```

THIS DATA SET GIVES SOME BASIC DATA ON THE ARRANGEMENT OF THE GAUGE NETWORK. THE RAIN GAUGES ARE ARRANGED IN A RECTANGULAR GRID OF 11 ROWS AND 10 COLUMNS. EACH GAUGE THUS

HAS A PAIR OF NOMINAL COORDINATES - THE COLUMN NUMBER AND THE ROW NUMBER OF THE SQUARE IN WHICH IT LIES. FOR EXAMPLE, THE NOMINAL COORDINATES OF GAUGE 10 ARE (3,2). THE X-COORDINATE IS GIVEN FIRST (RUNNING WEST TO EAST) AND THE Y-COORDINATE IS GIVEN SECOND (RUNNING NORTH TO SOUTH). FOR SIMPLICITY THE GAUGES ARE INDEXED BY THEIR NOMINAL COORDINATES, ALTHOUGH NOT ALL OF THE 110 SQUARES CONTAIN GAUGES.

THERE IS ONE CARD FOR EACH SQUARE. CARDS FOR SQUARES WITHOUT GAUGES CONTAIN ZEROS IN SOME FIELDS AS A DEFAULT VALUE.

1	1-3	XXX.	NOMINAL X-COORDINATE
2	4-6	XXX.	NOMINAL Y-COORDINATE
3	7-9	XXX.	GAUGE NUMBER
4	10-12	XXX.	SCANNING ORDER
5	13-18	XXXXXX.	TRUE X-COORDINATE FROM WEST TO EAST (FEET)
6	19-24	XXXXXX.	TRUE Y-COORDINATE FROM SOUTH TO NORTH (FEET) (NOTE THE REVERSAL OF DIRECTION OF THE Y-COORDINATE.)

FORMAT IS (4F3.0,2F6.0) N = 110

VAR	COL	FORMAT	DESCRIPTION OF STORM DATA
***	***	******	*********** ** ***** ****

THIS DATA SET GIVES 30-SECOND AVERAGES OF RECORDED RAIN RATES DURING SELECTED STORMS. IT CONSISTS OF SEVERAL BLOCKS OF CARDS, ONE BLOCK FOR EACH STORM SELECTED.

THE FIRST CARD IN THE BLOCK CONTAINS THE FOLLOWING INFORMATION -

1	3-6	XXXX.	DATE (MONTH, DAY)
2	7-12	XXXXXX.	NUMBER OF RECORDS IN THIS BLOCK
			THE NEXT TWO ITEMS REFER TO THE LOCATION OF THE STORM IN THE ISOMETRIC PLOTS.
3	13-18	XXXXXX.	STARTING TIME (MINUTES)
4	19-24	XXXXXX.	FINAL TIME (MINUTES)

FORMAT IS (2X,F4.0,3F6.0)

THE NEXT FIVE CARDS TELL WHICH GAUGES ARE OPERATIVE FOR THIS STORM. THE GAUGES ARE INDEXED ACCORDING TO THEIR NOMINAL COORDINATES AS FOLLOWS. THE SQUARES ON THE MAP ARE THOUGHT OF AS BEING STRUNG OUT IN LINEAR ORDER, STARTING IN THE UPPER LEFT CORNER, THEN MOVING EASTWARD ACROSS THE TOP ROW, THEN WEST TO EAST ACROSS THE SECOND ROW, AND SO ON, DOWN TO THE BOTTOM RIGHT CORNER. THE ENTRIES FOR THE GAUGES ARE PUNCHED CONSECUTIVELY, IN THE ABOVE ORDER,

```
                                 24 TO A CARD (14 ON THE LAST CARD).  THE
                                 ENTRIES PUNCHED ARE 1 IF THE GAUGE IS
                                 OPERATIVE, 0 OTHERWISE.

  1     1-3       XXX.          GAUGE 1
...     ...       ...           ...
 24    70-72      XXX.          GAUGE 24

 25     1-3       XXX.          GAUGE 25
...     ...       ...           ...
 48    70-72      XXX.          GAUGE 48

 49     1-3       XXX.          GAUGE 49
...     ...       ...           ...
 72    70-72      XXX.          GAUGE 72

 73     1-3       XXX.          GAUGE 73
...     ...       ...           ...
 96    70-72      XXX.          GAUGE 96

 97     1-3       XXX.          GAUGE 97
...     ...       ...           ...
110    40-42      XXX.          GAUGE 110

FORMAT IS (4(24F3.0/),14F3.0)

                                 FOLLOWING THESE CARDS ARE A NUMBER OF
                                 GROUPS OF FIVE CARDS.  EACH GROUP OF FIVE
                                 CARDS WILL BE CALLED A RECORD.  THE NUMBER
                                 OF RECORDS FOR THIS STORM IS GIVEN ON THE
                                 FIRST CARD DESCRIBED ABOVE.  EACH RECORD
                                 COVERS A PERIOD OF 30 SECONDS.

                                 FOR VARIABLES 1 TO 110 ZERO INDICATES THAT
                                 THE GAUGE WAS NOT OPERATING OR THAT THERE
                                 WAS NO RAINFALL.  A -1 INDICATES AN APPARENT
                                 GAUGE MALFUNCTION.  RAIN RATES ARE IN MM.
                                 PER HOUR.

  1     1-3       XXX.          FIRST RAIN RATE
...     ...       ...           ...
 24    70-72      XXX.          TWENTY-FOURTH RAIN RATE

 25     1-3       XXX.          TWENTY-FIFTH RAIN RATE
...     ...       ...           ...
 48    70-72      XXX.          FORTY-EIGHTH RAIN RATE

 49     1-3       XXX.          FORTY-NINTH RAIN RATE
...     ...       ...           ...
 72    70-72      XXX.          SEVENTY-SECOND RAIN RATE

 73     1-3       XXX.          SEVENTY-THIRD RAIN RATE
...     ...       ...           ...
 96    70-72      XXX.          NINETY-SIXTH RAIN RATE

 97     1-3       XXX.          NINETY-SEVENTH RAIN RATE
...     ...       ...           ...
110    40-42      XXX.          ONE HUNDRED AND TENTH RAIN RATE
111    43-48      XXX.XX        ATTENUATION (A 3-READING AVERAGE IN DECIBELS
```

```
                               OF THE ATTENUATION IN THE MICROWAVE SIGNAL)
112   49-54     XXX.XX         TIME (HOURS)

FORMAT IS (4(24F3.0/),14F3.0,2F6.2)


VAR    COL      FORMAT         DESCRIPTION OF ATTENUATION DATA
***    ***      ******         *********** ** *********** ****

                               THIS DATA SET GIVES MORE DETAILED DATA ON
                               THE MICROWAVE ATTENUATION AND THE RAIN
                               RATES AT THE FOUR GAUGES CLOSEST TO THE
                               MICROWAVE PATH - GAUGES 9, 17, 25 AND 33.
                               AS IN THE STORM DATA, THIS DATA SET CONSISTS
                               OF ONE BLOCK OF CARDS FOR EACH SELECTED
                               STORM.

                               THE FIRST CARD IS IDENTICAL IN FORM TO THE
                               FIRST CARD IN EACH BLOCK OF STORM DATA.
                               HOWEVER, THE MEANING OF THE TERM ''RECORD''
                               WILL BE DIFFERENT HERE.

  1     3-6     XXXX.          DATE (MONTH, DAY)
  2     7-12 XXXXXX.           NUMBER OF RECORDS IN THIS BLOCK

                               THE NEXT TWO ITEMS REFER TO THE LOCATION OF
                               THE STORM IN THE ISOMETRIC PLOTS.
  3   13-18 XXXXXX.            STARTING TIME (MINUTES)
  4   19-24 XXXXXX.            FINAL TIME (MINUTES)

FORMAT IS (2X,F4.0,3F6.0)


                               EACH RECORD COVERS A PERIOD OF TEN SECONDS.
                               AS BEFORE, A -1 INDICATES AN APPARENT GAUGE
                               MALFUNCTION.

  1     1-6     XXX.XX         TEN-READING AVERAGE OF THE MICROWAVE
                               ATTENUATION (DECIBELS)
  2     7-9     XXX.           RAIN RATE RECORDED AT GAUGE 9 (MM. PER HOUR)
  3    10-12    XXX.           RAIN RATE RECORDED AT GAUGE 17 (MM. PER HOUR)
  4    13-15    XXX.           RAIN RATE RECORDED AT GAUGE 25 (MM. PER HOUR)
  5    16-18    XXX.           RAIN RATE RECORDED AT GAUGE 33 (MM. PER HOUR)

FORMAT IS (F6.2,4F3.0)


LOCATION OF DATA

THE DATA ARE AVAILABLE IN CARD IMAGE FORM ON A STANDARD LABELED
9-TRACK TAPE LABELED DSLIBI.  THE DATA MUST BE REARRANGED IF ONE
WISHES TO USE THE USUAL PACKAGE PROGRAMS.

  DATA                         DSNAME     LABEL
  ****                         ******     *****
  GAUGE                        GAUGE        1
  STORM 7/21                   ST0721       2
  STORM 7/25                   ST0725       3
  STORM 7/28A                  ST0728A      4
  STORM 8/9                    ST0809       5
```

```
STORM 9/28              ST0928        6
STORM 10/25             ST1025        7
STORM 11/3              ST1103        8
ATTENUATION 7/14B       AT0714B       9
ATTENUATION 7/28B       AT0728B      10
ATTENUATION 8/25A       AT0825A      11
ATTENUATION 9/21        AT0921       12
ATTENUATION 10/10       AT1010       13
```

LISTING OF GAUGE DATA (THE FIRST TWO LINES ARE COLUMN GUIDES)

FOR COMPACT PRINTING TWO RECORDS ARE LISTED ON ONE LINE.

```
0         1         2    ...7 8
123456789012345678901234...890

  1  1  1  1  2200 45680     1
  2  1  2  3  6200 47000     2
  3  1  3  5 11140 46730     3
  4  1  4  7 14920 45330     4
  5  1  5  9 20520 45680     5
  6  1  6 11 24720 45270     6
  7  1  7 13 28170 45490     7
  8  1  0  0     0     0     8
  9  1  0  0     0     0     9
 10  1  0  0     0     0    10
  1  2  8 55  2650 42300    11
  2  2  9 57  6950 41510    12
  3  2 10 59 11370 41280    13
  4  2 11 61 15830 42090    14
  5  2 12 63 20000 40880    15
  6  2 13 65 23850 42320    16
  7  2 14 67 29200 41090    17
  8  2  0  0     0     0    18
  9  2  0  0     0     0    19
 10  2  0  0     0     0    20
  1  3 15 15  2780 38000    21
  2  3 16 17  6800 38100    22
  3  3 17 19 12250 37690    23
  4  3 18 21 16000 37520    24
  5  3 19 23 19850 37000    25
  6  3 20 25 24200 37340    26
  7  3 21 27 28230 37460    27
  8  3 71 29 33900 37250    28
  9  3 72 31 37060 38150    29
 10  3 73 33 41750 37800    30
  1  4 22 69  1650 32310    31
  2  4 23 71  5970 33060    32
  3  4 24 73 11250 33640    33
  4  4 25 75 16300 34340    34
  5  4 26 77 18610 34050    35
  6  4 27 79 24260 33230    36
  7  4 28 81 28340 34270    37
  8  4 74 83 34050 33640    38
  9  4 75 85 37430 31800    39
 10  4 76 87 40740 33800    40
  1  5 29 35  2150 29410    41
  2  5 30 37  6000 29540    42
  3  5 31 39 11140 28660    43
```

```
0         1         2    ...7 8
123456789012345678901234...890

  6  6 40 99 24550 24510    56
  7  6 41  2 29030 24430    57
  8  6 82  4 32600 24370    58
  9  6 83  6 37750 24100    59
 10  6 84  8 40430 22830    60
  1  7 36 48  3590 18900    61
  2  7 42 50  6080 18934    62
  3  7 43 52 10110 19860    63
  4  7 44 54 15960 19280    64
  5  7 45 56 19240 19510    65
  6  7 46 58 24030 19570    66
  7  7 47 60 29090 18400    67
  8  7 85 62 33760 20820    68
  9  7 86 64 37310 19490    69
 10  7 87 66 40572 19430    70
  1  8 48 10  3500 14130    71
  2  8 49 12  5730 15640    72
  3  8 50 14 10910 15200    73
  4  8 51 16 16300 15060    74
  5  8 52 18 20000 15580    75
  6  8 53 20 24610 16040    76
  7  8 54 22 28510 14600    77
  8  8 88 24 33270 16270    78
  9  8 89 26 35940 15030    79
 10  8 90 28 41540 14940    80
  1  9 55 68  4000 11400    81
  2  9 56 70  6720  9790    82
  3  9 57 72 11080 11580    83
  4  9 58 74 15660 10950    84
  5  9 59 76 20060 11530    85
  6  9 60 78 23910 11810    86
  7  9 61 80 28700 11440    87
  8  9 91 82 32520 10050    88
  9  9 92 84 36740 10800    89
 10  9 93 86 41490 10150    90
  1 10  0  0     0     0    91
  2 10 62 30  5790  7070    92
  3 10 63 32 11540  6780    93
  4 10 64 34 14530  6370    94
  5 10 65 36 19830  7130    95
  6 10 66 38 24200  5330    96
  7 10 67 40 27770  6750    97
  8 10 94 42 32120  5330    98
```

```
  4   5 32 41 16170 29300     44           9 10 95 44 36800   5890     99
  5   5 33 43 20230 29820     45          10 10 96 46 41980   6550    100
  6   5 77 45 25130 27800     46           1 11  0  0     0      0    101
  7   5 78 47 29570 29530     47           2 11  0  0     0      0    102
  8   5 79 49 32920 29390     48           3 11  0  0     0      0    103
  9   5 80 51 37030 28250     49           4 11 68 88 16000   1570    104
 10   5 81 53 42000 27500     50           5 11 69 90 19990   2610    105
  1   6  0  0  3420 25250     51           6 11 70 92 24660   2030    106
  2   6  0  0  7130 24780     52           7 11 97 94 29400   1970    107
  3   6 37 93 12160 24900     53           8 11 98 96 34170   2640    108
  4   6 38 95 15320 24380     54           9 11  0  0 38610   1660    109
  5   6 39 97 19100 23680     55          10 11  0  0  4160    249    110
```

LISTING OF STORM DATA FOR STORM 7/28A

(THE FIRST FOUR LINES ARE COLUMN GUIDES)

A SAMPLE OF THE DATA FILE, CONSISTING OF THE FIRST FIVE MINUTES OF THE STORM, IS LISTED BELOW. THE ENTIRE FILE FOR THIS ONE STORM IS 16 TIMES AS LONG AS THE SAMPLE.

```
0         1         2         3         4         5         6
123456789012345678901234567890123456789012345678901234567890123456

                                                               7         8
                                                      78901234567890

  728   160   146   225
                                                             7280001
 1  1  0  1  1  0  1  0  0  0  1  1  1  1  1  1  1  0  0  0  1  1
                                                         1  1 7280002
 1  0  1  1  1  1  1  1  1  1  1  1  1  1  1  1  1  1  1  1  1  1
                                                         1  1 7280003
 1  1  0  0  1  1  1  1  1  0  1  1  1  1  0  1  1  1  1  1  1  1
                                                         1  0 7280004
 0  0  1  1  0  0  1  1  0  1  1  1  1  0  1  1  1  1  0  1  0  1
                                                         1  1 7280005
 1  1  1  0  0  0  0  1  1  1  1  1  0  0
                                                             7280006
21 15  0  3  4  0  8  0  0  0  4  6  1 17  0 12  0  0  0  0 19 11
                                                         4  4 7280007
15  0 16  4  1  8 20  0  7 16  5  8  1  1  0  0  0 22  0  4 16  5
                                                         2 12 7280008
 8  0  0  0  1  1  0  0  0  0  6  1  0  0  0  0  0  5  4  2  5  8
                                                         8  0 7280009
 0  0  7  0  0  0  4  3  0  0  8  2  7  0  2  0  5 16  0  6  0  1
                                                         3 24 7280010
14 15  9  0  0  0  0  1  6  3 15 14  0  0  0.56 16.85
                                                             7280011
28 20  0  3  4  0  8  0  0  0  9  6  1 17  0 12  0  0  0  0 18 16
                                                         4  4 7280012
15  0 16  4  1  9 28  0  7 16  5  8  0  1  0  0  0 22  0  0 16  9
                                                         0 14 7280013
15  0  0  0  1  1  0  0  0  0  6  5  0  0  0  0  0  5  3  2  5  8
                                                         8  0 7280014
 0  0  7  0  0  0  4  3  0  0  8  2  7  0  2  0  5 19  0  6  0  1
                                                         3 24 7280015
14 28 19  0  0  0  0  1  6  3 11 50  0  0  0.91 16.85
                                                             7280016
```

```
26 21  0  3  4  0  8  0  0  0  5  6  1 17  0 12  4  0  0  0 24 12
                                                        4  4 7280017
15  0 16  4  1  7 31  5  7 16  5  8  0  1  0  0  4 22  0  0 16  9
                                                        0  9 7280018
 8  0  0  0  1  1  0  0  0  0  6  8  0  0  0  0  0  5  3  1  5  8
                                                        8  0 7280019
 0  0  7  0  0  0  4  3  0  0  8  5  9  0  0  0  5 14  0  5  0  1
                                                        3 24 7280020
13 26 11  0  0  0  0  1  6  3 11 31  0  0 1.34 16.86
                                                             7280021
21 14  0  3  4  0  8  0  0  0  3 11  5 17  0 12  0  0  0  0 30 15
                                                        4  4 7280022
15  0 16  4  1  6 51 11  7 16  5  8  0  6  0  0 11 26  0  0 16  9
                                                        0 12 7280023
15  0  0  0  1  1  0  0  0  0  6  4  0  0  0  0  0 10  4  1  4  8
                                                        8  0 7280024
 0  0  7  0  0  0  4  3  0  0  8  2 11  0  0  0  5 13  0  5  0  1
                                                        3 24 7280025
13 30 10  0  0  0  0  1  8  8  7 25  0  0 1.63 16.87
                                                             7280026
22 17  0  3  4  0  8  0  0  0  4 16  6 17  0 12  0  0  0  0 22 12
                                                        4  4 7280027
15  0 16  4  1  7 48  8  7 15  5  8  0  1  0  0  4 49  0  0 16  9
                                                        1  9 7280028
11  1  0  0  1  1  0  0  0  0  6  1  0  0  0  0  0  5  5  4  5  7
                                                        8  0 7280029
 0  0  7  0  0  0  4  3  0  0  8  2  7  0  0  0  5 12  0  4  0  1
                                                        3 25 7280030
13 27 10  0  0  0  0  1 10 14 16 21  0  0 2.03 16.88
                                                             7280031
20 16  0  3  4  0  8  0  0  0  5 20  5 17  0 12  0  0  0  0 21 12
                                                        4  4 7280032
15  0 16  4  1  5 46  9  7 15  5  8  0  1  0  0 24 55  0  0 16  9
                                                        1  7 7280033
 9  0  0  0  1  1  0  0  0  0  5  2  0  0  0  0  0  4  3  2  5  7
                                                        8  0 7280034
 0  0  7  0  0  0  6  3  0  0  8  3  7  0  0  0  5 11  0  3  0  5
                                                        3 22 7280035
12 22 36  0  0  0  0  1  9  7 13 26  0  0 2.10 16.89
                                                             7280036
25 20  0  3  7  0  8  0  0  0  3 23  3 17  0 12  0  0  0  0 41 20
                                                        4  4 7280037
15  0 16  4  1  4 47 13  7 15  5  8  0  3  0  0 36 91  0  0 16  8
                                                        0 10 7280038
10  0  0  0  1  1  0  0  0  0  5  2  0  0  0  0  0  4  3  2  5  7
                                                        8  0 7280039
 0  0  7  0  0  0  4  3  0  0  8  2  7  0  0  0  5 12  0  9  0  6
                                                        3 16 7280040
12 22 25  0  0  0  0  1  7  5 16 25  0  0 2.47 16.90
                                                             7280041
23 19  0  4 11  0  8  0  0  0  3 34  3 17  0 12  0  0  0  0 46 23
                                                       19  4 7280042
15  0 16  4  1  3 33 25 22 15  5  8  0  2  0  0 58 86  0  0 16  8
                                                        1 13 7280043
13  0  0  0  1  1  0  0  0  0  5  8  0  0  0  0  0  4  3  2  5  7
                                                        8  0 7280044
 0  0  7  0  0  0  4  3  0  0  8  2  7  0  0  0 26 18  0 13  0  6
                                                        3 16 7280045
13 25 23  0  0  0  0  5 10  6 40 26  0  0 4.51 16.90
                                                             7280046
```

```
 35 19  0  6  4  0  8  0  0  0  3 47  3 17  0 12  5  0  0  0 44 27
                                                      54  4 7280047
 15  0 16  4  1  2 40 27 31 15 13  8  0  1  0  0 67 76 81  0 16  8
                                                       1 14 7280048
 11  0  0  0  1  1  0  0  0  0  5  8  0  0  0  0  0  4  3  2  5  7
                                                       8  0 7280049
  0  0  7  0  0  0  4 47  0  0  8  2  7  0  0  4 55 63  0 34  0  6
                                                       3 16 7280050
 14 24 19  0  0  0  0 45 25  5 34 28  0  0  8.01 16.91
                                                            7280051
 33 18  0  6  4  0  8  0  0  0  3 48  4 20  0 12  4  0  0  0 29 31
                                                      63  4 7280052
 15  0 16  4  1  1 31 46 32 15 12  8  0  1  0  0 87 63101  0 16  8
                                                       0 13 7280053
 11  4  0  0  1  1  0  0  0  0  5  8  0  0  0  0  0  4  3  2  5  7
                                                       8  0 7280054
  0  0  7  0  0  0  4 73  0  1  8  2  7  0  0  7 53 59  0 32  0  6
                                                       3 16 7280055
 14 25 32  0  0  0  0 43 33  7 27 26  0  0 11.36 16.92
                                                            7280056
```

LISTING OF ATTENUATION DATA FOR STORM 9/21

(THE FIRST TWO LINES ARE COLUMN GUIDES)

EXCEPT FOR THE FIRST RECORD, TWO RECORDS ARE LISTED PER LINE
FOR COMPACT PRINTING.

```
0         1         2                 0         1         2
12345678901234567890123456            12345678901234567890123456

   921    240       1    40
  4.64 26  2  4  7 9218  1              10.33 19123 34 12 9218121
  3.67 21  2  4  7 9218  2               9.43 16117 34 12 9218122
  3.06 23  2  4  7 9218  3               9.84 24 69 55 12 9218123
  2.49 23  2  4  7 9218  4              12.09 15 74 66 12 9218124
  1.70 21  2  4  7 9218  5              12.92 16122 48 12 9218125
  1.03 18  2  4  7 9218  6              13.99 33145 71 12 9218126
  0.72 20  2  4  7 9218  7              13.85 19119 63 12 9218127
  0.94 13  2  4  7 9218  8              13.12 19176 75 12 9218128
  1.43 14  2  4  7 9218  9              10.49 18150 44 12 9218129
  1.83 19  2  4  7 9218 10               8.20 18 51 48 13 9218130
  1.68 18  2  4  7 9218 11               6.74 27 65 64 12 9218131
  2.43 14 15  4  7 9218 12               6.55 20 74 56 12 9218132
  2.42 17  8  4  7 9218 13               5.69 31 90 36 13 9218133
  2.43 12  5  4  7 9218 14               4.47 23 64 49 13 9218134
  2.62 19  9  4  7 9218 15               3.17 19 71 31 12 9218135
  4.09 18  6  4  7 9218 16               2.84 16 94 24 13 9218136
  5.39 19  5  4  7 9218 17               2.04 21 85 21 13 9218137
  7.64 22  5  4  7 9218 18               1.83 17 49 22 13 9218138
 10.78 31  6  4  7 9218 19               1.37 16 42 26 13 9218139
 15.11 29 17  4  7 9218 20               1.32 20 60 27 13 9218140
 19.01 46 12  4  7 9218 21               1.54 14 38 22 13 9218141
 22.83 24 21  4 24 9218 22               2.33 16 70 29 24 9218142
 24.95 40 59 75 13 9218 23               2.42 14 27 30 24 9218143
 26.83 37 76 31 15 9218 24               2.35 22 31 29 13 9218144
 23.86 30102 32 10 9218 25               2.47 14 25 29 13 9218145
 19.87 24129 25 11 9218 26               2.96 17 36 25 13 9218146
 19.40 34102 29 14 9218 27               3.62 18 37 40 18 9218147
```

16.70	37	111	28	13	9218	28
13.20	37	76	28	9	9218	29
9.95	27	65	23	11	9218	30
9.12	29	92	22	9	9218	31
10.87	33	79	21	9	9218	32
9.96	39	92	22	13	9218	33
8.12	24	78	25	12	9218	34
6.55	26	119	21	9	9218	35
6.30	29	116	14	10	9218	36
5.66	26	78	14	17	9218	37
5.41	26	113	20	13	9218	38
5.85	30	104	18	10	9218	39
6.43	17	83	17	10	9218	40
9.23	32	89	16	10	9218	41
15.35	24	92	24	9	9218	42
19.92	26	108	52	9	9218	43
20.97	31	140	49	10	9218	44
20.78	19	141	32	11	9218	45
19.69	17	112	55	18	9218	46
18.23	20	157	41	20	9218	47
19.27	51	164	32	17	9218	48
19.68	52	108	39	18	9218	49
19.11	65	74	36	24	9218	50
17.77	36	119	37	15	9218	51
15.91	39	83	39	19	9218	52
14.29	30	113	41	18	9218	53
13.10	25	88	35	10	9218	54
13.36	24	102	37	10	9218	55
13.75	21	67	24	11	9218	56
14.49	25	93	39	13	9218	57
15.10	19	98	25	11	9218	58
12.86	17	154	23	16	9218	59
9.23	14	141	32	20	9218	60
6.93	13	111	28	20	9218	61
3.66	15	162	19	12	9218	62
1.74	15	87	17	12	9218	63
1.16	15	102	27	12	9218	64
0.94	22	47	25	12	9218	65
0.73	13	39	16	12	9218	66
0.78	14	27	19	12	9218	67
0.62	13	29	21	8	9218	68
0.62	10	24	19	9	9218	69
0.51	10	23	17	9	9218	70
0.58	14	26	15	8	9218	71
0.75	12	35	22	8	9218	72
0.86	14	21	26	9	9218	73
1.03	12	22	15	9	9218	74
0.98	20	38	16	9	9218	75
0.97	14	34	22	9	9218	76
0.74	15	33	26	9	9218	77
0.69	17	47	15	9	9218	78
0.58	15	46	20	8	9218	79
0.54	13	45	17	8	9218	80
0.31	9	40	16	19	9218	81
0.31	11	34	16	22	9218	82
0.50	9	31	16	16	9218	83
0.65	14	35	19	18	9218	84
0.87	11	33	18	9	9218	85
0.93	14	30	18	17	9218	86
0.99	13	22	21	9	9218	87

3.79	24	54	33	16	9218	148
4.10	16	41	42	16	9218	149
4.99	26	53	29	14	9218	150
5.06	23	76	30	13	9218	151
5.32	16	59	31	11	9218	152
5.09	16	65	33	9	9218	153
5.13	15	70	31	9	9218	154
5.10	18	70	30	19	9218	155
4.19	21	30	29	8	9218	156
3.70	14	44	42	9	9218	157
3.55	13	37	32	9	9218	158
4.10	12	28	32	9	9218	159
4.21	14	22	30	9	9218	160
3.89	15	37	33	9	9218	161
4.07	13	59	34	8	9218	162
3.37	12	36	31	9	9218	163
2.93	13	26	35	9	9218	164
2.65	12	35	27	9	9218	165
2.16	16	36	25	9	9218	166
1.87	15	32	30	9	9218	167
1.50	19	31	31	9	9218	168
1.47	10	37	22	9	9218	169
1.35	12	43	27	18	9218	170
1.25	11	32	28	9	9218	171
1.31	15	34	22	21	9218	172
1.21	13	57	24	8	9218	173
1.28	16	34	20	9	9218	174
1.78	14	40	19	9	9218	175
2.06	15	36	18	8	9218	176
2.41	12	38	19	8	9218	177
2.74	13	51	26	8	9218	178
2.88	10	33	24	8	9218	179
3.04	10	36	21	8	9218	180
2.89	12	56	20	8	9218	181
2.76	10	93	18	8	9218	182
2.84	19	80	21	9	9218	183
3.08	16	71	24	8	9218	184
2.97	21	38	27	8	9218	185
3.02	13	65	24	8	9218	186
2.30	11	54	28	8	9218	187
2.27	13	47	29	8	9218	188
1.24	10	50	22	8	9218	189
0.94	13	25	22	8	9218	190
0.41	10	32	19	8	9218	191
0.19	9	30	19	8	9218	192
0.04	11	55	19	8	9218	193
0.03	10	31	18	8	9218	194
0.01	10	30	18	8	9218	195
0.06	8	17	14	8	9218	196
0.17	17	25	16	8	9218	197
0.02	14	17	14	8	9218	198
0.04	12	32	11	8	9218	199
0.08	6	23	12	8	9218	200
0.44	7	21	15	8	9218	201
0.08	17	29	11	8	9218	202
0.09	6	14	14	8	9218	203
0.02	8	21	13	8	9218	204
0.26	6	25	11	8	9218	205
0.01	14	13	12	8	9218	206
0.07	6	17	11	8	9218	207

1.01	16	31	20	9	9218 88
0.95	13	24	20	9	9218 89
1.17	13	35	22	9	9218 90
1.01	12	32	18	9	9218 91
0.83	13	47	16	24	9218 92
0.78	9	52	16	13	9218 93
0.60	11	45	17	12	9218 94
0.52	9	38	17	12	9218 95
0.63	19	47	18	13	9218 96
0.76	16	36	19	12	9218 97
1.08	12	23	17	12	9218 98
1.72	18	22	17	12	9218 99
1.70	13	22	16	12	9218100
1.63	15	25	16	12	9218101
1.60	24	33	18	13	9218102
1.33	20	23	19	12	9218103
1.31	14	26	20	12	9218104
1.72	16	36	17	12	9218105
2.09	18	27	19	12	9218106
2.72	11	59	19	12	9218107
2.76	14	33	17	12	9218108
2.70	15	64	18	12	9218109
2.74	20	58	29	12	9218110
3.72	16	71	28	12	9218111
4.34	16	44	24	12	9218112
5.10	17	41	21	12	9218113
6.29	24	53	26	12	9218114
6.92	16	77	22	12	9218115
7.79	17	63	32	12	9218116
9.05	17	98	28	12	9218117
9.58	19	128	37	12	9218118
9.92	16	122	25	12	9218119
9.74	17	108	25	12	9218120
0.08	6	13	12	8	9218208
0.04	6	20	12	8	9218209
0.03	6	17	10	8	9218210
0.08	8	24	9	8	9218211
0.09	6	17	14	18	9218212
0.04	6	16	11	9	9218213
0.15	5	19	10	20	9218214
0.04	7	17	12	10	9218215
0.33	6	16	12	18	9218216
0.08	13	18	11	12	9218217
0.01	7	17	10	15	9218218
0.06	6	30	10	16	9218219
0.01	8	18	12	21	9218220
0.03	6	17	14	12	9218221
0.03	6	16	16	16	9218222
0.01	6	19	18	18	9218223
0.01	12	17	13	16	9218224
0.	7	16	14	18	9218225
0.03	6	19	17	22	9218226
0.07	6	17	19	14	9218227
0.02	15	16	14	12	9218228
0.01	7	21	12	9	9218229
0.01	6	18	15	14	9218230
0.22	6	17	13	15	9218231
0.02	7	16	12	15	9218232
0.38	6	20	13	9	9218233
0.12	10	18	15	9	9218234
0.14	7	17	17	13	9218235
0.24	6	20	13	18	9218236
0.07	6	33	12	9	9218237
0.14	6	36	11	22	9218238
0.08	8	28	15	19	9218239
0.18	7	39	14	20	9218240

REFERENCES

Amorocho, J. and A. Brandstetter (1967). The representation of storm precipitation fields near ground level. J. Geophys. Res. 72, 1145.

Freeny, A. E. (Mrs.) (1969). Statistical treatment of rain gauge calibration data. Bell Sys. Tech. J. 48, 1757.

Freeny, A. E. (Mrs.) and J. D. Gabbe (1969). A statistical description of intense rainfall. Bell Sys. Tech. J. 48, 1789.

Marquardt, D. W. (1963). An algorithm for least squares estimation of nonlinear parameters. J. Soc. Indust. Appl. Math. 2, 431.

Medhurst, R. G. (1965). Rainfall attenuation of centimeter waves: comparison of theory and measurement. I. E. E. E. Trans. on Antennas and Propagation 550.

Semplak, R. A. (1966). Gauge for continuously measuring rate of rainfall. Rev. Sci. Instr. 37, 1554.

Semplak, R. A. and H. E. Keller (1969). A dense network for rapid measurement of rainfall rate. Bell Sys. Tech. J. 48, 1745.

Semplak, R. A. and R. H. Turrin (1969). Some measurements of attenuation by rainfall at 18. 5 GHz. Bell Sys. Tech. J. 48, 1767.

DISCUSSION OF THE JAECKEL AND GABBE CHAPTER

R. Moore

In reading the preliminary draft of this presentation and associated references prior to the Conference, I was reminded that communication between the researcher and the data analyst is often hampered by ill-designed designation systems. For example, in this work the gauge numbering system seems to have no special ordering within the network (as shown in figure 1). Nor is the sampling order conveniently identified with the gauge or location orders. Now, this type of condition is seldom the fault of the data analyst. But it does point up the desirability of getting the analyst into the game early, so that systematic designations can be established before the data are actually taken. Such designations most certainly permit more convenient reporting of the experiment and its results, thus improving the information transmittal to the reader.

This paper was well-received by the Conference attendees, as indicated by the high ranking accorded it on the rating forms distributed by the Chairman. The problem statement, the presentation, and the use of computing capabilities were deemed exemplary.

Still, these very virtues give me pause. We have completed a decade of data analysis (stemming from Tukey's article in The Annals of Mathematical Statistics, March 1962). Is a new profession developing, or are we seeing a revival of the 19th century phenomena that gave rise to statistics in the first place? Moreover, I have some concern about the use of some statistical methods (such as F tests and factor analysis) whose basic validity is highly dependent upon distributional assumptions, which are used without so much as a caveat. Should we be training our students to ignore fundamental assumptions, or just not to worry about them? If these be adjudged personal hangups, so be it; but enough of my statistical friends express similar concerns to convince me that our profession should examine its own criteria for membership.

J. Dickey

We have, in this paper, an example of statisticians letting their life's blood in an analysis of data that has been handed over to them. The statisticians survived and the resulting analysis is a thing of beauty. They seem to have first asked themselves, "What is it possible to do?" There were, for instance, all kinds of uncalibrated, nonlinear-malfunctioning rain rate gauges (cleaned each six weeks, but calibrated each two years). The gauge response time was under one second but the sampling frequency was once per ten seconds for each gauge, except for gauge 33 which was sampled once per second but which usually malfunctioned. They take more than one notice of a handful "of gauges, at which something peculiar must have been happening."

The things possible to analyze turned out to be:

- drift of storm (as defined by authors);
- relation of rainfall rates to attenuation of microwave radio transmission; and
- compact description of a high-rain-rate event.

The authors decided to ignore higher-frequency variation by smoothing to one observation per 30 seconds for each gauge, in order to distinguish "true rainfall" from "noise." I wonder whether the "noise" was in the actual rain rates (seems likely) or in the gauges (seems likely too). I feel that the authors' smoothing was done merely to help the broader patterns emerge from the analysis, especially for the isometric plots.

Rain rate power spectra and cospectra (perhaps after a log transformation) would be of interest in the higher frequencies;

and I would imagine such have been studied in the meteorological literature. They would certainly be of some interest in the attenuation problem; and also, a falling off of cospectra in the higher frequencies might narrow the range of possible propagational mechanisms for drift.

Surely a lot more can be said about the possible causes of drift from general meteorological knowledge of wind-speed orders of magnitude, variation and covariation, as functions of altitude, over time in single storms and from storm to storm. The "real" location differences in whole storm drift from quadrant to quadrant are puzzling, but perhaps nonuniformity of gauges is somehow to blame.

It took me a while to see that α does <u>not</u> "represent the time it takes for the storm to move one unit of distance eastward" (that is what $1/v_x$ represents). But, α is the time lag of greatest correlation for two places differing in only their x coordinates by one unit; the quotient of their difference-vector's component parallel to (v_x, v_y) divided by the storm speed; or the time it takes for the storm to travel from Place 1 to the point on its path nearest to Place 2 = Place 1 + (1,0).

I can't help thinking that treatment of a storm as a rain-rate-valued random function of a three-dimensional (x, y, t) variable would be fruitful. The authors have studied one aspect of the correlation function.

On the attenuation regression, I would just like to say it might still be of interest to fit also the theoretical A(t) curve after empirical linear calibration of the gauges.

J. Tukey

(Ed. Note: Professor Tukey was not present during the discussion of the Jaeckel and Gabbe analysis. His brief comment was prompted by a preconference outline of the analysis.)

I do not have the longer accounts with me at this writing, so I will confine my remarks to suggesting two things:

- Might not using log rain rate simplify some analyses?
- In assessing the microwave absorption, would it not make sense to include slopes and curvatures of rain aslant the measured path?

CHAPTER 4

EEG FREQUENCY DISTRIBUTION DATA

M.R. MICKEY
Department of Biomathematics, University of California, Los Angeles

These analyses are based on a study that attempted to make quantitative and objective measurements of drug induced alterations in spontaneous electrical activity of the brain. The general goal of the study was to contribute to the understanding of the effects of psychotropic drugs in man through studying these effects in experiments based on the cat.

The two drugs used in the study were atropine and physostigmine; they were chosen because of their action on the cholinergic mechanisms in the central nervous system.

The following discussion is self-contained. One reference, Jenden et al (1972), is appropriate since it presents a general review of the study to which the present discussion relates and contains detailed references on various points that may be of interest to the reader.

Electrical activity of the brain was assessed from chronically implanted electrodes consisting of two lengths of 38 gauge, enamel insulated, stainless steel wire affixed together with epoxylite insulator. The electrodes were placed in four morphologically defined regions: prepyriform cortex, ventral hippocampus, lateral geniculate and the midbrain reticular formation. Implants in the first three regions were made using a "mirror-image" evoked potential method, a technique that results in accurate emplacement.

Since drug effects may last for hours, the data were reduced to manageable proportions by means of a laboratory designed and constructed frequency analyzer. The analyzer is a broad-band instrument consisting of eight tuned amplifiers covering the frequency range between 2.5 and 40 Hz in equal, half-octave steps with sharp attenuation (50 db/octave) outside the band. The outputs are connected through rectifying circuits to integrators with essentially infinite time constants.

Digital information concerning the mean amplitude in each channel was obtained for every one-minute period of an experiment; the output was in the form of typewriter printing and punched cards. The reduced data thus consist of values of 32 variables (average amplitude in each of eight frequency bands in each of four brain regions) for each minute of an experiment except for some intentional gaps during the recovery portion of some of the experiments.

Over a period of time, six experiments were conducted on each of three cats; each cat received two dose levels of each drug and two normal saline controls. The drug experiments were alternated with control studies in which no injections were made. Experiments were done weekly so the drug experiments were separated by intervals of at least two weeks.

In contrasting data between pre- and post-drug periods, we realized that there was a possibility that the pattern of pre-drug data was poorly defined — it might depend strongly on the behavioral state of the animals. For example, recordings from an alert animal might be characteristically different from those of a sleeping animal. Therefore, a study of data patterns that might be associated with behavioral states was given top priority.

BEHAVIORAL STATES

Three behavioral states were recognized in the study: slow

sleep (S), fast sleep (F), and alert (A). Classification of state during an experiment was based on visual observation of the animal (through a one-way mirror), and monitoring electrical activity from neck and eye muscles and cortical EEG leads as well as monitoring the leads from the brain areas under study. Not all minutes are classified as one of these three states; minutes for which the animal is not clearly classifiable as S, F or A throughout the minute are classified as U (unclassified); and minutes for which there is a clear artifact (such as that resulting from some movement) are classified as R (reject). The classification process thus also accomplished some data screening as well.

We were interested in knowing whether the points representing data cases classified as S, F or A formed clusters; and if clusters were formed, if they were stable over time. Would different cats give essentially the same clusters? Would the plane formed by the cluster averages, the SFA plane, contain the transitional periods as well?

To investigate these questions we made use of linear discriminant analysis — in particular, the version used in computer program BMD07M (Dixon, 1971). A high fraction of "correct" classification from the three-group discriminant analysis would indicate clustering of the data corresponding to behavioral state. We were also interested in questions concerning the stability of the discriminant functions. (Would the functions developed from one experiment apply successfully in classifying the states of subsequent experiments from the same cat? And to what extent could the states of a cat be successfully predicted from using the discriminant functions computed from other cats?) A substantial stability was found within the same animal and from cat to cat.

To illustrate the analysis, the data base was shortened by restricting it to the eight frequency channels from the first brain area (the prepyriform cortex). These restricted data are listed in the first data set of the Data Set Description at the end of this chapter. The data are listed in the observed order and include the unclassified minutes (U) as well as those classified as S, F or A.

The restriction of data limits the discrimination information and in particular omits the structure resulting from contrasting frequency band amplitudes between the brain regions; the illustrative value of the analysis is not handicapped by this limitation. To illustrate the discrimination stability, the data set was further restricted to the first and last control experiments for each of the three cats.

The experiments were spread out over the better part of a year. Table 1 shows this dispersion and is included here to emphasize that a substantial time period elapsed between the first and last experiments. Prediction of state over a longer time lapse would be expected to be more difficult than over a shorter time span.

Table 1. Experiments in first data set.

Cat No.	Exp. No.	Date
2	2	24 Jan 69
2	40	17 Oct 69
8	27	25 July 69
8	42	31 Oct 69
9	14	18 Apr 69
9	26	18 July 69

The clustering of data by states is illustrated in figure 1; it is shown as a multigroup discriminant analysis from output of BMD07M program (Dixon, 1971). In addition to computing discriminant functions and classifying cases (both those on which the analysis is based and additional cases to be classified, if desired), the program computes a canonical correlations analysis in which a "second set" of dummy variables corresponds to groups. The generated canonical variables form an orthogonal (with respect to the within groups covariance matrix) set of

FIGURE 1. PLOT FROM PROGRAM BMDO7M SHOWING CLUSTERING OF EEG AMPLITUDE PROFILES

KEY

S = SLOW SLEEP
F = FAST SLEEP
A = ALERT
* = GROUP MEANS
$ = OVERLAP

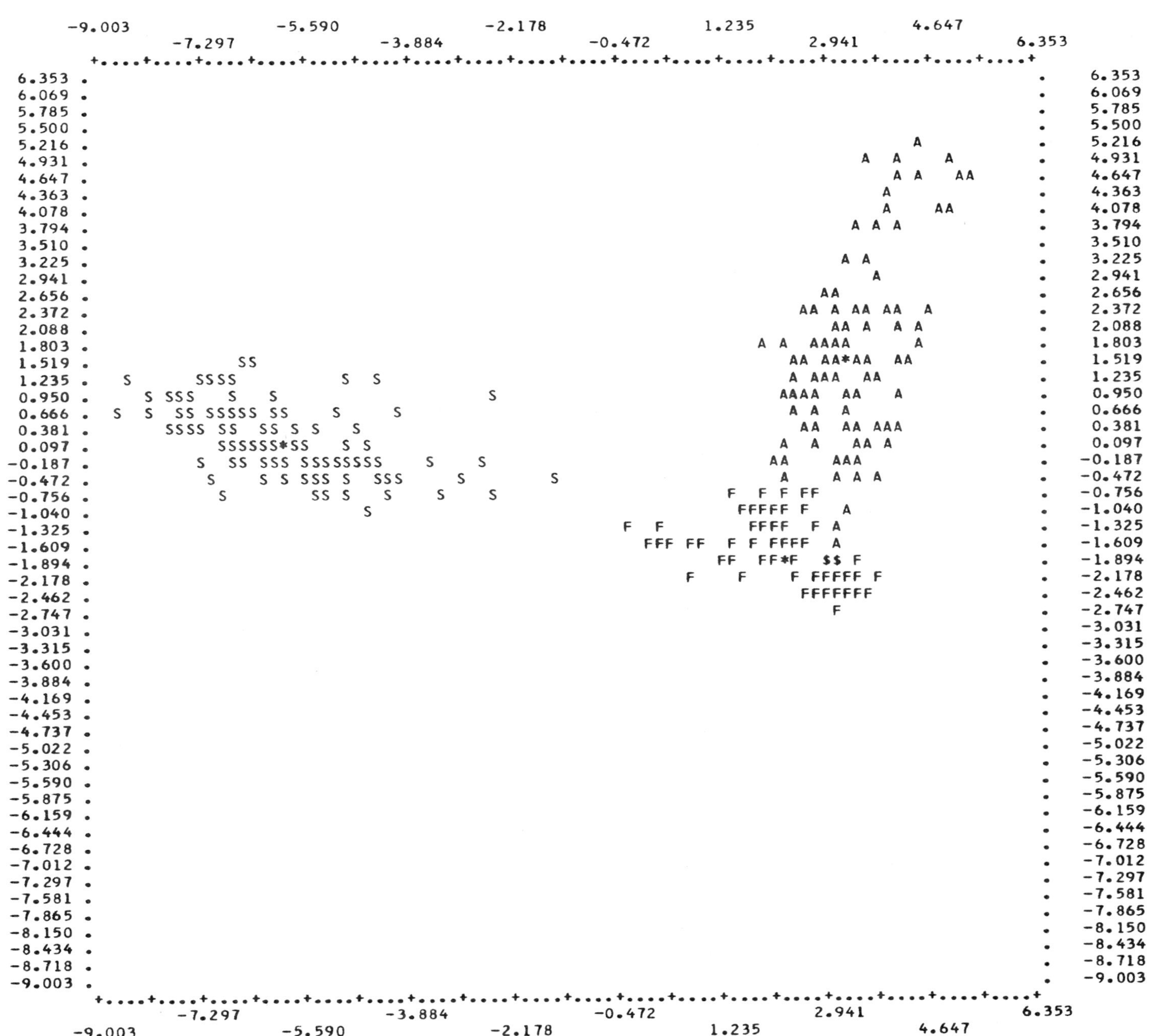

variables in which the first (g - 1) contain all of the information for linear discrimination among the g groups and the remaining canonical variables are residuals that do not linearly discriminate among the groups. The canonical variables are a useful linear transformation of the input data that can be written out and used as input for further analysis. The plot in figure 1 shows the first two canonical variables containing the linear discrimination information. Data on all six experiments are pooled for the analysis. The results show a fairly clear separation among the behavioral states of S, F and A.

Table 2. Discriminant analysis classification of last experiment on basis of first and first on basis of last.

		Class. of Last on basis of First			Class. of First on basis of Last		
Cat No.	Group	S	F	A	S	F	A
2	S	20	0	0	20	0	0
	F	0	20	0	0	20	0
	A	0	0	20	0	3	17
8	S	20	0	0	20	0	0
	F	0	20	0	0	20	0
	A	0	1	19	0	1	19
9	S	13	0	0	20	0	0
	F	0	20	0	0	20	0
	A	0	7	13	0	0	20
		95.4% correct classification			97.8% correct classification		

Key: S = slow sleep; F = fast sleep; A = alert

Table 2 shows a considerable predictive stability and substantial differences among the patterns of spontaneous electrical activity corresponding to the behavioral states of slow sleep, fast sleep and alert. The predictability over the fairly long time periods also reflects stability of the (calibrated) frequency analyzer.

The same data set can be used to demonstrate the degree of consistency of the activity patterns among cats. For this purpose we have used data from two of the cats to predict the state of the third. Results are given in table 3.

Table 3. Discriminant analysis classification of one cat on basis of data from two other cats.

Cat No.	Group	S	F	A
2	S	30	10	0
	F	0	40	0
	A	0	8	32
8	S	40	0	0
	F	0	40	0
	A	0	14	26
9	S	33	0	0
	F	0	40	0
	A	0	0	40
		90.9% correct classification		

Key: S = slow sleep; F = fast sleep; A = alert

The percentage of minutes correctly classified on the basis of other cats is somewhat less than on a "within" cat basis, but the percentage is impressively high. The amplitude profiles that correspond to the observable behavioral states are sufficiently different that they transcend differences among cats.

To some extent it is not surprising that the states are well separated by the discriminant analysis, since the initial classification was based in part on the EEG tracings. The slow sleep pattern is quite different from either fast sleep or alert; fast sleep and alert are very difficult to distinguish by visual inspection of the EEG, and leads to indicate neck muscle electrical activity and eye movement are both required for visual separation of these two states.

The fact that our objective classification of state on the basis of average amplitude data is fairly stable over both time and animals is important. The correlation of electrical activity with behavioral states is of particular interest because it provides a base for assessing drug effects apart from behavioral state effects.

DRUG EFFECTS

The time course of drug effects is expected to follow a pattern of increased activity after administration of the drug and then a dropping off. The time course could be quite complicated if various metabolites of the drug are active and if different tissues respond differentially to the various products. In any event the observed data can be represented formally as

$$a(t) = s(t) + g(t - t_D)$$

in which

$a(t)$ is the observed average amplitude

$s(t)$ is a component appropriate to a no-drug situation

$g(t)$ is the drug effect ($g(t) = 0,\ t < 0$)

and

t_D is the time of administration of drug.

The decomposition of a(t) into its component parts was affected indirectly by elimination of part of the spontaneous component s(t) and possibly also part of the drug component, by means of the canonical analysis associated with the discriminant analysis of BMD07M (Dixon, 1971). In this analysis, a discrimination is made between the S, F and A classified minutes with the drug minutes included as data to be classified. As a result of the calculations, a set of (within groups, SFA) uncorrelated canonical variables are generated as linear transforms of the initial data in which the first two variables contain the information for discriminating between the S, F and A groups. Therefore, it is assumed that the remaining variables do not contain the portion of s(t) that corresponds to the behavioral state. The drug effects are calculated in terms of these remaining canonical variables.

The data used to illustrate the computation of drug effects are listed as Data Set 2 in the Data Set Description at the end of this chapter; they consist of the initial data and the generated canonical variables for the high dose of physostigmine for cat 9.

We considered two ways of proceeding. The first was to fit functions of particular parametric form to the data; the second was to form the linear discriminant function between the post-drug and pre-drug minutes (using the residual canonical variables only). The motivations for the latter were that if a strong departure existed it was quite likely to be revealed, and that a simple calculation would be preferred if it yielded appropriate results.

In the first procedure, the functional form g(t), in our earlier considerations was a weighted sum of negative exponentials together with the restriction that g(0) = 0. We were led to the degenerate forms

$$g(t) = \gamma t^{\alpha} e^{-\beta t}$$

for integral α. It is convenient to scale and parameterize g so its maximum value is unity and is assumed at $t = t_m$. This results in the expression

$$g(t;\alpha,t_m) = \left(\frac{t}{t_m}\right)^{\alpha} \exp\left\{-\alpha\left(\frac{t}{t_m} - 1\right)\right\} \qquad t \geq 0$$

If we discard the first two canonical variables (that correspond to the SFA plane) and denote the residual canonical variables as $y_i(t)$, $i = 3,\ldots,8$, then the curve fitting problem was assumed to fit

$$y_i(t) = \beta_i g(t;\alpha, t_m), \quad i = 3,\ldots,8.$$

Since for the S, F and A minutes the y_i are uncorrelated, have zero means and unit variance (in the numerical or sample sense), the curve fitting was done using the nonlinear least squares program BMDX85 (Dixon, 1969) with the six variables for a given time used as different cases for purposes of analysis; in this way, a multivariate nonlinear curve fitting program was avoided.

For an illustrative case, only the high dose experiment of physostigmine on cat number 9 was used. The parameters of best fit for the above equations representing the $y_i(t)$ are given in table 4.

Table 4. Results of fitting response curve to high dose physostigmine, cat No. 9.

	Canonical Variable					
	3	4	5	6	7	8
Coefficients $\hat{\beta}_i$	-.565	-1.338	-4.267	-1.993	.887	-.666
	$\hat{\alpha} = 1.783$		$t_m = 30.4$ minutes			
	residual mean square = .65					

The calculations for table 4 included pre-drug minutes (which contributed to the mean square) simply because it was easier for this calculation to include them. Since the mean square for the pre-drug minutes would be approximately 1., the residual for the drug minutes only would be even smaller than the given low value of 0.65. The data for analysis included 172 minutes from the two control experiments for cat 9, 44 minutes prior to drug administration and 169 minutes post-drug administration. (See Data Set 2 in Data Set Description at the end of this chapter.)

A single drug response variable was computed as

$$z_1(t) = c \Sigma \beta_i y_i(t)$$

The response variable has unit standard deviation in the control and pre-drug data. The results were averaged over 10 minute intervals (although not all intervals contained 10 data points since some pre-drug minutes were not classified) and plotted as figure 2. The fitted curve is also shown.

The second procedure was to form a discriminant function based on the residual canonical variables. By taking the covariance estimate as that based on the nondrug minutes, the coefficients of the discriminant function are simply the averages of the variables over the drug minutes (averages for the nondrug minutes are zero and the covariance estimate is a multiple of the identity matrix). A drug response variable was obtained from the discriminant function by shifting the origin and rescaling so the average over nondrug minutes was zero and the standard deviation over the same data base was unity. The response variable $z_2(t)$ formed in this way is also plotted in figure 2.

The two ways of forming a variable corresponding to drug response are closely related as shown in figure 2; the curve fitting approach in this case was hardly worth the bother. After preliminary work some time ago, we settled on the discriminant function to represent the time course of the drug effect. However, the curve fitting approach might be substantially better to bring out local maximums that correspond to distinctive effects.

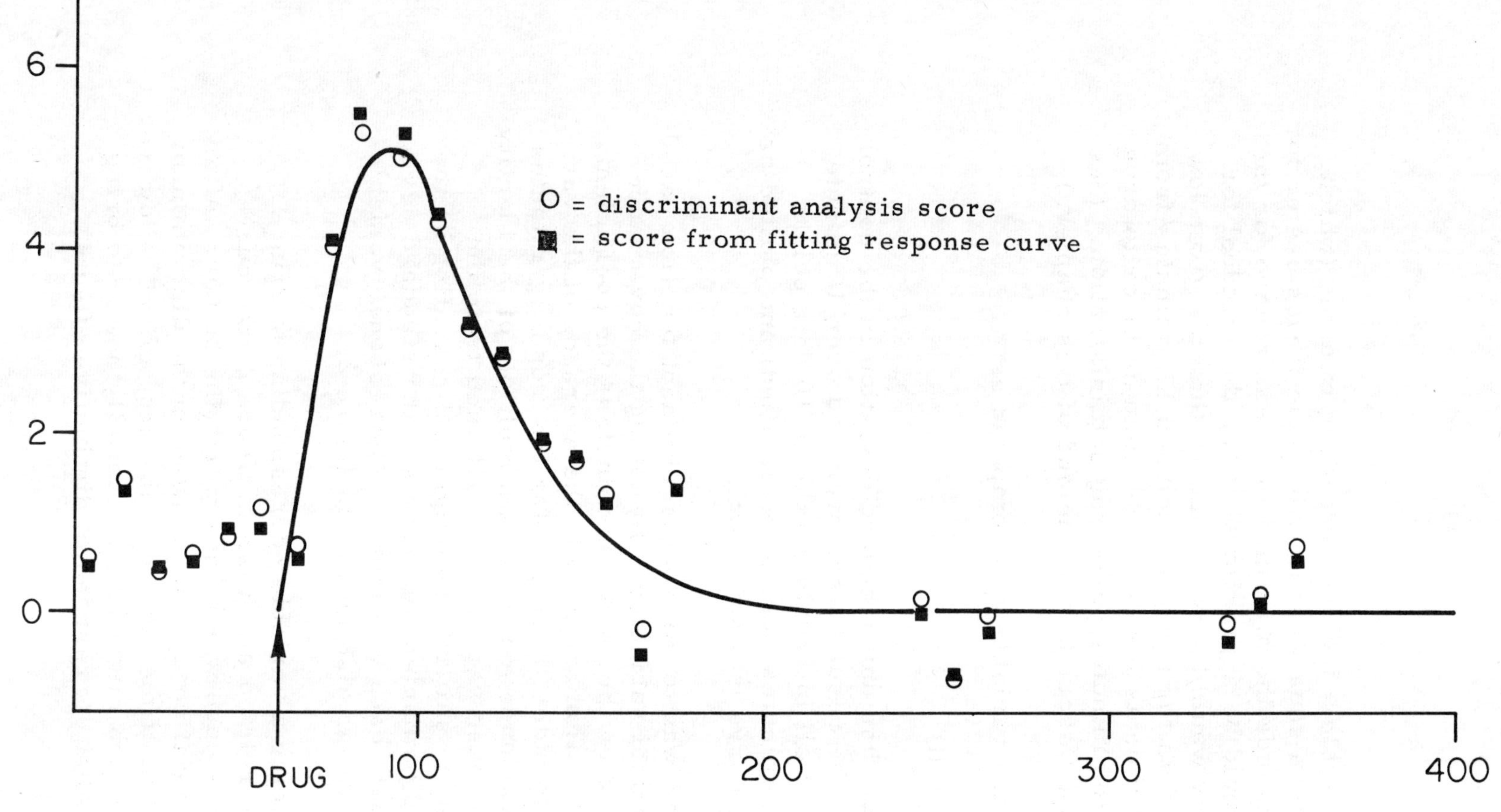

Figure 2. Response to high dose physostigmine, cat No. 9. Points plotted are averages, usually 10 minutes.

Note: Response curve is of the form

$$Z(t) = (\sqrt{\Sigma \hat{\beta}_i^2})\ g\ (t - t_D; \hat{\alpha},\ t_m)$$

in which

t_D is the time of drug administration,

$\hat{\beta}_i$, $\hat{\alpha}$ and t_m are fitted constants (table 4),

and g is of the form

$$g\ (x; \alpha,\ T) = (\frac{x}{T})^{\alpha} \exp\ \left\{ -\alpha\ (\frac{x}{T} - 1) \right\}$$

We constructed variables representing drug effects to see whether the response conformed to our expectations. We watched for artifacts that may be picked up from the method of analysis rather than being indications of some underlying physiological phenomenon. Previous and subsequent control experiments were included in the nondrug part of the data. This prevented the pre-drug portion of the drug experiment from dominating the nondrug data when checking the response before and after drug injection. In general we found it is possible in many cases to detect the buildup of the drug response before any response is noticeable either by direct observation or by scanning the EEG tracing.

"Characteristic profiles" of the alteration in frequency response were constructed as a dual to the construction of a drug response; either should determine the other and the pair should satisfy the reproducibility criteria,

$$P(H(p)) = p \quad \text{and}$$

$$H(P(h)) = h,$$

in which H(p) denotes the response function corresponding to profile (p) and P(h) denotes the profile corresponding to response function h. These properties will hold if, starting with the profile, the problem is formulated as an extreme value problem. Let m and S be the mean vector and (within groups) covariance matrix of the variables x in the SFA classified nondrug minutes. Determine the response function* $h(x) = \beta^{\tau}(x - m)$ so as to maximize $\beta^{\tau} p$ subject to the constraint $\text{var}(h) = \beta^{\tau} S \beta = 1$. The solution to the problem is:

$$\beta = \lambda S^{-1} p$$

$$p = \lambda^{-1} S \beta$$

where λ is chosen to satisfy the variance constraint.

*In the above notation, τ designates the transpose.

For any given response function, the corresponding profile is readily computed from the within groups covariance matrix of the response function with the original variables since

$$\text{cov}\,(x,\ (x-m)^{T}\beta\,) = \text{ave}_{\omega}\,(x(x-m)^{T}\beta\,) = S\beta$$

where ave_{ω} denotes average within groups. The calculations are easily carried out by adding the computed values of the response function to the data set and recomputing the SFA discrimination using BMD07M (Dixon, 1971) printing the covariance matrix.

Characteristic profiles corresponding to the two response functions for the high dose physostigmine (cat 9) are shown in figure 3. As expected from the response functions, the two profiles are quite similar.

Production of the profile is one of the main goals of the experiment. It represents a physiological fact and serves as a data basis for discussion, theorizing, possible correlation with other findings, and so on.

DOSE RESPONSE

For the analysis, the drug experiments were reduced to characteristic profiles, and then the profiles were further analyzed.

In the complete data, profiles based on discrimination between drug and nondrug minutes were computed for the drug experiments and for the saline (nondrug) controls. In the restricted data set (based on the first brain region only), a different procedure was followed: a profile was constructed as the vector perpendicular from the plane defined by S, F and A group means to the mean of the drug minutes.

Thus the data set consists of 18 nominal profiles, each made up of 8 variables corresponding to amplitude alterations in the 8 frequency bands. For each of the three cats there are six profiles: two saline controls, a low and a high atropine, and a low and a high physostigmine. Data for the profiles are listed as

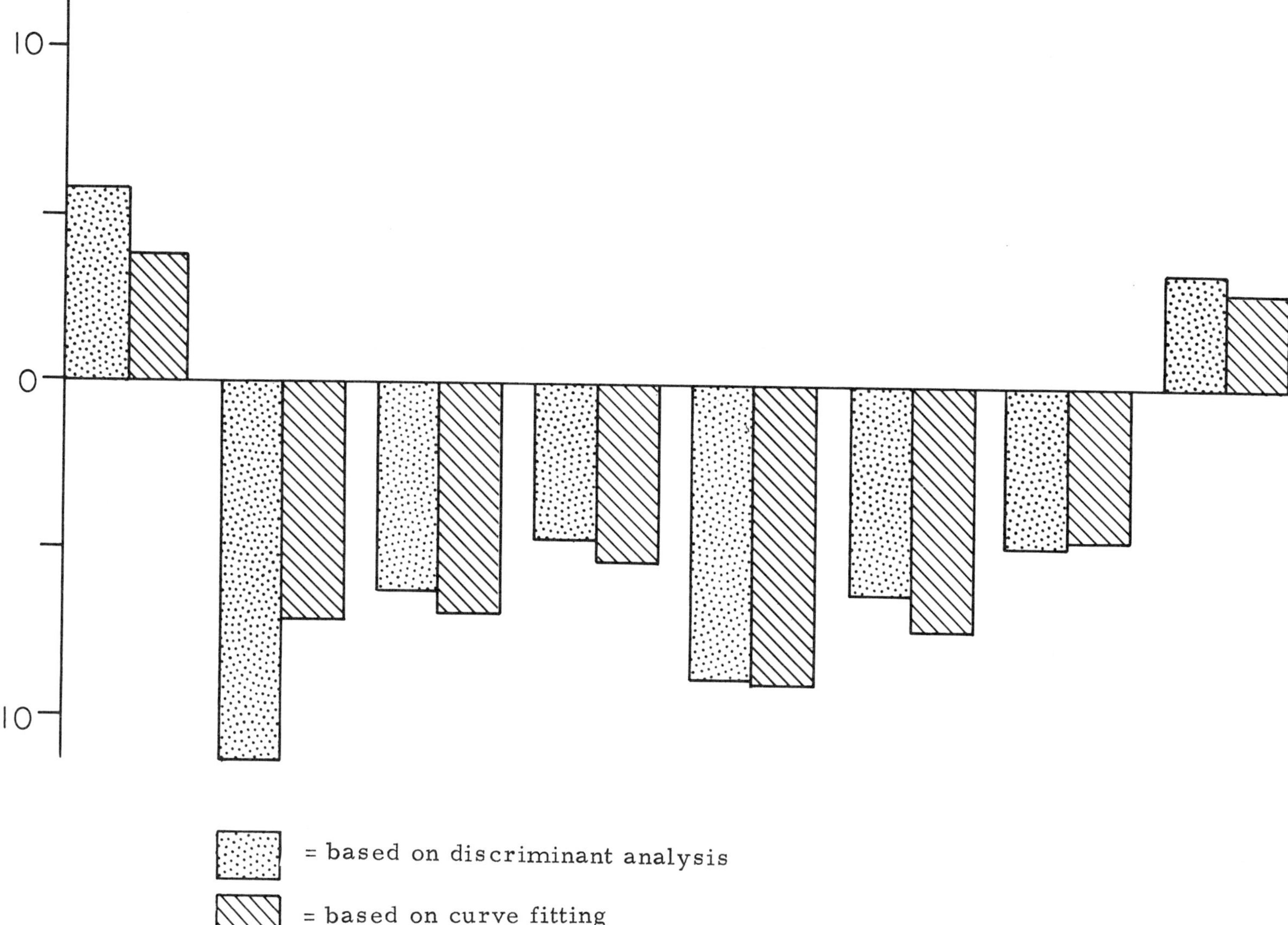

Figure 3. Characteristic profiles of high dose physostigmine, cat No. 9. The characteristic profile can be thought of as a set of additive components, one for each frequency band, that corresponds to the alteration in band amplitude.

Data Set 3 in the Data Set Description at the end of this chapter.

A simple expression of the content of the results was contemplated by expressing the profiles in terms that would reduce them to two dimensional representation. Each experiment would be represented by a point in the plane, and it was hoped that points corresponding to the same drug would be collinear with a dose dependent distance from the origin; saline controls should plot near the origin.

If such a simplified model would hold, several aspects should be discernible from a principal component/factor analysis. To test the model, such an analysis was constructed using the program BMDX72 (Dixon, 1969) and requesting the covariances to be computed about the origin.

The numerical results are short of ideal, and we found them difficult to interpret. The characteristic values (or eigenvalues) obtained were the following in decreasing order:

1701.81	309.30	148.01	89.88
32.83	13.60	4.35	3.49

The largest value stands in isolation, but the next five are related by factors of about two. The two smallest are presumably small enough to reflect that the profiles were constructed as perpendicular to a plane. It is difficult to judge whether a two-dimensional representation would summarize the data adequately. It is quite possible that there is a "saline effect," although it is more likely that the SFA plane is not an adequate description of nondrug data. We considered this aspect in greater detail and found that a quadratic surface seemed to account well for the transitional periods between states as well as for the well-defined states. We did not apply these results to the drug studies, partly because we thought that if the drug effects were fairly strong, we would not be seriously misled.

The problems of using a principal component analysis at this stage are more severe when the entire set of variables is used.

The analysis would then involve 32 variables and 18 cases, and the effects of sampling variation would be enhanced accordingly.

A two dimensional representation can be developed using a canonical correlation/multivariate regression analysis. We used program BMDX75 (Dixon, 1969). One set of variables consists of the eight variables of the characteristic profile; a second set consists of the two variables giving the dosages of the two drugs, atropine and physostigmine. A compact printout of the analysis is shown in exhibit 1.

The results are plotted in figures 4 and 5. Figure 4 shows the averages of the canonical variables for the various treatments and the "fitted" response. Also shown are characteristic profiles corresponding to the two axes. These profiles were computed as the covariance (about the origin) of the canonical variables with the initial profile data.

Figure 5 shows the canonical variables for each experiment in a similar plot. The drugs appear to be well separated, and although there is a substantial variability, there is enough consistency to make the results pharmacologically interesting.

As in factor analysis, including all 32 variables improves the results. It is not clear how to make appropriate allowance for the small sample sizes. In a somewhat similar study we used large sample multivariate normal results to obtain some nominal confidence ellipsoids; the regions seemed large enough to be reasonable yet small enough to be useful. The adequacy was not studied further.

SUMMARY

The study investigated alterations in spontaneous brain electrical activity in response to particular drugs. A quantitative summary of the electrical activity is given by average amplitudes of frequency bands of the recorded activity. While this particular summary discards information that may be useful, it contains a substantial amount of information that is useful and is feasible to

EXHIBIT 1. CANONICAL CORRELATION - MULTIVARIATE REGRESSION ANALYSIS USING PROGRAM BMDX75

BMDX75 - CANONICAL ANALYSIS - REVISED DECEMBER 19, 1972

PROBLEM CODE

NUMBER OF VARIABLES 10

NUMBER OF CASES 18

INPUT FORMAT
(8F6.2,F3.1,F3.2)

VARIABLE	MEAN	STANDARD DEVIATION	VARIABLE	MEAN	STANDARD DEVIATION	VARIABLE	MEAN	STANDARD DEVIATION	VARIABLE	MEAN	STANDARD DEVIATION
1	9.984437	23.677917	4	-8.511102	23.288422	7	5.642216	12.002718	10	0.041667	0.075245
2	0.988329	10.726023	5	-0.458889	17.507355	8	1.585548	11.189098			
3	-6.798882	15.118748	6	5.362219	10.821854	9	0.416666	0.752446			

THE COVARIANCE MATRIX ABOUT THE ORIGIN IS USED IN THE FOLLOWING CALCULATION

CANONICAL CORRELATIONS

1	2
0.90582	0.72097

VARIABLE COEFFICIENTS FOR CANONICAL VARIABLES OF THE FIRST SET

VARIABLE	1	2
1	-0.04536	0.02892
2	-0.07223	0.07170
3	0.00176	0.01472
4	-0.03636	0.08278
5	0.10090	-0.16185
6	0.00292	-0.02671
7	-0.04948	0.11690
8	0.18834	-0.28315

VARIABLE COEFFICIENTS FOR CANONICAL VARIABLES OF THE SECOND SET

VARIABLE	1	2
9	-1.15945	-0.25970
10	2.59696	-11.59452

CANONICAL VARIABLES

CASE NO.	1		2	
1	-0.11941	0.0	1.13406	0.0
2	0.14869	0.0	0.31705	0.0
3	-0.78284	-0.57972	-0.62825	-0.12985
4	-2.46890	-2.31890	-0.37997	-0.51939
5	1.13713	0.12985	-0.38164	-0.57973
6	0.11895	0.51939	-2.00423	-2.31890
7	0.31405	0.0	-0.80249	0.0
8	0.53234	0.0	-1.02926	0.0
9	-1.34822	-0.57972	-0.22448	-0.12985
10	-1.90941	-2.31890	-0.71301	-0.51939
11	0.04817	0.12985	-0.32856	-0.57973
12	0.78071	0.51939	-2.41762	-2.31890
13	0.26962	0.0	1.55863	0.0
14	0.21924	0.0	0.62989	0.0
15	-0.19843	-0.57972	0.58624	-0.12985
16	-1.73468	-2.31890	-0.03400	-0.51939
17	0.54178	0.12985	-0.01746	-0.57973
18	0.16820	0.51939	-0.72529	-2.31890

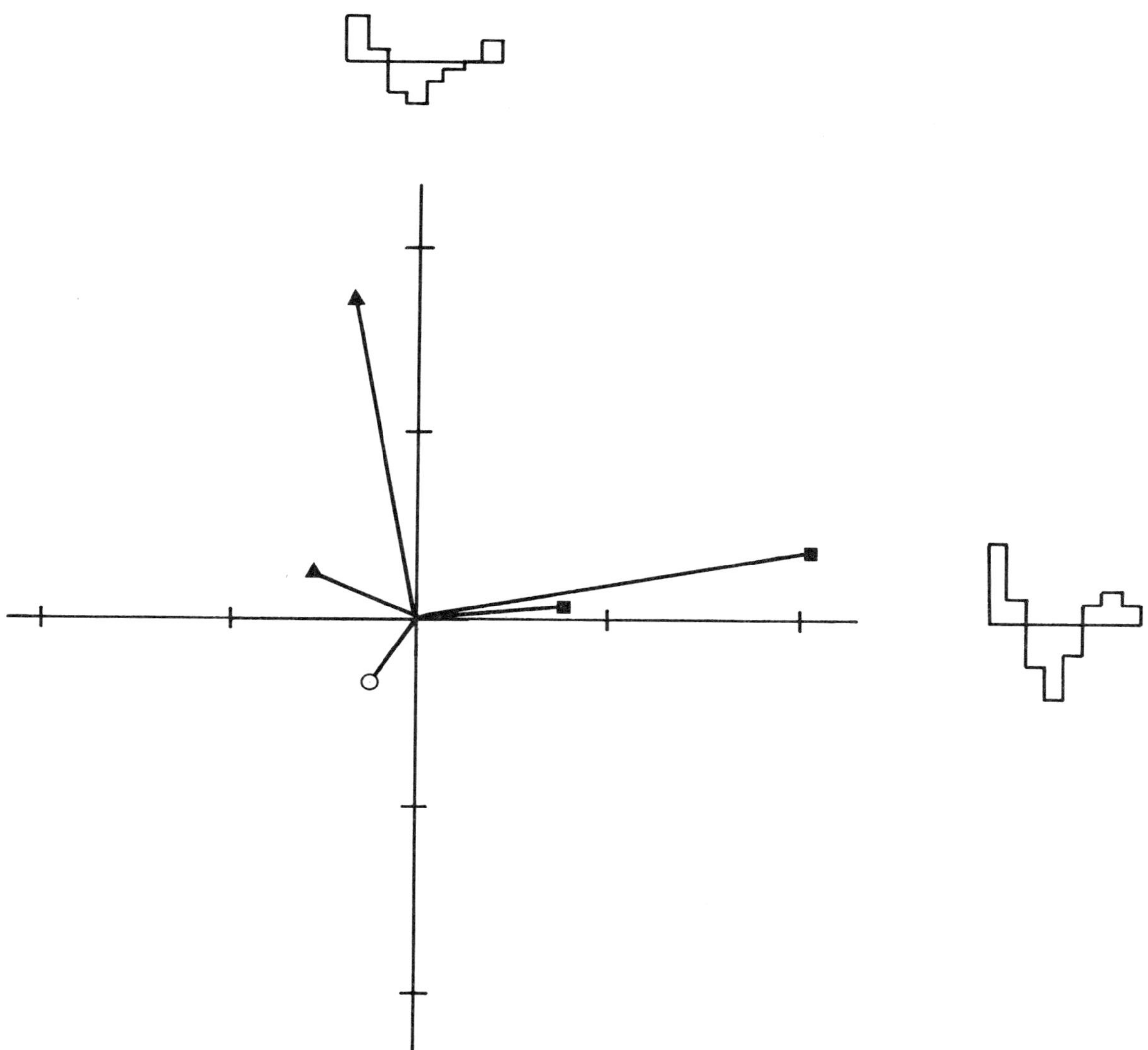

O = saline
■ = atropine
▲ = physostigmine

Figure 4. Averages of canonical variables for dose response, and characteristic profiles corresponding to the canonical axes. Points in plane correspond to profiles obtained as weighted sums of the characteristic profiles corresponding to the axes, with weights equal to the coordinates of the points.

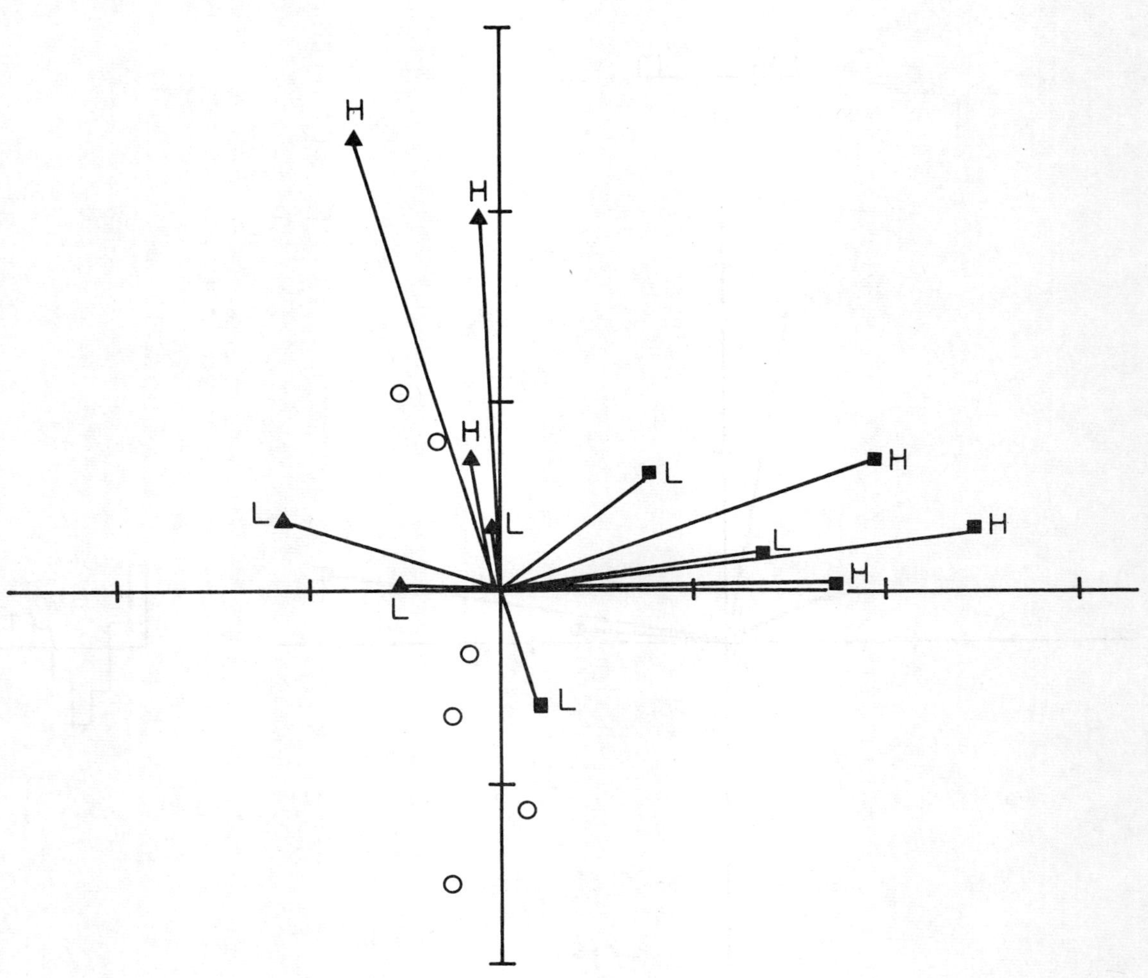

O = saline

■ = atropine
▲ = physostigmine
L = low dose
H = high dose

Figure 5. Canonical variables for dose response. Profiles corresponding to axes are those given in figure 4.

obtain and process.

We illustrate the main analysis steps that we have been following using only a portion of the data — that from one brain region only and from selected experiments for different parts of the analysis. Three somewhat distinct aspects of the analysis are illustrated.

It is apparent that distinct states of slow sleep, fast sleep and alert are readily distinguishable on the basis of the amplitudes in the frequency bands filtered out by the frequency analyzer. This separation was developed using discriminant analysis. The distinctions are great enough to be stable for a cat examined at widely different times, and also fairly stable among animals. The similarity among the animals is perhaps especially noteworthy in view of the technical problems of electrode emplacement, to say nothing of the natural variations among animals; at the very least, it reflects a high degree of technical competence in the execution of the experiment.

It is important to allow for this distinction between behavioral states in comparative drug studies. For example, one drug dose may result in the appearance of sleep while another may result in an appearance of alertness. While this is an important comparative distinction, we were particularly interested in components of electrical activity above and beyond that corresponding to the behavioral state. Consequently, we based much of our analysis on these additional components. We chose to identify them with residual canonical variables as generated in the course of the discriminant analysis calculations of BMD07M (Dixon, 1971).

Our conceptual representation is in terms of normal activity on an SFA plane with the drug effects of interest being those out of the plane. Even in our terms this conception is simplified because transitional periods between the states do not lie quite on the plane but tend to be distributed about a quadratic surface. This curvature probably accounts for the "saline effects" that appear in subsequent parts of the analysis. We were not overly concerned about this, because the drug effects tend to be substantially stronger, and we were reluctant to complicate both the analysis and its exposition by taking the curvature into account.

We characterized the drug effects in terms of the time course of action and as a profile of changes from normal. We tried curve fitting of various sorts to generate a response variable, but came to the tentative conclusion that the simple discriminant function of drug. vs. nondrug on the residual canonical variables is adequate for this purpose.

We formulated a resolution of the problem of constructing characteristic profiles. This problem is not well defined and there may be alternative methods for making the construction. Perhaps the choice should depend mostly on the performance for subsequent analysis.

Finally, we wanted to use the profiles to compare the responses of various drugs given in various doses. Principal components analyses seemed appropriate, but we had difficulties in arriving at interesting conclusions or consequences. We found canonical correlation/regression analyses more directly useful. But one of the difficulties with these analyses is that the number of cases is modest in comparison with the number of variables; we did not cope particularly well with the problems of evaluating uncertainties in the results.

Some additional complications of experimental design may be related to problems of changes in the animals with successive treatments of some of the psychotropic drugs. We did not feel that this was a problem in the investigation under discussion, but it appears to be a factor in some current work.

DATA SET DESCRIPTION

DATA - FAIRCHILD, EEG RESPONSE TO DRUG (R. MICKEY)

CASES AND VARIABLES

DATA SET 1
473 CASES (WITHIN EXPERIMENT TEMPORAL SUBINTERVALS)
14 VARIABLES - 8 MEASUREMENT, 2 CATEGORICAL, 4 IDENTIFICATION

DATA SET 2
385 CASES (WITHIN EXPERIMENT TEMPORAL SUBINTERVALS)
19 VARIABLES - 17 MEASUREMENT, 2 CATEGORICAL

DATA SET 3
18 CASES (EXPERIMENTS)
12 VARIABLES - 8 MEASUREMENT, 3 CATEGORICAL, 1 IDENTIFICATION

BRIEF DESCRIPTION

VAR	NAMES OF VARIABLES IN DATA SET 1
***	***** ** ********* ** **** *** *
1-3	DATE OF EXPERIMENT
4	CAT IDENTIFICATION NUMBER
5	EXPERIMENT
6	BEHAVIORAL STATE
7-14	AVERAGE AMPLITUDES

VAR	NAMES OF VARIABLES IN DATA SET 2
***	***** ** ********* ** **** *** *
1	BEHAVIORAL STATE
2	TYPE OF TRIAL
3	TIME
4-11	AVERAGE AMPLITUDES
12-19	CANONICAL VARIABLES

VAR	NAMES OF VARIABLES IN DATA SET 3
***	***** ** ********* ** **** *** *
1-8	PROFILES
9	DOSE OF ATROPINE
10	DOSE OF PHYSOSTIGMINE
11	CAT IDENTIFICATION NUMBER
12	EXPERIMENT NUMBER WITHIN CAT

PURPOSE OF STUDY

THE PURPOSE OF THE STUDY IS TO ATTEMPT TO MAKE QUANTITATIVE AND OBJECTIVE MEASUREMENTS OF DRUG INDUCED ALTERATIONS IN SPONTANEOUS BRAIN ELECTRICAL ACTIVITY. THE GENERAL GOAL IS TO CONTRIBUTE TO THE STUDY OF EFFECTS OF PSYCHOTROPHIC DRUGS IN MAN. THE APPROACH IS TO STUDY EFFECTS IN CAT EXPERIMENTS AND TO ATTEMPT TO CORRELATE FINDINGS WITH EFFECTS IN MAN.

REFERENCE

JENDEN, DONALD J., M. DAVID FAIRCHILD, MAX R. MICKEY, ROBERT W. SILVERMAN, AND CORALEE YALE. A MULTIVARIATE APPROACH TO THE ANALYSIS OF DRUG EFFECTS ON THE ELECTROENCEPHALOGRAM. BIOMETRICS 28, 1972, 73-80.

DETAILED DESCRIPTION

VAR	COL	FORMAT	DESCRIPTION OF DATA SET 1
***	***	******	*********** ** **** *** *
			THIS DATA SET CONSISTS OF THE FIRST AND THE LAST CONTROL EXPERIMENTS AND NOT MORE THAN 20 MINUTES OF EACH BEHAVIORAL STATE FOR EACH OF THREE CATS.
1	1-2	XX.	MONTH OF EXPERIMENT
2	3-4	XX.	DAY OF EXPERIMENT
3	5-6	XX.	YEAR OF EXPERIMENT
4	7-8	XX.	CAT IDENTIFICATION NUMBER
5	9	X.	EXPERIMENT 1 = FIRST CONTROL 2 = LAST CONTROL
6	10	X.	BEHAVIORAL STATE 1 = SLOW SLEEP 2 = FAST SLEEP 3 = ALERT 4 = UNCLASSIFIED
			VARIABLES 7 THROUGH 14 ARE AVERAGE AMPLITUDES IN FREQUENCY BANDS. THE AMPLITUDES ARE AVERAGED OVER ONE-MINUTE EPOCHS. THE FREQUENCY BANDS COVER A RANGE OF 2.5 TO 40 HZ IN HALF-OCTAVE STEPS.
7	11-14	XXXX.	FIRST
8	15-18	XXXX.	SECOND
9	19-22	XXXX.	THIRD
10	23-26	XXXX.	FOURTH
11	27-30	XXXX.	FIFTH
12	31-34	XXXX.	SIXTH
13	35-38	XXXX.	SEVENTH
14	39-42	XXXX.	EIGHTH

FORMAT IS (4F2.0,2F1.0,8F4.0)   N = 473

VAR	COL	FORMAT	DESCRIPTION OF DATA SET 2
***	***	******	*********** ** **** *** *
			THIS DATA SET CONSISTS OF DATA FROM ONE DRUG EXPERIMENT - HIGH DOSE PHYSOSTIGMINE FOR ONE CAT.
1	1	X.	BEHAVIORAL STATE 1 = SLOW SLEEP 2 = FAST SLEEP 3 = ALERT 7 = FOLLOWING DRUG

```
 2      2        X.     TYPE OF TRIAL
                          1 = CONTROL TRIAL
                          2 = DRUG TRIAL
 3     3-5      XXX.    TIME
                          0 = CONTROL TRIAL

                        VARIABLES 4 THROUGH 11 ARE AVERAGE AMPLITUDES
                        IN FREQUENCY BANDS.  THE AMPLITUDES ARE
                        AVERAGED OVER ONE-MINUTE EPOCHS.  THE
                        FREQUENCY BANDS COVER A RANGE OF 2.5 TO
                        40 HZ IN HALF OCTAVE STEPS.

 4     6-8      XXX.    FIRST
 5     9-11     XXX.    SECOND
 6    12-14     XXX.    THIRD
 7    15-17     XXX.    FOURTH
 8    18-20     XXX.    FIFTH
 9    21-23     XXX.    SIXTH
10    24-26     XXX.    SEVENTH
11    27-29     XXX.    EIGHTH

                        VARIABLES 12 THROUGH 19 ARE CANONICAL
                        VARIABLES GENERATED FROM A DISCRIMINANT
                        ANALYSIS OF BEHAVIORAL STATES.  ADDITIONAL
                        DATA WERE USED TO SET THE BEHAVIORAL STATE
                        PLANE.

12    30-33     XX.XX   FIRST
13    34-37     XX.XX   SECOND
14    38-41     XX.XX   THIRD
15    42-45     XX.XX   FOURTH
16    46-49     XX.XX   FIFTH
17    50-53     XX.XX   SIXTH
18    54-57     XX.XX   SEVENTH
19    58-61     XX.XX   EIGHTH

FORMAT IS (2F1.0,9F3.0,8F4.2)      N = 385


VAR    COL     FORMAT   DESCRIPTION OF DATA SET 3
***    ***     ******   *********** ** **** *** *

                        THIS DATA SET CONSISTS OF PROFILES FOR THE
                        DRUG-SALINE EXPERIMENTS.

                        VARIABLES 1 THROUGH 8 ARE THE PROFILES.  THEY
                        WERE DETERMINED BY PROJECTION FROM THE
                        BEHAVIORAL STATE PLANE.  THERE ARE EIGHT
                        PROFILES FOR EACH CAT.  THE PROFILES WERE
                        GENERATED FROM THE CHANNEL AVERAGES TAKEN
                        FROM PREVIOUS OUTPUT.

 1     1-6     XXXX.XX  FIRST
 2     7-12    XXXX.XX  SECOND
 3    13-18    XXXX.XX  THIRD
 4    19-24    XXXX.XX  FOURTH
 5    25-30    XXXX.XX  FIFTH
 6    31-36    XXXX.XX  SIXTH
 7    37-42    XXXX.XX  SEVENTH
 8    43-48    XXXX.XX  EIGHTH
```

```
 9   49-51       XX.X   DOSE OF ATROPINE
10   52-54        X.XX  DOSE OF PHYSOSTIGMINE
11   55-56       XX.    CAT IDENTIFICATION NUMBER
12   57-58       XX.    EXPERIMENT NUMBER WITHIN CAT

FORMAT IS (8F6.2,F3.1,F3.2,2F2.0)      N = 18
```

```
LOCATION OF DATA

  CARD IMAGE - FS.C073.CFAIR1 (DATA SET 1)
  CARD IMAGE - FS.C073.CFAIR2 (DATA SET 2)
  CARD IMAGE - FS.C073.CFAIR3 (DATA SET 3)
```

```
LISTING OF DATA SET 1 (THE FIRST TWO LINES ARE COLUMN GUIDES)

  0              1          2           3              4
  12 34 56 78 9 0 1234 5678 9012 3456 7890 1234 5678 9012

   1 24 69 02 1 1  802  732  672  576  456  347  288  171
   1 24 69 02 1 1  607  559  482  450  348  330  281  162
   1 24 69 02 1 1  777  829  720  609  443  360  295  170
   1 24 69 02 1 1  502  497  454  484  369  329  279  157
   1 24 69 02 1 1  552  548  525  454  378  331  288  170
   1 24 69 02 1 1  807  711  606  562  475  334  274  166
   1 24 69 02 1 1  751  744  675  586  477  340  279  168
   1 24 69 02 1 1  737  713  690  609  476  343  283  162
   1 24 69 02 1 1  773  700  645  576  439  342  282  157
   1 24 69 02 1 1  871  893  767  697  490  377  284  175
   1 24 69 02 1 1  888  842  735  639  479  356  284  168
   1 24 69 02 1 1  813  795  718  608  424  363  279  167
   1 24 69 02 1 1  829  814  789  659  466  363  294  178
   1 24 69 02 1 1  724  773  677  619  503  344  275  168
   1 24 69 02 1 1  854  849  793  704  471  398  310  179
   1 24 69 02 1 1  776  683  611  498  383  352  288  167
   1 24 69 02 1 1  686  587  567  529  421  357  285  171
   1 24 69 02 1 1  510  552  521  498  435  338  282  166
   1 24 69 02 1 1  608  603  552  512  403  340  290  175
   1 24 69 02 1 1  916  850  744  684  485  367  297  179
  10 17 69 02 2 1  545  554  549  491  400  357  307  227
  10 17 69 02 2 1  715  656  641  587  418  356  291  206
  10 17 69 02 2 1  719  702  623  579  449  364  286  190
  10 17 69 02 2 1  741  720  651  581  419  352  296  205
  10 17 69 02 2 1  925  909  775  680  466  373  296  189
  10 17 69 02 2 1  913  757  692  627  442  358  284  214
  10 17 69 02 2 1  665  633  570  533  425  352  286  219
  10 17 69 02 2 1  895  805  732  684  476  377  292  197
  10 17 69 02 2 1  658  611  585  571  398  280  218  136
  10 17 69 02 2 1  788  769  701  587  430  298  232  133
  10 17 69 02 2 1  784  734  648  587  410  296  244  139
  10 17 69 02 2 1  705  730  700  633  459  320  247  161
  10 17 69 02 2 1  718  718  614  617  433  334  275  167
  10 17 69 02 2 1  870  725  675  591  465  386  286  166
  10 17 69 02 2 1  634  617  623  555  446  364  278  162
  10 17 69 02 2 1  880  878  800  730  465  366  264  153
  10 17 69 02 2 1  902  877  758  612  449  363  275  175
  10 17 69 02 2 1  747  726  686  614  425  361  266  165
  10 17 69 02 2 1  884  921  840  688  499  383  280  169
  10 17 69 02 2 1  757  692  561  512  406  372  285  186
```

7	25	69	08	1	1	612	617	606	536	296	187	144	81
7	25	69	08	1	1	621	589	566	473	281	216	161	92
7	25	69	08	1	1	646	669	598	532	312	202	156	80
7	25	69	08	1	1	786	810	772	617	345	241	172	85
7	25	69	08	1	1	738	824	669	579	343	231	165	88
7	25	69	08	1	1	693	685	618	511	297	214	159	81
7	25	69	08	1	1	747	796	705	564	318	218	156	79
7	25	69	08	1	1	788	762	717	636	371	235	162	87
7	25	69	08	1	1	682	685	711	586	327	223	161	82
7	25	69	08	1	1	708	748	746	605	327	231	166	86
7	25	69	08	1	1	835	859	804	633	347	241	176	85
7	25	69	08	1	1	788	857	844	644	367	258	174	89
7	25	69	08	1	1	847	863	830	634	339	233	159	85
7	25	69	08	1	1	743	796	699	541	317	210	161	87
7	25	69	08	1	1	777	814	732	612	343	224	166	81
7	25	69	08	1	1	750	839	796	595	331	222	158	81
7	25	69	08	1	1	703	775	710	623	338	234	164	82
7	25	69	08	1	1	579	590	574	440	267	192	153	85
7	25	69	08	1	1	763	819	733	605	325	222	160	88
7	25	69	08	1	1	674	702	602	477	293	195	155	83
10	31	69	08	2	1	715	726	689	533	331	229	180	106
10	31	69	08	2	1	752	759	729	577	350	239	189	108
10	31	69	08	2	1	690	702	626	514	283	218	167	83
10	31	69	08	2	1	796	778	729	595	345	232	165	89
10	31	69	08	2	1	734	805	694	554	327	226	171	89
10	31	69	08	2	1	767	837	785	547	315	236	170	86
10	31	69	08	2	1	623	664	600	469	276	208	164	87
10	31	69	08	2	1	631	576	586	433	282	198	171	89
10	31	69	08	2	1	715	777	753	603	322	218	158	87
10	31	69	08	2	1	721	745	705	542	322	215	176	87
10	31	69	08	2	1	803	828	782	612	349	223	165	91
10	31	69	08	2	1	795	924	844	638	360	248	170	90
10	31	69	08	2	1	691	724	699	515	306	212	164	88
10	31	69	08	2	1	736	791	676	549	339	234	171	92
10	31	69	08	2	1	818	892	809	642	357	259	178	92
10	31	69	08	2	1	719	843	764	612	350	250	173	91
10	31	69	08	2	1	774	819	832	638	369	243	170	88
10	31	69	08	2	1	851	917	863	644	357	234	181	89
10	31	69	08	2	1	702	803	751	554	373	247	182	92
10	31	69	08	2	1	789	757	734	620	354	250	189	101
4	18	69	09	1	1	737	760	611	465	288	213	196	121
4	18	69	09	1	1	730	710	638	502	310	227	200	121
4	18	69	09	1	1	838	655	589	477	289	230	203	125
4	18	69	09	1	1	691	563	461	367	231	183	183	123
4	18	69	09	1	1	637	606	517	402	254	187	187	122
4	18	69	09	1	1	744	705	614	462	284	222	207	123
4	18	69	09	1	1	833	774	616	485	290	226	202	120
4	18	69	09	1	1	535	525	467	361	225	187	182	138
4	18	69	09	1	1	749	655	520	424	263	210	183	119
4	18	69	09	1	1	802	676	576	428	272	215	188	117
4	18	69	09	1	1	617	598	519	399	271	209	190	115
4	18	69	09	1	1	785	662	556	417	256	208	189	118
4	18	69	09	1	1	778	758	633	464	282	224	196	121
4	18	69	09	1	1	867	730	627	488	294	217	192	119
4	18	69	09	1	1	674	602	503	381	258	201	190	111
4	18	69	09	1	1	721	698	578	442	267	202	189	120
4	18	69	09	1	1	732	684	575	408	250	194	189	117
4	18	69	09	1	1	739	615	539	418	274	208	193	119
4	18	69	09	1	1	754	686	626	449	278	220	191	113
4	18	69	09	1	1	832	732	669	507	286	221	196	118

7	18	69	09	2	1	939	887	748	606	378	268	215	147
7	18	69	09	2	1	882	801	683	568	350	252	224	147
7	18	69	09	2	1	887	818	673	584	337	261	221	137
7	18	69	09	2	1	909	835	711	599	333	243	217	137
7	18	69	09	2	1	835	827	691	546	315	226	206	136
7	18	69	09	2	1	909	890	693	514	297	222	200	132
7	18	69	09	2	1	853	830	661	516	324	243	214	132
7	18	69	09	2	1	928	879	702	571	344	253	217	139
7	18	69	09	2	1	979	936	765	591	344	241	204	119
7	18	69	09	2	1	897	881	787	592	350	265	213	133
7	18	69	09	2	1	798	783	718	554	301	239	196	122
7	18	69	09	2	1	837	807	641	511	306	227	200	130
7	18	69	09	2	1	819	786	691	562	363	242	204	133
1	24	69	02	1	2	277	290	310	309	250	256	244	166
1	24	69	02	1	2	213	266	253	246	223	238	239	158
1	24	69	02	1	2	224	225	225	254	221	251	249	166
1	24	69	02	1	2	228	245	229	220	211	235	233	166
1	24	69	02	1	2	208	212	219	230	210	239	242	166
1	24	69	02	1	2	243	247	233	238	197	225	242	167
1	24	69	02	1	2	201	238	229	232	204	229	249	173
1	24	69	02	1	2	304	329	289	287	248	275	256	173
1	24	69	02	1	2	255	254	240	245	217	238	242	170
1	24	69	02	1	2	230	233	222	239	225	245	251	168
1	24	69	02	1	2	228	267	243	257	221	236	245	165
1	24	69	02	1	2	219	223	218	232	210	246	249	183
1	24	69	02	1	2	211	212	219	220	191	221	227	173
1	24	69	02	1	2	213	219	229	229	211	232	243	169
1	24	69	02	1	2	273	273	262	268	238	268	258	172
1	24	69	02	1	2	235	258	256	252	215	230	238	175
1	24	69	02	1	2	217	204	214	226	197	230	235	160
1	24	69	02	1	2	262	256	258	256	219	236	243	166
1	24	69	02	1	2	213	245	230	225	212	227	239	164
1	24	69	02	1	2	244	229	234	237	218	252	257	170
10	17	69	02	2	2	220	204	216	237	198	208	212	153
10	17	69	02	2	2	178	207	201	214	182	192	194	139
10	17	69	02	2	2	198	210	205	224	186	190	208	145
10	17	69	02	2	2	201	207	221	213	184	202	207	156
10	17	69	02	2	2	189	192	176	180	176	180	199	154
10	17	69	02	2	2	202	224	208	225	203	199	203	152
10	17	69	02	2	2	183	214	215	212	173	203	205	146
10	17	69	02	2	2	211	238	244	236	207	194	187	129
10	17	69	02	2	2	171	218	221	233	199	196	197	141
10	17	69	02	2	2	183	222	205	200	178	195	204	145
10	17	69	02	2	2	187	205	206	205	169	179	180	130
10	17	69	02	2	2	201	208	197	196	176	199	205	135
10	17	69	02	2	2	416	319	294	256	194	190	192	136
10	17	69	02	2	2	401	347	267	232	185	197	196	141
10	17	69	02	2	2	399	349	272	241	185	196	202	137
10	17	69	02	2	2	331	293	247	220	174	192	188	133
10	17	69	02	2	2	355	261	212	211	182	185	181	131
10	17	69	02	2	2	409	344	262	237	170	183	183	141
10	17	69	02	2	2	300	331	348	307	278	250	211	142
10	17	69	02	2	2	328	320	279	296	240	216	202	132
7	25	69	08	1	2	120	134	146	131	113	121	146	104
7	25	69	08	1	2	93	114	133	123	100	110	142	117
7	25	69	08	1	2	94	108	111	118	91	113	146	106
7	25	69	08	1	2	112	110	105	113	87	110	145	113
7	25	69	08	1	2	91	112	116	107	82	108	133	94
7	25	69	08	1	2	84	97	107	117	91	116	134	106
7	25	69	08	1	2	102	108	116	117	94	114	148	115

7	25	69	08	1	2	96	109	129	121	94	113	128	99
7	25	69	08	1	2	93	113	114	110	95	106	132	95
7	25	69	08	1	2	105	116	117	109	84	104	124	86
7	25	69	08	1	2	82	108	111	112	95	107	130	97
7	25	69	08	1	2	112	108	108	96	82	97	125	95
7	25	69	08	1	2	110	134	109	111	94	100	137	99
7	25	69	08	1	2	101	105	107	112	86	99	128	100
7	25	69	08	1	2	100	119	106	98	81	92	127	96
7	25	69	08	1	2	107	122	126	113	93	110	129	92
7	25	69	08	1	2	276	260	231	256	179	166	162	98
7	25	69	08	1	2	141	161	159	152	114	128	146	97
7	25	69	08	1	2	75	109	95	102	84	107	139	107
7	25	69	08	1	2	86	101	99	113	93	112	139	99
10	31	69	08	2	2	129	139	136	129	116	128	149	117
10	31	69	08	2	2	105	117	123	105	83	94	130	112
10	31	69	08	2	2	104	119	132	120	89	97	133	100
10	31	69	08	2	2	118	118	123	110	83	90	131	111
10	31	69	08	2	2	100	120	129	120	87	97	133	105
10	31	69	08	2	2	115	123	131	124	87	97	122	100
10	31	69	08	2	2	120	128	134	109	87	89	119	92
10	31	69	08	2	2	187	196	210	195	174	157	168	104
10	31	69	08	2	2	99	124	132	125	115	120	142	101
10	31	69	08	2	2	109	137	131	125	100	102	127	93
10	31	69	08	2	2	113	121	116	115	87	101	127	96
10	31	69	08	2	2	100	125	131	119	90	101	132	102
10	31	69	08	2	2	98	117	118	111	89	96	122	92
10	31	69	08	2	2	89	121	127	111	92	99	121	92
10	31	69	08	2	2	118	121	126	106	83	90	120	89
10	31	69	08	2	2	92	116	110	108	79	85	119	88
10	31	69	08	2	2	109	123	114	107	89	97	132	97
10	31	69	08	2	2	104	121	118	109	79	97	124	94
10	31	69	08	2	2	109	137	145	131	116	118	144	102
10	31	69	08	2	2	103	124	128	111	88	101	135	94
4	18	69	09	1	2	340	291	251	220	159	152	169	118
4	18	69	09	1	2	186	176	170	158	137	140	162	122
4	18	69	09	1	2	191	189	173	170	140	140	162	124
4	18	69	09	1	2	177	184	178	181	136	143	157	119
4	18	69	09	1	2	153	167	172	170	140	140	167	124
4	18	69	09	1	2	169	171	176	180	141	142	166	127
4	18	69	09	1	2	183	192	173	183	151	140	164	116
4	18	69	09	1	2	171	165	168	169	142	149	172	122
4	18	69	09	1	2	176	175	179	174	141	142	162	118
4	18	69	09	1	2	171	163	181	165	144	137	161	119
4	18	69	09	1	2	181	186	186	178	146	152	158	114
4	18	69	09	1	2	152	154	167	165	138	139	153	112
4	18	69	09	1	2	150	159	170	170	138	136	161	121
4	18	69	09	1	2	166	173	176	168	143	133	161	122
4	18	69	09	1	2	157	161	172	163	137	126	159	114
4	18	69	09	1	2	276	229	241	207	159	145	170	121
4	18	69	09	1	2	183	168	175	172	146	146	166	122
4	18	69	09	1	2	188	182	174	177	141	143	161	121
4	18	69	09	1	2	189	176	173	176	136	142	157	117
4	18	69	09	1	2	152	165	179	177	140	143	158	124
7	18	69	09	2	2	238	213	204	202	163	164	166	136
7	18	69	09	2	2	204	200	188	196	154	157	170	132
7	18	69	09	2	2	199	186	182	202	163	163	180	138
7	18	69	09	2	2	231	231	222	205	166	158	172	135
7	18	69	09	2	2	195	198	188	199	158	152	166	133
7	18	69	09	2	2	189	190	201	216	170	157	165	132
7	18	69	09	2	2	194	201	196	212	158	164	166	130

7	18	69	09	2	2	188	195	199	207	157	150	158	129
7	18	69	09	2	2	198	207	197	198	160	163	169	129
7	18	69	09	2	2	176	206	213	202	158	155	169	134
7	18	69	09	2	2	204	198	189	193	157	164	180	138
7	18	69	09	2	2	188	219	213	215	166	167	180	141
7	18	69	09	2	2	216	225	211	225	171	154	169	136
7	18	69	09	2	2	217	203	212	209	160	160	172	136
7	18	69	09	2	2	221	194	211	201	162	159	169	134
7	18	69	09	2	2	220	196	186	197	156	157	166	130
7	18	69	09	2	2	231	207	213	202	166	150	168	130
7	18	69	09	2	2	219	199	184	194	168	166	170	130
7	18	69	09	2	2	193	195	195	198	156	151	168	130
7	18	69	09	2	2	221	185	189	203	161	166	168	127
1	24	69	02	1	3	233	227	242	248	225	268	277	230
1	24	69	02	1	3	225	243	271	260	221	267	295	242
1	24	69	02	1	3	268	244	255	260	224	260	307	272
1	24	69	02	1	3	236	230	238	281	219	279	297	260
1	24	69	02	1	3	257	250	236	246	227	279	306	249
1	24	69	02	1	3	234	232	240	252	223	288	334	307
1	24	69	02	1	3	212	219	237	270	221	258	276	244
1	24	69	02	1	3	223	229	241	281	214	256	275	240
1	24	69	02	1	3	224	252	258	256	215	245	271	226
1	24	69	02	1	3	183	213	234	249	214	245	248	205
1	24	69	02	1	3	229	246	255	260	233	260	267	189
1	24	69	02	1	3	214	230	243	258	213	237	281	227
1	24	69	02	1	3	218	229	254	275	242	253	345	303
1	24	69	02	1	3	206	237	256	248	217	258	303	259
1	24	69	02	1	3	226	231	243	249	213	251	282	225
1	24	69	02	1	3	187	206	218	256	204	242	274	221
1	24	69	02	1	3	219	239	224	258	207	246	273	229
1	24	69	02	1	3	230	230	224	236	214	250	285	237
1	24	69	02	1	3	245	239	251	255	215	230	258	193
1	24	69	02	1	3	240	249	250	272	233	246	260	197
10	17	69	02	2	3	179	202	214	243	204	226	238	243
10	17	69	02	2	3	198	208	219	230	200	207	238	268
10	17	69	02	2	3	182	200	216	231	198	218	242	292
10	17	69	02	2	3	203	203	197	216	200	217	255	295
10	17	69	02	2	3	175	211	212	223	194	209	250	279
10	17	69	02	2	3	205	230	233	239	198	208	237	267
10	17	69	02	2	3	206	200	222	232	194	233	274	326
10	17	69	02	2	3	184	199	217	234	195	208	245	298
10	17	69	02	2	3	201	212	224	247	209	201	245	273
10	17	69	02	2	3	210	218	206	230	212	218	253	300
10	17	69	02	2	3	205	220	246	247	202	227	229	225
10	17	69	02	2	3	246	203	212	216	188	202	219	251
10	17	69	02	2	3	259	222	225	248	205	220	232	212
10	17	69	02	2	3	186	204	218	247	206	209	239	289
10	17	69	02	2	3	231	211	197	232	192	211	257	303
10	17	69	02	2	3	221	226	216	235	197	217	262	298
10	17	69	02	2	3	217	235	237	229	200	205	216	204
10	17	69	02	2	3	193	244	234	236	206	214	260	294
10	17	69	02	2	3	235	205	208	207	201	232	345	344
10	17	69	02	2	3	215	208	193	220	212	218	284	348
7	25	69	08	1	3	124	121	129	138	124	133	152	171
7	25	69	08	1	3	124	124	141	151	114	124	154	187
7	25	69	08	1	3	141	137	150	152	124	150	190	266
7	25	69	08	1	3	135	131	145	138	109	152	200	284
7	25	69	08	1	3	147	142	138	142	129	168	215	299
7	25	69	08	1	3	119	123	136	133	109	141	186	221
7	25	69	08	1	3	122	121	122	135	116	139	174	236

7	25	69	08	1	3	101	120	133	137	113	142	174	230
7	25	69	08	1	3	127	139	143	140	118	145	187	233
7	25	69	08	1	3	148	141	135	133	125	143	181	210
7	25	69	08	1	3	141	141	146	147	114	130	140	123
7	25	69	08	1	3	139	137	135	148	129	138	153	151
7	25	69	08	1	3	100	122	132	133	127	137	167	228
7	25	69	08	1	3	116	131	138	146	129	158	197	261
7	25	69	08	1	3	146	156	151	150	123	137	168	204
7	25	69	08	1	3	137	163	153	150	120	129	160	184
7	25	69	08	1	3	129	146	150	153	126	138	179	205
7	25	69	08	1	3	117	139	144	139	121	157	182	201
7	25	69	08	1	3	147	150	169	191	157	174	156	126
7	25	69	08	1	3	125	153	162	164	138	161	171	221
10	31	69	08	2	3	136	155	164	146	115	140	201	311
10	31	69	08	2	3	151	173	163	142	110	142	209	319
10	31	69	08	2	3	110	154	153	149	113	172	282	421
10	31	69	08	2	3	137	162	167	155	124	179	284	431
10	31	69	08	2	3	135	160	166	165	140	191	288	395
10	31	69	08	2	3	139	151	154	158	140	196	304	436
10	31	69	08	2	3	138	145	155	153	130	182	295	402
10	31	69	08	2	3	131	155	152	149	125	175	280	406
10	31	69	08	2	3	112	166	172	166	131	175	254	301
10	31	69	08	2	3	136	154	159	145	121	162	237	311
10	31	69	08	2	3	129	132	131	125	108	122	152	152
10	31	69	08	2	3	131	155	157	151	115	133	164	205
10	31	69	08	2	3	130	140	153	150	121	147	177	228
10	31	69	08	2	3	196	188	177	156	123	153	212	268
10	31	69	08	2	3	117	147	161	148	119	172	219	294
10	31	69	08	2	3	139	151	156	147	124	179	236	332
10	31	69	08	2	3	151	155	162	149	134	163	225	292
10	31	69	08	2	3	120	143	159	144	113	162	229	305
10	31	69	08	2	3	162	190	185	169	122	156	208	245
10	31	69	08	2	3	146	162	161	157	137	163	219	278
4	18	69	09	1	3	202	204	202	188	151	161	206	331
4	18	69	09	1	3	147	159	181	178	143	172	198	326
4	18	69	09	1	3	177	176	181	183	153	169	204	361
4	18	69	09	1	3	191	188	177	196	165	180	224	365
4	18	69	09	1	3	198	177	186	198	168	190	215	358
4	18	69	09	1	3	171	182	187	195	169	168	204	320
4	18	69	09	1	3	188	178	195	189	173	176	205	314
4	18	69	09	1	3	165	179	196	202	181	185	220	331
4	18	69	09	1	3	174	176	177	174	170	180	225	430
4	18	69	09	1	3	196	189	187	196	166	174	214	377
4	18	69	09	1	3	185	186	182	189	164	187	221	405
4	18	69	09	1	3	157	161	176	178	158	182	231	445
4	18	69	09	1	3	180	189	193	194	179	175	220	407
4	18	69	09	1	3	203	191	195	227	195	216	246	442
4	18	69	09	1	3	222	202	191	197	185	216	241	451
4	18	69	09	1	3	251	209	201	196	170	194	240	456
4	18	69	09	1	3	205	203	189	192	181	223	252	487
4	18	69	09	1	3	197	192	192	193	186	202	235	417
4	18	69	09	1	3	217	181	202	218	196	221	240	461
4	18	69	09	1	3	228	198	200	192	173	192	227	381
7	18	69	09	2	3	273	270	243	241	177	193	220	318
7	18	69	09	2	3	318	265	252	234	174	190	213	290
7	18	69	09	2	3	210	230	245	207	162	189	213	258
7	18	69	09	2	3	257	238	217	233	200	213	216	279
7	18	69	09	2	3	225	228	228	228	174	191	216	290
7	18	69	09	2	3	225	229	224	251	192	188	201	298
7	18	69	09	2	3	246	237	240	234	185	191	208	278

7	18	69	09	2	3	249	238	222	219	185	183	207	297
7	18	69	09	2	3	215	226	231	232	183	190	200	244
7	18	69	09	2	3	236	237	236	240	187	184	195	243
7	18	69	09	2	3	231	233	218	220	178	181	201	242
7	18	69	09	2	3	205	210	210	221	174	163	191	243
7	18	69	09	2	3	249	212	215	219	192	192	210	291
7	18	69	09	2	3	215	221	221	226	172	183	200	309
7	18	69	09	2	3	241	233	224	237	176	208	219	291
7	18	69	09	2	3	221	227	219	225	185	201	215	280
7	18	69	09	2	3	209	219	225	234	179	203	217	324
7	18	69	09	2	3	203	225	229	211	169	190	204	276
7	18	69	09	2	3	214	244	217	240	194	195	212	305
7	18	69	09	2	3	226	242	229	234	185	184	198	261
1	24	69	02	1	4	283	281	306	307	244	250	250	170
1	24	69	02	1	4	175	221	249	270	219	250	245	171
1	24	69	02	1	4	189	227	248	273	228	273	255	167
1	24	69	02	1	4	263	329	342	360	298	291	265	166
1	24	69	02	1	4	344	387	402	393	324	292	256	160
1	24	69	02	1	4	394	424	402	406	332	308	271	158
1	24	69	02	1	4	321	332	367	350	292	278	271	181
1	24	69	02	1	4	191	224	218	262	219	247	266	201
1	24	69	02	1	4	224	250	300	292	246	257	267	207
1	24	69	02	1	4	206	230	279	276	226	253	245	178
1	24	69	02	1	4	355	344	364	384	327	281	254	157
1	24	69	02	1	4	466	449	409	374	321	326	277	166
1	24	69	02	1	4	246	283	304	300	261	289	262	163
1	24	69	02	1	4	450	464	474	436	342	303	264	155
1	24	69	02	1	4	374	331	357	382	335	304	268	158
1	24	69	02	1	4	224	218	225	269	231	268	277	191
1	24	69	02	1	4	189	218	238	258	216	246	257	181
1	24	69	02	1	4	193	225	229	257	211	249	263	184
1	24	69	02	1	4	246	240	258	258	216	244	252	179
1	24	69	02	1	4	259	269	279	288	232	254	248	164
10	17	69	02	2	4	411	397	443	425	321	338	283	226
10	17	69	02	2	4	438	449	416	407	318	323	276	212
10	17	69	02	2	4	541	515	469	471	376	385	301	212
10	17	69	02	2	4	502	502	463	426	342	291	245	168
10	17	69	02	2	4	298	340	303	360	290	353	273	205
10	17	69	02	2	4	348	402	396	401	352	313	285	212
10	17	69	02	2	4	773	704	660	607	475	370	300	200
10	17	69	02	2	4	388	371	365	385	318	364	289	204
10	17	69	02	2	4	691	678	664	627	412	350	280	198
10	17	69	02	2	4	503	491	499	495	370	344	284	202
10	17	69	02	2	4	402	414	353	337	285	305	273	264
10	17	69	02	2	4	151	188	214	232	221	243	266	322
10	17	69	02	2	4	179	187	224	254	224	239	234	282
10	17	69	02	2	4	151	162	201	245	220	241	252	320
10	17	69	02	2	4	179	230	252	267	230	245	295	410
10	17	69	02	2	4	194	211	222	246	216	227	260	378
10	17	69	02	2	4	256	235	256	294	254	257	248	215
10	17	69	02	2	4	295	281	290	310	247	253	239	175
10	17	69	02	2	4	304	306	386	353	275	277	235	148
10	17	69	02	2	4	293	319	330	317	260	259	227	145
7	25	69	08	1	4	638	661	601	473	295	205	165	98
7	25	69	08	1	4	599	585	515	426	237	199	159	95
7	25	69	08	1	4	110	134	140	141	101	109	138	100
7	25	69	08	1	4	195	182	204	182	140	134	140	109
7	25	69	08	1	4	134	153	178	163	117	121	143	114
7	25	69	08	1	4	133	158	160	170	132	140	159	131
7	25	69	08	1	4	166	148	139	147	118	144	179	243

7	25	69	08	1	4	129	119	123	127	107	123	142	153
7	25	69	08	1	4	135	135	158	165	130	133	138	130
7	25	69	08	1	4	175	177	177	177	142	153	140	123
7	25	69	08	1	4	169	185	183	182	150	145	138	120
7	25	69	08	1	4	250	219	210	189	147	149	143	113
7	25	69	08	1	4	274	275	251	213	158	144	146	116
7	25	69	08	1	4	127	149	139	160	128	139	153	121
7	25	69	08	1	4	185	236	217	219	166	147	142	87
7	25	69	08	1	4	483	429	366	315	213	170	140	73
7	25	69	08	1	4	501	460	434	356	218	169	145	78
7	25	69	08	1	4	529	525	454	390	252	184	148	79
7	25	69	08	1	4	665	638	540	462	284	189	150	82
7	25	69	08	1	4	371	366	344	293	195	156	151	88
10	31	69	08	2	4	582	540	440	361	258	214	188	126
10	31	69	08	2	4	174	194	186	185	164	181	185	137
10	31	69	08	2	4	263	236	259	235	191	187	174	120
10	31	69	08	2	4	480	495	465	379	250	196	163	106
10	31	69	08	2	4	357	377	341	273	212	177	177	129
10	31	69	08	2	4	121	153	148	137	104	111	139	92
10	31	69	08	2	4	251	225	211	202	149	164	156	94
10	31	69	08	2	4	228	231	228	199	156	152	167	103
10	31	69	08	2	4	360	345	300	276	208	171	162	101
10	31	69	08	2	4	427	424	400	309	213	181	163	90
10	31	69	08	2	4	289	305	250	240	188	178	181	116
10	31	69	08	2	4	231	260	233	240	189	184	182	114
10	31	69	08	2	4	277	299	274	266	201	171	170	105
10	31	69	08	2	4	379	376	360	302	222	183	170	108
10	31	69	08	2	4	433	512	435	353	236	180	161	92
10	31	69	08	2	4	224	197	184	170	128	148	181	157
10	31	69	08	2	4	191	175	197	191	152	160	170	112
10	31	69	08	2	4	320	291	275	239	188	162	169	114
10	31	69	08	2	4	455	440	413	316	205	178	162	93
10	31	69	08	2	4	319	339	333	295	200	165	162	101
4	18	69	09	1	4	159	181	193	194	162	147	169	145
4	18	69	09	1	4	175	221	218	209	150	153	175	126
4	18	69	09	1	4	377	371	346	308	206	173	182	133
4	18	69	09	1	4	175	169	175	168	137	138	179	226
4	18	69	09	1	4	193	192	192	180	152	157	178	212
4	18	69	09	1	4	205	242	229	218	189	167	175	156
4	18	69	09	1	4	302	306	295	267	196	162	182	124
4	18	69	09	1	4	336	322	310	267	188	166	180	147
4	18	69	09	1	4	394	380	364	283	212	187	191	145
4	18	69	09	1	4	396	343	317	273	189	170	177	132
4	18	69	09	1	4	505	493	401	314	221	185	188	129
4	18	69	09	1	4	235	246	234	221	164	153	172	137
4	18	69	09	1	4	466	448	396	324	215	183	184	121
4	18	69	09	1	4	653	570	488	374	249	196	180	114
4	18	69	09	1	4	303	323	300	259	195	177	187	123
4	18	69	09	1	4	178	174	173	178	148	153	175	170
4	18	69	09	1	4	173	193	201	211	185	172	199	316
4	18	69	09	1	4	204	194	202	203	165	164	187	219
4	18	69	09	1	4	210	198	208	195	162	162	179	183
4	18	69	09	1	4	224	223	231	221	195	173	181	149
7	18	69	09	2	4	217	195	206	224	174	170	197	209
7	18	69	09	2	4	260	264	258	231	174	187	190	201
7	18	69	09	2	4	272	259	254	236	175	181	192	170
7	18	69	09	2	4	207	233	217	229	175	172	190	167
7	18	69	09	2	4	300	303	289	276	191	180	187	139
7	18	69	09	2	4	475	492	445	384	244	203	197	137
7	18	69	09	2	4	609	527	478	418	263	225	203	144

```
7 18 69 09 2 4   423   421   403   350   238   201   192   143
7 18 69 09 2 4   598   553   483   360   241   221   206   152
7 18 69 09 2 4   206   255   243   248   196   187   194   168
7 18 69 09 2 4   348   315   298   306   219   203   200   185
7 18 69 09 2 4   454   435   422   389   261   204   209   180
7 18 69 09 2 4   744   669   539   463   274   220   211   151
7 18 69 09 2 4   747   723   636   520   335   255   212   145
7 18 69 09 2 4   561   493   450   352   237   216   202   168
7 18 69 09 2 4   221   248   272   279   185   187   185   145
7 18 69 09 2 4   469   485   452   397   269   218   206   149
7 18 69 09 2 4   446   412   423   373   244   201   201   152
7 18 69 09 2 4   800   753   659   553   315   242   208   139
7 18 69 09 2 4   806   723   622   518   311   248   219   155
```

LISTING OF DATA SET 2 (THE FIRST TWO LINES ARE COLUMN GUIDES)

```
0          1      2            3       4         5         6
1 2 345 678901 234567 890123 456789 0123456789012345678901234567890 1

1 1   0 653649 564421 279213 208138  377  44 111 -22 155  41 171  87
1 1   0 802806 689538 318243 223146  651  75 -30 -92  75 183  -7   8
1 1   0 744759 572462 304232 220159  454 112-275  76 153  43 144 147
1 1   0 673713 603488 277211 209136  438   6-122-241 -44 160  91 -88
1 1   0 784742 682516 316228 209140  629  50 203 -43  33  25  98-119
1 1   0 723751 652520 296222 217138  563  45 -46-219  -8 227  94-101
1 1   0 833784 689491 300236 218131  725  43 193 -37  89 142  39 239
1 1   0 901913 726534 304229 212128  797 -19-140-214 -33 -29 100 189
1 1   0 927915 756577 355247 219134  847  37 -50  -1 115 -32 155 -37
1 1   0 932948 763584 335255 218132  803 -25-179-128  50   7-108 100
1 1   0 874886 727545 310231 207125  733 -53 -92-207 -29 -59  22  58
1 1   0 802803 669512 303238 211130  586 -29 -59 -88  59  53 -84 105
1 1   0 857890 700488 289221 211129  726  -8 -75-251  57-100 175 322
1 1   0 865856 685513 288225 214135  745  16-144-185 -95  87  79 212
1 1   0 904859 692513 313218 209129  823  25 -57 -31 -48 -62 327  93
1 1   0 867855 714523 306236 217129  764  10   8-131  38  89  42 193
1 1   0 828806 639508 299235 212129  614 -25-213  -4 -20 114 -34 140
1 1   0 854940 772591 327245 220127  745 -34-162-347  83  69 -45  -6
1 1   0 863833 709550 343241 211126  715  -9  10  53 103 -26  98 -96
1 1   0 876843 710492 300236 214127  761   3 130 -80  96   4  38 330
1 1   0 952914 790588 336244 215123  927 -12  68-117 -25  50  31   4
1 1   0 875866 732548 340247 209129  698 -30  33 -14 146-131 -13   0
1 1   0 897927 760553 343246 212131  750 -18  -9-111 197-202  63  67
1 1   0 851843 713534 347254 218128  680   3  62  79 279 -20  19  59
1 1   0 880840 687554 337239 201121  663 -83-153  84  -7-110   1-127
1 2   1 747689 541435 275215 197128  418 -51-173 120 -53 -20  46 131
1 2   2 890778 678515 321227 203122  767 -11 115 148 -59 -17 161 -19
1 2   3 929863 715566 338245 207134  766 -16-105  84 -59 -58 -25 -59
1 2  14 810700 671499 301228 205129  672  13 318  46 -57 110   1 -18
1 2  15 831821 687503 303227 202130  630 -38  22-117  25-145  34  90
1 2  16 945802 702536 296228 196127  819 -52  54  28-323   2 -93  41
1 2  17 968870 770542 308237 196123  859 -79 214 -83-144-175-133 144
1 2  18 896778 617489 293216 193125  682 -59-167 148-240 -95 104  71
1 2  19 942850 687577 318240 210131  809 -21-285  69-270 175 -71 -50
1 2  20 932885 714530 331223 203126  825 -10 -46  17 -12-215 304  12
1 2  21 976867 745595 345241 210129  923   7 -33 101-173  68  76-157
1 2  22 933884 715554 339261 208129  702 -67-128 128  75-101-219 114
1 2  23 978930 776595 344261 208130  823 -64 -73   1 -21 -95-234  31
1 2  24 982878 664529 306246 207135  778 -36-333 162-194   7-150 271
1 2  25 768706 562468 299226 202124  463 -51-187 200   0  42  36  28
```

1	2	26	867782	669524	328240	203127	657	-36	19	169	7	-51	-17	-37
1	2	27	876777	652483	293223	196123	679	-59	39	70	-112	-100	27	150
1	2	29	945829	727558	334239	202128	827	-22	92	118	-115	-65	9	-77
1	2	30	874812	663536	321237	215133	728	25	-154	95	-53	143	75	6
1	2	31	919792	668491	327249	205132	705	-7	115	307	88	-122	-50	213
1	2	32	842764	645520	318230	203124	643	-36	-58	130	-60	23	58	-82
1	2	47	707619	549435	264208	201129	458	0	59	60	-121	175	78	47
1	2	48	865795	642465	275214	205120	733	-26	-28	-35	-112	41	161	300
1	2	49	722689	602430	264202	185120	427	-103	147	-121	-44	-218	59	81
1	2	50	753723	649498	306231	201134	490	-27	119	-22	50	-54	-68	-73
1	2	51	923792	646501	306216	191127	739	-44	-69	164	-227	-182	164	5
1	2	52	810761	695507	309233	197126	590	-63	236	-45	38	-139	-93	-13
1	2	56	773685	623473	306231	198127	505	-38	196	154	56	-56	-43	-8
1	2	57	812654	577454	256206	188117	585	-88	42	97	-353	93	-35	80
1	2	59	858716	666492	285208	192120	744	-38	255	10	-261	-7	123	-30
1	2	60	672529	462367	236194	187118	328	-67	63	215	-185	98	75	157
1	1	0	728722	638508	305233	215148	505	64	20	-66	50	147	-13	-80
1	1	0	914880	764582	331242	211136	822	5	31	-123	-68	-1	-25	-73
1	1	0	904897	784597	358265	218144	761	25	68	-45	145	-40	-163	-76
1	1	0	961950	780602	360253	221138	896	42	-105	-27	68	-13	91	-50
1	1	0	884864	775603	356267	221147	747	46	90	-30	110	75	-210	-128
1	1	0	882883	732536	323241	214151	731	60	-3	-125	67	-122	33	105
1	1	0	794891	738590	359261	223138	579	0	-156	-143	301	9	-90	-160
1	1	0	977929	798627	370266	222144	908	54	-21	42	32	48	-77	-127
1	1	0	885849	741587	336247	214142	760	27	-15	-62	-57	72	-75	-148
1	1	0	953833	716561	365259	214141	809	54	63	331	83	-38	17	-58
1	1	0	914930	759598	395280	228142	735	48	-100	187	395	-60	-41	-64
1	1	0	735683	642478	278211	198135	536	0	231	-145	-108	12	40	-64
1	1	0	866849	767601	335253	208138	698	-32	58	-152	-46	6	-262	-196
1	1	0	965952	795587	343249	206137	836	-30	-12	-126	6	-237	-65	3
1	1	0	858801	719542	315244	212137	718	14	137	-50	-20	87	-121	23
1	1	0	863845	702575	356248	211141	658	13	-140	85	73	-71	27	-225
1	1	0	769684	661474	284214	199137	607	26	365	-82	-84	-11	68	-16
1	1	0	697644	583451	283218	204140	432	31	132	23	9	73	46	-8
1	1	0	735756	624473	291218	207143	489	28	-48	-135	59	-56	128	45
1	1	0	865811	722532	322246	211138	710	17	172	-9	57	-9	-89	60
1	1	0	884807	712551	315241	207136	733	-6	53	-13	-119	53	-125	-26
1	1	0	793793	683520	289227	211130	639	-14	-6	-221	-60	139	-34	53
1	1	0	784709	589495	301227	213149	564	81	-143	132	-100	188	82	-40
1	1	0	860889	729569	318244	214137	692	-18	-175	-207	-14	33	-113	8
1	1	0	912875	778610	346260	212147	762	10	34	-73	-41	-3	-250	-162
1	1	0	760703	609474	295227	208138	525	26	42	62	11	84	27	44
1	1	0	911852	725553	359254	222141	806	85	67	177	179	18	125	-2
1	1	0	922928	778611	358246	216141	836	34	-97	-108	17	-52	94	-203
1	1	0	898814	686540	348245	203136	682	-7	-6	225	47	-168	27	-102
1	1	0	877811	694571	354249	216139	726	47	-59	177	34	85	52	-197
1	1	0	903871	746606	375258	225144	811	88	-50	114	121	102	94	-236
1	1	0	800652	623485	295227	197132	579	-10	241	172	-143	70	-84	-61
1	1	0	804743	641502	303238	216139	619	45	12	59	3	196	-35	84
1	1	0	922895	707587	366252	208141	698	-8	-281	157	44	-175	25	-202
1	1	0	843778	713554	319237	213142	733	54	136	-51	-84	141	-18	-123
1	1	0	923850	771600	340244	214140	875	47	113	-55	-122	102	-8	-189
1	1	0	899830	668503	323254	214147	669	40	-60	185	103	-57	-100	254
1	1	0	741716	628520	303228	198131	450	-64	-62	-38	-73	13	-119	-207
1	1	0	677713	617510	309232	198138	304	-64	-85	-89	81	-97	-155	-237
1	1	0	758733	660560	322245	219143	563	46	-50	-13	-20	306	-129	-230
1	1	0	854899	749540	342248	209131	652	-37	50	-115	271	-258	-2	53
1	1	0	915913	793599	349261	211141	753	-18	45	-117	75	-123	-220	-69
1	1	0	978978	838642	379276	227141	918	39	1	-50	161	30	-139	-73
1	1	0	954920	787600	364256	214133	859	2	32	18	87	-89	2	-101

1	1	0	886753	678507	315241	204132	713	0	201	186	-39	1	-76	78
2	1	0	157195	187195	151155	175127	-597	-111	-36	-37	62	154	-7	40
2	1	0	174189	177187	159152	163123	-637	-161	-26	60	57	-50	4	-21
2	1	0	197209	205201	166160	171134	-569	-98	26	48	80	7	-10	10
2	1	0	136185	178182	152150	170127	-650	-123	-9	-40	101	39	41	7
2	1	0	166162	161176	159165	176134	-633	-83	7	155	128	134	-64	101
2	1	0	215227	223214	174168	174129	-537	-112	38	69	122	46	-61	29
2	1	0	156178	176190	146145	166127	-617	-132	-34	-39	-26	55	33	-39
2	1	0	160214	191200	156143	164126	-619	-145	-82	-83	17	-55	102	-93
2	1	0	147175	178185	147150	175124	-588	-105	6	-27	55	186	51	40
2	1	0	144177	166173	165151	169120	-648	-134	10	87	181	-5	125	-3
2	1	0	141161	177189	151146	164124	-645	-145	30	-16	13	28	25	-86
2	1	0	159187	173179	148145	163123	-638	-157	-33	-15	26	-32	45	0
2	1	0	176184	183196	148155	169127	-600	-135	-31	4	-14	110	-77	22
2	1	0	182187	173188	152150	163123	-618	-161	-55	47	-6	-5	-8	-5
2	1	0	152162	165182	150140	168121	-587	-119	-2	16	0	100	165	-59
2	1	0	146169	186199	161147	172123	-580	-104	41	2	57	130	133	-109
2	1	0	148155	163179	142148	165123	-647	-151	-2	9	-11	81	-33	2
2	1	0	157173	179189	156149	168132	-617	-102	17	15	33	29	56	-54
2	1	0	198241	210203	162155	178125	-507	-96	-50	-28	101	116	108	71
2	1	0	194220	217208	160162	169123	-578	-158	20	-15	72	33	-108	35
2	1	0	160195	187195	152142	161129	-630	-143	-36	-67	-20	-76	68	-101
2	1	0	156187	180177	146140	165121	-600	-141	10	-53	26	0	129	0
2	1	0	166200	182184	138145	160118	-639	-200	-60	-86	-29	-27	-48	36
2	1	0	135167	187190	153149	162124	-673	-162	61	-40	51	-30	-28	-85
2	1	0	160172	178176	149147	167126	-607	-121	53	7	43	27	60	21
2	1	0	153176	170177	146144	160120	-657	-178	-13	-12	13	-50	19	-16
2	1	0	162182	182181	152149	165125	-625	-141	29	3	54	-15	25	5
2	1	0	140161	180166	147152	170135	-644	-88	123	-9	105	23	2	75
2	1	0	160171	163175	153152	163133	-677	-131	-19	64	46	-66	-25	2
2	1	0	150185	179181	147144	168126	-612	-122	-3	-53	35	38	89	1
2	1	0	149185	171181	142142	159122	-669	-182	-60	-65	-17	-61	5	-28
2	1	0	148156	160180	149145	170121	-603	-121	-7	39	20	141	102	-20
2	1	0	188226	228220	155152	169124	-538	-142	9	-121	-8	66	-15	-45
2	1	0	173190	199199	153149	159126	-630	-169	29	-38	-17	-81	-50	-78
2	1	0	163174	181182	154146	165118	-600	-151	52	25	50	19	83	-22
2	1	0	167177	187187	148134	163117	-550	-141	51	-47	-40	25	200	-86
2	1	0	160175	169190	147145	163117	-625	-177	-58	0	-28	53	16	-47
2	1	0	156173	174187	141141	162120	-619	-166	-26	-52	-58	39	30	-45
2	1	0	169181	170184	143144	161126	-632	-157	-57	-16	-55	-17	2	-17
2	1	0	175184	170172	138137	156117	-619	-193	-26	-30	-61	-82	59	9
2	2	5	203198	207195	152150	159118	-571	-188	71	0	-19	-61	-44	-4
2	2	6	203188	198198	140149	158129	-589	-167	17	-38	-135	-30	-130	-6
2	2	7	180176	167183	148147	153125	-677	-198	-48	46	-58	-147	-75	-50
2	2	8	158173	187182	138140	159126	-623	-156	53	-91	-63	-41	7	-31
2	2	9	237196	189188	153158	161123	-558	-171	14	132	-36	-33	-112	87
2	2	10	205181	175183	140145	159126	-582	-159	-23	29	-118	-19	-30	26
2	2	34	184206	181173	134151	155128	-683	-204	-54	-59	-36	-159	-183	118
2	2	35	187189	175179	149147	157124	-637	-179	-23	36	-16	-122	-17	4
2	2	36	184172	163169	138139	154121	-630	-188	-16	20	-87	-106	16	10
2	2	37	198192	190198	144143	155126	-600	-179	-29	-33	-136	-88	-40	-70
2	2	38	194179	188180	135141	158125	-575	-159	51	-42	-116	-35	-12	27
2	2	39	182166	158167	134137	153126	-639	-175	-26	9	-125	-116	11	2
2	2	40	182181	190177	140140	154120	-611	-191	72	-44	-62	-125	0	-4
2	1	0	187224	213201	154157	166129	-603	-150	0	-66	45	-40	-90	36
2	1	0	168201	185187	152148	165130	-625	-131	-34	-31	30	-45	34	-4
2	1	0	191201	186177	143154	167128	-595	-138	-2	-1	20	13	-64	139
2	1	0	178181	187184	141147	162131	-609	-137	29	-39	-56	-19	-43	19
2	1	0	165201	188186	147153	166138	-650	-117	-28	-58	25	-45	-65	41
2	1	0	202179	192194	151151	165130	-555	-119	51	44	-56	35	-18	-5

```
2 1   0 171171 199192 156155 169127 -582-116  123   22   52   69  -21   -4
2 1   0 195163 168177 155143 160125 -579-133   42  121  -32  -54  110  -45
2 1   0 207188 195195 156161 168139 -580  -97   41   75   -3   16-109   43
2 1   0 201202 203203 165161 172143 -567  -65   27   57   43   17  -29    0
2 1   0 185186 179199 155151 169139 -586  -82  -55   33  -42   52   25  -47
2 1   0 189199 171189 155162 173141 -620  -85-104   78   38   63  -80   86
2 1   0 204222 212228 163173 187144 -515  -32  -78   14   37  302-130   66
2 1   0 183200 202199 155170 176147 -615  -66   16    8   69   89-198  104
2 1   0 182201 185196 155146 163145 -619  -86  -62  -22  -47-106   51  -82
2 1   0 194167 173195 145153 173138 -547  -68  -28   55  -97  194  -26   25
2 1   0 187199 189200 156159 173145 -593  -64  -46   18    5   63  -52   19
2 1   0 187194 174201 157156 163142 -656-118-117   51  -43  -63  -93  -59
2 1   0 224214 189197 170165 174146 -558  -50  -47  150   66    3  -13   58
2 1   0 179175 180188 159149 164135 -609-103   28   63    6  -50   52  -63
2 1   0 185197 181183 155151 168125 -585-128  -16   44   53   14   53   44
2 1   0 150164 155169 135131 136115 -770-291  -56  -48-127-359  -62-137
2 1   0 261238 240212 169159 171129 -428  -95  110   61   23   11   41   50
2 1   0 186202 204203 161154 171137 -563  -80   25   -3   30   30   34  -32
2 1   0 195189 204185 139147 166132 -541-109   96  -62  -62   34  -12   70
2 1   0 169190 185179 140137 160124 -597-152   10  -84  -50  -57   91  -10
2 1   0 181194 191174 134144 163132 -591-128   38  -89  -50  -33  -11   94
2 1   0 171164 179167 139143 158126 -625-155   98   -9  -29  -79   -5   36
3 1   0 158177 174178 163165 195265 -654  418    4   -9    4  -93   89   11
3 1   0 164174 178192 155162 194264 -629  408  -27  -61-108  -19   47  -30
3 1   0 194186 178214 176177 208360 -653  758-125    0-201-166   32-144
3 1   0 143180 197207 186194 223356 -685  786   28  -18   73  -19  -32  -25
3 1   0 196198 188191 162174 206329 -618  653  -38  -50-123-127   25   41
3 1   0 164173 190191 175160 197247 -568  398   87   18   34   -7  245  -80
3 1   0 190192 192191 164171 191215 -586  232   26   44   40   42  -20   82
3 1   0 198204 206209 160161 184203 -554  173   -3  -39  -67   23   14  -25
3 1   0 178194 198195 175159 190196 -537  202   64   38   92   44  216  -32
3 1   0 181207 204217 165161 180155 -562   -1  -38  -13   25  123    2  -48
3 1   0 181206 191186 155182 211332 -651  657  -46-101  -68  -71  -92  153
3 1   0 187207 207199 168182 205302 -629  546   16  -51   -9  -83  -73   68
3 1   0 163202 189189 168162 202292 -614  540  -22  -87  -19-118  210  -36
3 1   0 179186 185199 166158 188211 -569  230  -16   14  -23   35  139  -59
3 1   0 138181 203195 155160 190184 -584  138   86-104   72  166   48   25
3 1   0 174180 189195 156162 178164 -603   19   22   18   17   81  -59   25
3 1   0 188236 215232 179165 191189 -535  148-101  -36   46  108  105  -96
3 1   0 212225 215200 163156 180145 -479  -10   37    2   61   93  121   65
3 1   0 170214 201214 176158 173150 -617  -41  -50    8   91  -42   60-126
3 1   0 177220 205209 170161 173141 -608  -79  -35    0  107   -9   -5  -37
3 1   0 252234 206208 179180 197216 -488  256  -46  147   52   80    3  129
3 1   0 209210 209224 182187 205273 -577  452  -36   56    0   50  -84   -5
3 1   0 212211 208213 177174 195282 -596  460  -14    6  -72-153    6  -67
3 1   0 190189 214205 176170 189221 -572  257  121   34   46  -30   25  -41
3 1   0 205208 199204 153169 192237 -570  294  -53  -53-106   44  -75   70
3 1   0 217217 203210 166181 197246 -577  330  -60   25  -31   43-131   93
3 1   0 213230 215214 192208 206279 -665  434  -19  131  175  -67-302  138
3 1   0 200206 198211 164169 184199 -590  142  -54   17  -29   31  -80    0
3 1   0 165185 202199 173167 187183 -588  125   89   38  126   69   42  -12
3 1   0 237219 216215 181187 202282 -557  475    7   78  -19  -57  -87   45
3 1   0 193199 214205 171177 212327 -564  679   64  -71  -71  -44   66    1
3 1   0 196199 219217 165183 198232 -569  289   50  -13    0  140-183   71
3 1   0 221207 229227 183189 202302 -590  534   63   21  -56-100-146  -47
3 1   0 219213 233242 180180 192234 -548  281   20    8  -57   33-128-102
7 2  62 250227 209239 196206 224250 -447  443  -98  234  114  421-103  150
7 2  63 231197 211206 180215 239374 -532  892   63  111   25  208-200  270
7 2  64 209200 185209 198228 253425 -6161090  -80  185  105  179-180  231
7 2  65 215180 193208 176189 225396 -575  944   -1   26-176  -35   15   -1
```

```
7 2   66 197180 177199 159169 205344 -617 705 -79 -50-246-118  47 -54
7 2   67 177180 185181 153169 198333 -680 634  12-102-164-242 -26  13
7 2   68 178182 181182 141163 185294 -703 447 -34-133-218-250-122  15
7 2   69 201166 165193 138162 185278 -650 401-107 -48-321 -87-136  -7
7 2   70 166147 161179 139174 208365 -694 764 -40-104-278 -75-122  58
7 2   71 173166 161174 132156 191339 -704 629 -86-172-347-265 -26 -12
7 2   72 141143 156173 138175 209386 -760 826 -47-150-260-156-143  44
7 2   73 182171 175181 150197 236510 -7311312 -55-158-314-242-177 125
7 2   74 175150 155172 145179 228480 -6981220 -63-135-368-219   6  45
7 2   75 179157 156164 140171 212392 -675 881 -40 -95-278-167  -4 104
7 2   76 234172 164176 134162 216429 -5551040 -87-128-502-170 123  63
7 2   77 335269 216208 154176 224390 -347 935-152 -48-400 -28 140 223
7 2   78 295201 184180 140162 221479 -4661238 -66-118-606-312 222  71
7 2   79 222163 158172 148186 221412 -626 969 -58  22-287 -89 -94 182
7 2   80 223202 168173 144173 215437 -6501029-140-129-367-330  19 108
7 2   82 234173 178186 147164 216462 -5811150 -38-136-510-327 150 -55
7 2   83 181149 155165 148163 216423 -6401028 -11 -93-326-232 193  -5
7 2   86 234180 163167 130158 196413 -660 895 -97-146-516-454   0  31
7 2   90 180162 174181 140166 203384 -684 819 -23-163-348-250 -40 -12
7 2   91 206160 167167 133149 197374 -605 804   0-151-436-296 157 -20
7 2   93 196171 165173 140153 192365 -671 736 -60-139-383-378  86 -60
7 2   94 253188 169183 152151 191300 -508 551 -72  34-330-172 221 -17
7 2   95 180166 160186 142147 192346 -654 683-113-155-400-273 162-154
7 2   96 242195 192191 158161 213386 -492 910  -6 -72-340-175 263 -19
7 2   98 184181 165177 140143 186398 -729 823-117-240-477-606 161-203
7 2   99 220174 168181 144151 182325 -642 573 -63 -64-389-375  61 -92
7 2  100 216187 164184 139146 183279 -587 432-131 -78-350-181 123 -40
7 2  101 205160 157181 136135 186331 -585 640 -90-135-489-275 275-172
7 2  102 184137 169176 134147 179271 -626 397  60 -81-328-159  28 -67
7 2  103 185155 166181 140148 175297 -696 451 -28-100-355-355   3-140
7 2  104 202158 173168 133144 170238 -617 256  54 -59-281-231  25  -6
7 2  105 199175 163184 151150 180289 -655 450 -88 -36-285-307 100-121
7 2  106 165163 167180 143155 181331 -760 564 -52-150-314-428 -43-131
7 2  107 217195 180186 158157 188327 -634 600 -57 -50-269-379 111 -86
7 2  108 183161 156180 139145 177266 -657 364 -88 -75-311-213  72-100
7 2  109 179184 168179 147148 179270 -669 379 -78 -96-217-286 103 -80
7 2  110 192178 184184 149144 165214 -640 155  22 -52-189-291  61-114
7 2  111 157172 160174 130136 166226 -698 187 -94-170-268-262  62 -96
7 2  112 169159 166179 136139 169231 -659 226 -26-115-273-217  75-119
7 2  113 195157 160175 143145 171241 -643 268 -21  -8-262-239  67 -83
7 2  114 154160 154174 133144 169229 -719 198 -87-113-235-214 -15 -70
7 2  115 158158 169173 141155 178236 -688 254  17 -70-134-141 -45   2
7 2  116 168153 163177 138141 161207 -696 110 -12 -65-234-259  -3-122
7 2  117 175171 159172 146153 172234 -707 225 -60 -18-150-257 -22 -25
7 2  118 173162 156170 138146 174218 -650 196 -47 -40-182-122  53  -7
7 2  119 182170 167186 137146 169224 -664 185 -77 -86-267-184 -25 -80
7 2  120 185176 176190 149143 172208 -603 167 -31 -50-188-133 126-117
7 2  121 195164 176193 144140 156175 -635  -5 -11 -20-246-208   0-162
7 2  122 196166 168184 145148 169165 -583  10 -18  42-136   3  25 -10
7 2  123 182152 167187 146151 177220 -616 217  -4  -1-200 -47  26 -71
7 2  124 202186 167190 152144 169196 -601 113-100  12-183-150 118-100
7 2  125 191163 176178 138140 158201 -658  85  28 -60-250-295  -2-103
7 2  126 184148 169193 135135 164186 -590  73 -15 -66-313 -56  75-160
7 2  127 195181 168179 133135 153177 -658 -16 -75 -78-266-284   5 -88
7 2  128 156151 161180 133133 155162 -676 -50 -28 -87-223-162  30-143
7 2  129 150136 160171 135133 158173 -671   8  46 -65-195-170  78-135
7 2  130 162160 174182 138151 158163 -714 -68   9 -45-122-156-180 -37
7 2  131 180146 174171 144139 156162 -636 -31 114  14-137-209  50 -91
7 2  132 174178 181190 136146 162171 -657 -19 -35-101-179-116 -89 -50
7 2  133 147157 171180 133131 160153 -634 -50   9-128-172 -75 105-117
```

7	2	134	164163	174178	130138	160146	-628	-87	12	-96	-159	-44	-3	-25
7	2	135	185178	166174	132131	160158	-594	-32	-51	-78	-206	-115	133	-41
7	2	136	194167	184182	139132	150162	-623	-56	56	-63	-235	-283	57	-146
7	2	137	204198	192192	138138	169144	-500	-44	-18	-82	-158	65	132	4
7	2	138	162161	167183	144151	166129	-634	-132	-9	26	-26	75	-57	11
7	2	139	178164	179178	138140	170128	-523	-85	65	-20	-73	147	123	36
7	2	140	184157	174176	136138	162130	-560	-111	61	-4	-126	35	71	-1
7	2	141	162159	160175	133142	173141	-572	-46	-25	-30	-94	173	79	51
7	2	142	173162	181174	136141	158150	-634	-79	78	-56	-122	-128	-16	-21
7	2	143	196182	162165	138142	168165	-585	12	-39	5	-111	-65	102	73
7	2	144	172147	163175	124133	160142	-592	-86	8	-70	-228	27	39	-26
7	2	145	177172	175177	137136	170163	-554	26	12	-72	-143	9	169	-15
7	2	146	187175	167173	128140	162153	-608	-60	-41	-60	-181	-45	-6	43
7	2	147	175148	174174	129139	162162	-608	-18	71	-66	-200	-46	-2	-17
7	2	148	180165	163178	126141	162177	-646	8	-63	-86	-251	-92	-57	-11
7	2	149	197169	172190	143152	176189	-575	107	-35	6	-175	47	-1	1
7	2	150	234181	174183	156156	187202	-475	216	0	123	-107	98	154	79
7	2	151	167171	170176	131134	163161	-612	-18	-23	-111	-180	-71	93	-39
7	2	152	168147	157183	138144	149144	-723	-154	-40	12	-181	-186	-152	-114
7	2	153	180165	172188	137133	160135	-570	-104	-17	-47	-188	5	110	-96
7	2	154	164166	181190	137136	161136	-595	-106	13	-92	-150	3	61	-92
7	2	155	183172	180177	134143	162141	-597	-96	33	-47	-118	-23	-18	33
7	2	156	164162	185184	128143	165130	-586	-125	53	-102	-125	112	-53	31
7	2	157	181181	196192	137146	166129	-560	-125	45	-78	-92	88	-32	23
7	2	158	164168	186201	145144	166122	-575	-140	10	-44	-80	127	22	-81
7	2	159	188192	178192	138141	165128	-560	-128	-79	-53	-126	72	46	-1
7	2	160	163172	207198	145155	169136	-597	-101	115	-73	-17	89	-113	0
7	2	161	186172	203208	147153	167125	-554	-138	69	-20	-80	140	-92	-37
7	2	162	189193	222210	153152	166122	-540	-147	112	-57	-23	65	-43	-44
7	2	163	212200	204204	146150	169124	-501	-127	11	-19	-86	134	-7	28
7	2	164	184172	195187	146149	170120	-528	-125	107	4	-7	147	29	46
7	2	165	180184	208203	157156	164117	-590	-179	93	6	31	33	-87	-34
7	2	166	192201	211203	156160	167122	-580	-162	57	1	40	42	-118	29
7	2	167	202176	201208	151153	162118	-556	-178	57	30	-84	75	-100	-51
7	2	168	206218	217209	169166	172119	-548	-152	43	63	121	71	-76	44
7	2	169	204217	232238	171166	168127	-565	-153	17	0	12	51	-161	-111
7	2	170	185193	211198	147152	159118	-602	-200	82	-54	-21	-36	-125	-3
7	2	171	171170	160180	122139	151115	-667	-245	-101	-73	-191	-29	-142	9
7	2	172	169178	165184	136140	146122	-713	-242	-94	-45	-142	-201	-119	-81
7	2	173	181171	168174	137137	148114	-652	-237	-9	-2	-112	-159	-22	-37
7	2	174	190174	165165	137137	149118	-640	-215	0	21	-95	-184	7	8
7	2	175	199188	179175	139139	158134	-583	-126	1	-18	-108	-106	55	19
7	2	176	167190	171179	132136	153120	-654	-210	-81	-97	-115	-109	-9	-17
7	2	177	181175	167179	140141	152115	-646	-225	-39	11	-92	-102	-35	-26
7	2	178	189173	176178	133139	148114	-638	-243	5	-23	-142	-134	-83	-14
7	2	179	203177	184177	138143	151115	-606	-225	53	14	-103	-119	-73	23
7	2	180	191179	171177	142147	163160	-628	-38	-25	3	-110	-110	-17	14
7	2	241	155151	172194	141146	163178	-682	13	-12	-68	-186	-92	-72	-141
7	2	242	163186	187196	146153	172164	-630	-5	-30	-78	-60	16	-50	-19
7	2	243	185183	184184	141149	164155	-622	-56	3	-40	-97	-71	-59	15
7	2	244	193196	215206	149153	159119	-590	-200	69	-50	-46	-26	-137	-29
7	2	245	207237	254236	174163	170127	-522	-133	90	-58	67	12	-67	-89
7	2	246	248241	227229	163173	177129	-485	-120	-50	50	0	184	-192	98
7	2	247	256244	244243	181170	170131	-486	-125	11	81	12	19	-115	-68
7	2	248	333334	311268	208171	184128	-243	-47	81	68	130	55	186	-22
7	2	249	393346	345301	202186	185134	-168	-54	102	66	-26	172	-104	21
7	2	250	240241	238241	185170	183143	-446	-23	3	80	66	158	20	-51
7	2	251	594517	464382	233198	198129	250	-14	28	54	-125	262	20	88
7	2	252	487404	410311	224195	192146	25	36	300	135	51	84	5	125
7	2	253	397438	397330	236193	179136	-228	-100	45	-22	175	-182	-73	-105

```
7 2 254 485428 406351 233192 194141   37  16  75  97 -34 185  56 -58
7 2 255 514480 463385 252191 196130  151   1 141  20  11 166 179-159
7 2 256 484479 457387 252202 208135  121  40  96   3 103 332 101 -84
7 2 257 497488 422375 246202 194133    7 -47 -89  69  40 126 -31 -64
7 2 258 476414 376356 229199 191144  -52 -16 -78 151 -93 204-120 -76
7 2 259 634519 462384 254205 193127  256 -32  48 243 -75 131  25  47
7 2 260 397360 320269 208194 194151 -194  30  52 175 130 123 -39 205
7 2 261 203189 180188 147153 173187 -590  83 -39  -1-128 -53  -5  17
7 2 262 156162 195199 136149 167146 -615 -76  58-117-121  86-111 -43
7 2 263 255258 255245 165169 178131 -429-101  -3 -36 -30 188-134  37
7 2 264 399427 405347 224200 188132 -173 -93  40 -67  96  94-210 -48
7 2 265 486451 406333 230199 191123   -7 -75  64 108  89 119 -48 103
7 2 266 604545 495418 255204 202132  286   5  17  44 -86 268  52 -53
7 2 267 344335 311267 183166 167130 -318-134  30 -43 -66-110 -50 -14
7 2 268 181172 182193 147142 161123 -579-150  12   0 -87  17  45 -75
7 2 269 147158 162186 134143 161119 -652-185 -57 -48 -96  84 -59 -32
7 2 270 168172 185164 136145 155123 -654-188 108 -40  -8-134 -76  71
7 2 331 695610 520439 283225 200136  327 -16 -61 229 -17  96 -78   4
7 2 332 466438 387359 229186 193140   -1   0 -98  42 -89 202  89-126
7 2 333 563441 431377 252190 207137  293 107 138 234 -93 417 323-103
7 2 334 603611 557489 296223 208134  280  -4 -48 -17  50 242 -32-292
7 2 335 285274 242242 177165 171153 -454 -48 -89  46 -73 -60 -40 -44
7 2 336 171182 204201 152158 172145 -595 -63  70 -38   0  75 -87  -3
7 2 337 165162 187201 147153 164153 -651 -69  25 -33-102 -14-128 -92
7 2 338 188213 236213 153155 173132 -516 -94 114-118  10 102 -29   5
7 2 339 438370 351321 215190 190134  -80 -29  23 153 -54 244 -43  -3
7 2 340 550491 432375 247192 188127  116 -53 -25 126 -62  70 100 -94
7 2 341 290282 282262 211178 189140 -325   4  98 148 171 145 120 -65
7 2 342 596610 539408 270210 205132  273   6 114 -53 178  63 119  56
7 2 343 767680 608483 304223 202133  535   7  92 133 -32  42  75 -93
7 2 344 462445 465352 217200 187136  -20 -63 293-135  17  70-242  44
7 2 345 209196 190184 154153 157125 -613-179  28  70  -2-141 -73  27
7 2 346 168174 176174 145144 160119 -623-173  37   5   3 -42  25  13
7 2 347 188171 174190 142143 159116 -586-188 -21  19-110  36  -8 -34
7 2 348 194175 166183 144143 153122 -633-197 -51  50-111-107 -34 -42
7 2 349 214173 182182 136136 155122 -538-165  39   5-183 -40  47 -15
7 2 350 190182 179185 146153 157118 -641-212  -8  45 -30 -57-138  29
7 2 351 180180 181167 131132 149112 -616-229  50 -73-102-160  37   8
7 2 352 190190 187186 140143 155113 -604-221   0 -31 -79 -62 -44   2
7 2 353 192181 178175 148153 163150 -634 -75  25  44 -26-115 -60  51
7 2 356 145186 181179 139152 174175 -665  33 -18-125 -25 -22 -42  51
7 2 357 175175 208196 155150 169155 -576 -17 138 -41 -25 -26  32 -69
7 2 358 187194 190195 153153 164149 -627 -80 -16  -5 -42 -88 -56 -32
7 2 359 165178 202200 149161 164155 -680 -82  59 -56 -31 -73-230 -29
7 2 360 201222 238220 163155 170146 -530 -59  87 -87  -6 -25   0 -75
```

LISTING OF DATA SET 3 (THE FIRST TWO LINES ARE COLUMN GUIDES)

```
0        1         2         3         4         5
1234567890123456789012345678901234567890123456789012345678

 -1391 -1305  1240   426  1593  1959   449 -1594 00 00 2 1
  -175  -395   375   572  -071  -447  -016   033 00 00 2 2
  3480   666 -2512 -4085 -1442  3195  4607  1846 05 00 2 3
  6122  2130 -1833 -6052 -5991   469  1445  3411 20 00 2 4
 -1456  -804  1194  1630   585  -280  -492  -190 00 05 2 5
   874  2312 -1690 -2479  -019  -488   589   870 00 20 2 6
   415  -103  -007  -487   131   051  -100   036 00 00 8 1
   531  -611  -718   479   827   587  -096  -202 00 00 8 2
```

3501	-448	-2997	-1524	1570	1656	1026	-908	05	00	8	3
3831	1800	-2305	-4971	022	505	1171	-051	20	00	8	4
081	-403	-574	096	1257	1541	403	-677	00	05	8	5
1751	148	-720	-897	-732	-938	-960	881	00	20	8	6
-3649	651	1580	2028	1724	1135	835	-831	00	00	9	1
-1143	-245	562	1374	829	672	401	-342	00	00	9	2
816	-881	-338	287	340	004	698	-187	05	00	9	3
3160	726	-3552	-2591	-769	1085	885	279	20	00	9	4
-468	-584	444	1186	499	-187	-009	-091	00	05	9	5
1692	-875	-387	-312	-1179	-867	-680	571	00	20	9	6

REFERENCES

Dixon, W. J. (1971) BMD Biomedical Computer Programs. Berkeley, Los Angeles, London, University of California Press.

Dixon, W. J. (1969) BMD Biomedical Computer Programs. Berkeley and Los Angeles, University of California Press.

Jenden, D. J., M. D. Fairchild, M. R. Mickey, R. W. Silverman and C. Yale (1972) A multivariate approach to the analysis of drug effects on the electroencephalogram. Biometrics 28, 73-80.

DISCUSSION OF THE MICKEY CHAPTER

D. Martin

The data collection system actually performs most of the analysis because it reduces the huge volume of complex data down to 8 or 32 observations. I would be interested in more discussion of the alternatives and the basis for choices at this stage of analysis. While I realize that a frequency analyzer approach is standard, I feel that the statistical analysis should consider this preliminary analysis as a part of the overall analysis. (This may be given in earlier reports.) As an example, there was no motivation for the use of the integral of the absolute value of the amplitude function instead of the squared amplitude.

The curve fitting and discriminant analysis of the dose response were interesting but seemed to raise a few unanswered questions. The one example given showed that both methods led to essentially the same results with a remarkable consistency.

Much of the discussion was about the definition of the behavioral states. The identification of the two sleep states was a special problem since this could have been confounded with EEG measurements. I am not sure, but I suspect that these states used frequency analysis or empirical analysis of EEG to arrive at these definitions. (Not in this study but in the literature.)

I don't think that the aim of the initial discriminant analysis was to predict behavioral states from the EEG. The usual assumptions are clearly invalid but the results are quite good. The basic question is why should a linear discriminant analysis be used at all? It does establish a relation between the frequency data and the state for these three cats and hints that the between-cat variation may be small.

A basic problem that recurs is the problem of variation between cats, variation within cats over long time spans and short term variation. This problem was considered but there are no very solid results, and no good alternatives other than running more cats were suggested. This is probably the fundamental problem in this analysis. The within cat variability data is available from the experiment although there was not much use of this except in the predictive validation studies in the first section.

The discriminant procedure was used to produce canonical variables rather than to perform a classification. It seems to me that the discrimination was used to reduce the dimensionality of the observation vector. It is quite possible that much drug effect was lost by removing the first two canonical variables. This was a basic concern of several of the discussants.

I would describe the use of the discriminant function based on the remaining canonical variates as an index of drug effect. I am somewhat unsure that this is really a good measure of drug effects. The elimination of the first two canonical variables may be a relatively poor way to adjust the drug effect index for state variables. Given time, I would like to try a more direct covariance approach using a linear model. This would tend to combine the drug effect analysis and dose response analysis, which seems reasonable.

The functional form used to fit the dose response was not identified as a biological model or as simply a convenient curve (empirical model?).

Are the canonical variables interpretable? They are far removed from the raw data. There was much discussion of their utility to medical practitioners.

J. Tukey

I hope that now that the half-octave analysis is in good shape some thought will be given to what might be done to improve it.

As an easy first step, it might be sensible to compare different "integrations" of the output over each minute. If the present device integrates a mean square, it might be natural to try integrating either a mean absolute value or a mean fourth power.

One naturally also wonders how much better two minutes of data would be (this can easily be checked) and how much worse 50 seconds would be.

Some analysts take the view that discriminant analysis is a non-arbitrary approach. This is true in a very important sense, but there are a number of arbitrarinesses that remain and deserve to be pointed out:

- The choice of the half-octave bands is arbitrary, though for all I know, it may have been guided by long experience.
- The choice of volts, $(\text{volts})^2$, or db as the measure of the activity on individual channels is subject to the same remark.
- So too is the restriction to linear discrimination, as contrasted to linear and quadratic or linear and discontinuous, etc.

- There must be others, if we think hard and long enough.

The general approach in this study is pleasing and appears sound. This is no reason against asking how it might be polished up, however, Even if this study is adequately techniqued, the next one may require better technique for adequate success.

It seems to me that one question that deserves careful attention is how does one "tighten up" one or more discriminant functions.

The present data, with their externally judged S, F and A, and their time sequences seem to offer special opportunities for such a study. Let us begin with either one or two discriminant functions, and hold them fixed. A question that seems natural to me is the following: If we subdivide one kind of minutes, say S, in accordance with the characteristics of the previous minute (or of the following minute), does this make the clustering tighter? Let us suppose we have examined this, and other possible ways to tighten up the S, F and A clusters, and that we are now going to work only with these tightened clusters.

Take all such clusters, and one discriminant function, and separate the data points into three classes (-, 0, +) in accord with the relation of the function values to the center of the cluster. For each input separately, ask if the +'s are separated from -'s (a linear sort of separation is unlikely but not impossible; a quadratic sort of separation is more likely.) Do the same for each pair of inputs together. (Here a cross product of deviations from means is the more likely pattern.) By such approaches we can gain insight into the possibilities of modifying the discriminant function.

For those who want it all nicely mechanized, the natural thing to try is a stepwise regression function (about three such using three different cutoffs) for discriminant function minus median of same for cluster with carriers

- quadratic orthogonal to linear for each single input, and

- product of the two deviations from means for each pair of inputs.

Since there is a question of serendipitous gain, we may well want an unbiased estimate of the improvement due to such a modification. To do this most simply, divide the data being analyzed into 10 parts, perhaps randomly, and do the stepwise 10 times, once for each 9/10th of the data formed by leaving one tenth out, trying both original and modified discriminants on the observations of the left out 10th. The result will be to find original and modified values for each minute analyzed. It should then be easy to observe any tightening of clustering.

Once we know what the various discriminant functions look like, other questions are likely to arise naturally.

J. Gabbe

This paper contains analyses of subsets of data chosen to illustrate methods. Is this enough, or should there be some results of analyses of the entire body of data?

The paper describes a successful analysis and makes the approach seem a natural one. Examples of success are given, but leave unanswered such questions as:

- Does this analysis work on other subsets of the data? If not, why not? What similarities should I look for in another body of data before trying this method?

- What techniques were tried and rejected, and in what way did they prove unsuitable?

The example is lovely but can you offer some general guidance, particularly as it is apparent that the distributional assumptions on which the classical theory is based are badly violated.

The data base is a good one. It is extensive, well described, simple in design and thus easy to understand in a superficial sense.

The key analytical ideas are discriminant analysis and canonical correlation. Because they worked well in the examples given, they seem natural to the analysis. A discussion of examples for which the method does not work so well, and of alternate (rejected) techniques and the reasons for which they were rejected might offer valuable guidance to those with somewhat similar problems.

CHAPTER 5

ROBUSTNESS STUDY ANALYSIS
(A Further Analysis of the First Phase of the Princeton Robustness Study; Examples of Less Standard Two-Way Table Analysis)

JOHN W. TUKEY
Bell Telephone Laboratories Incorporated and Princeton University

To follow the spirit of the workshop, I should give an account of a successful data analysis, hopefully proceeding without overt difficulties. Under other circumstances I might have done just that - perhaps. As it is I shall illustrate many aspects of how not to do a data analysis - aspects that appear, however, in very many data analyses accomplished under the pressures of real life.

The format of a workshop is somewhat more forgiving than that of a conventional publication. Both, of course, tacitly urge omission from mention of dead ends and clue sources, two of the most important aspects of data analysis. It seems that we have not yet learned how to describe the process of data analysis in a form that exposes to learners its true character. It may well be that there is no alternative to going back to Robert Burns' "A chid's among ye taking notes" and putting informed witnesses into instances of the actual process of analyzing data.

Prepared in part in connection with research at Princeton University sponsored by the Army Research Office (Durham).

After all, this never seems to be the best of all possible worlds – we must face reality in writing about data analysis, as well as in doing it.

THE PROBLEM

The data discussed here come from the first phase of the Princeton robustness study. The general results of this phase have been discussed in book form (Andrews et al, 1972), to which we refer the interested reader for details of the Monte Carlo approach and other background information. The entire study was confined to

- point estimates of location, and
- samples from symmetric distributions or polysamples from symmetric situations.

A simple example of the last alternative is a set of 20 values, 19 from a Gaussian (often miscalled normal) distribution with one variance and 1 from a Gaussian distribution with the same mean but ten times the variance. We call both kinds of specifications situations. Where necessary we distinguish the set of values obtained in a situation of the second kind – which though independently distributed are not identically distributed – as a polysample.

Our considerations were further limited in sample size and measures of variability of our estimates to

- samples of 5, 10, 20 and 40, with emphasis on samples of 20; and
- variances, and pseudovariances at 0.1%, 0.5%, 1%, ?, 10%, 25% and 50%.

The concept of a pseudovariance is related to that of an "effective standard deviation" as used by Bahadur (1960). For a Gaussian distribution, the ratio

$$\left[\frac{\text{estimated (one-sided) p\% point}}{\text{(one-sided) p\% point of unit Gaussian}}\right]^2$$

estimates the variance of the distribution. If the distribution is symmetric (so that either side may be used) but not necessarily Gaussian, we define the ratio or its estimand (what it estimates) as the pseudovariance at p%.

In the first phase we acquired variances and pseudovariances for 65 estimates in each of a number of situations. Our concern here is only with the situations for n = 20.

The offer of having computations done far away was tempting; the pressures of other activities were unwisely strong (another week-long meeting that coincided with the workshop, for one). As a result the data were sent West from other places than where the writer was writing down what was supposed to be done. The data were sent as printout, gathered up from various runs – and, alas, different titles were used from time to time to describe the same situation. As a result, those who volunteered to do the computing got 32 "situations" which later turned out to consist of 14 pairs and 4 singles as shown in exhibit 1.

The data for the 18 distinct situations are on file at the Health Sciences Computing Facility, University of California at Los Angeles. Situation numbers are given in exhibit 1; where two numbers appear the lesser identifies the situation in the file.

The descriptive material on file at HSCF and the data for situations 2, 16 and 18 are included as a Data Set Description at the end of this chapter. These situations are used to illustrate the less standard two-way table analyses that were performed at HSCF on all 18 situations. The text summarizes the calculations on the illustrated situations. Computer output for some steps of the analysis on these situations is included for the reader who wishes to examine the analysis approach in depth.

Exhibit 1. The data as submitted – and later put in order.

Situation Number(s)	Rather More Explicit Title
16, 23	75% of N(0, 1) and 25% of N(0, 1)/U(0, 1/3)
14, 21	N(0, 1)/U(0, 1)
2	Cauchy
9, 20	5 from N(0, 1) and 15 from N(0, 9)
8, 24	10 from N(0, 1) and 10 from N(0, 9)
17	Double exponential
12, 27	15 from N(0, 1) and 5 from N(0, 100)
3, 31	75% of N(0, 1) and 25% of N(0, 1)/U(0, 1)
15, 22	90% of N(0, 1) and 10% of N(0, 1)/U(0, 1/3)
13, 29	90% of N(0, 1) and 10% of N(0, 1)/U(0, 1)
10, 19	19 from N(0, 1) and 1 from N(0, 100)
18	Student's t with 3 degrees of freedom
11, 28	18 from N(0, 1) and 2 from N(0, 100)
7, 25	15 from N(0, 1) and 5 from N(0, 9)
5, 26	18 from N(0, 1) and 2 from N(0, 9)
6, 32	17 from N(0, 1) and 3 from N(0, 9)
4, 30	19 from N(0, 1) and 1 from N(0, 9)
1	N(0, 1)

Notes: $N(\mu, \sigma^2)$ is a Gaussian distribution with mean μ and variance σ^2.

U(a, b) is a (uniform) rectangular distribution between a and b.

a "/" indicates a ratio of independent deviates.

"20-k from ... and k from ..." implies drawing of fixed numbers from each distribution.

"(100-p)% of ... and p% of ..." implies 20 from a mixed distribution in which the component percentages are as stated.

Previous examinations of the data had indicated that

- the pseudovariances were similar in value to the variances;
- there were trends, which seemed to follow a reasonably consistent pattern for the well-behaved estimates in each situation;
- generally speaking, the well-behaved estimates in a given situation were those that performed well there;
- sampling fluctuations mattered, so that an attempt to "borrow strength" across percent levels (for pseudovariances) seemed worthwhile;
- it seemed convenient to express performance by deficiency = 1 - efficiency;

where the best choice we could make for "efficiency" was efficiency relative to the corresponding variance or pseudovariance for the estimate that performed best for the situation in question.

The task set for the next stage of data analysis was to try to improve the observed variances and pseudovariances by borrowing strength across p-values.

This last statement is undoubtedly not as clear to the reader as it is to the writer. So let us try to give more detail from varied directions.

All our results are subject to sampling error. We have no absolute standards – except for pure Gaussian situations. So we must assess each estimate relatively. We chose to do this by comparing (in each situation) with the estimate that does best in that situation. If we can produce a somewhat hypothetical behavior (in that specific situation) for the best-performing estimate <u>that is less in error</u> than the behavior we have actually observed, we will improve the quality of all the comparisons we make with that reference standard.

To obtain better figures (for a specific estimate in a specific situation) than we actually observe, we must borrow strength – take information from the behavior of other estimates in that situation, or from the behavior of estimates in other situations. To do this we must recognize regularities; we must recognize them not as qualitative facts but rather as quantitative structures that can be used to guide us in replacing actual behavior by some form of fitted behavior.

If we can replace the 8 degrees of freedom involved in the 7 pseudovariances and 1 variance (for the apparently best estimate in a given situation) by 2 or 3 degrees of freedom – for example 2 or 3 regression constants for carriers whose quantitative nature has been borrowed – we can hope to reduce the total variability involved. If the carriers are well enough chosen, we can convert hope into reality.

This is one step more flexible (and perhaps more sophisticated) than when we have (x, y) data, fit a straight line, decide that it fits well enough, and then use $a + bx_i$ instead of y_i when we want to know what y is at an x where we have data. In that situation the two carriers involved (which we can choose, equivalently, as 1 and x, or x-3 and 3x+2, or 1 and $x-\bar{x}$, among many) came from outside. In our case, we will pull the quantitative nature of the carriers out of the data – out of the rest of the data, out of the data bank from which we can borrow.

The main effort of the analysis here is the structuring of appropriate carriers. Once we have them, our problem is much closer to a nearly familiar one.

THE HASTY SPECIFICATION

The text of what was written down for the study follows. Some typographical errors have been corrected, with the original left in [ ]. Remember that "32[37]" really refers to 14 pairs of duplicates and 4 singles.

(direct quote)

We are all used to fitting "rows-PLUS-columns" to a two-way table of responses by <u>means</u>. The purpose of the examples covered here is to show how two modifications/extensions of this type of fitting
- fitting "rows-PLUS-columns" by medians (as an example of more resistant fitting)
- fitting a rank-one "rows-TIMES-columns" term

(since [some] resistant fits have not been thought through, we will use some conventional fit) can be combined to give useful analyses of moderately complex sets of data.

<u>The robustness data</u>

The data here can be thought of as 32 [37] tables, each 65 x 8 (the 8 being a 7 + 1, a distinction we shall postpone as long as we can). (The 32 [37] are situations, the 65 estimates, the 8 percents.) We eventually want to find analyses of the moderately few "best performing" estimates out of the 65 (which may involve different estimates for different situations). The analyses will be then used as specifying the nearest-to-ideal behavior for an estimate that we have realized.

Since all numbers are based on Monte Carlos, all results are subject to fluctuations, which are far from independent within situation.

The plan of analysis to be tried first proceeds as follows:

<u>Step 1</u>: Analyze each 65 x 8 table by MEDIANS, fitting "rows-PLUS-columns". (Note that a few "infinites" and very oddly performing estimates do not prevent the analyses from coming out reasonably well.)

Step 2: For each of the 32 [37] situations, select the 20 estimates whose fitted values are smallest, and repeat step 1 for the 20 x 8 tables.

Step 3: Take the tables of residuals from step 2 and fit "rows-TIMES-columns." (In particular, this will give 32 [37] 8-vectors, one per situation. Some of these will be poorly determined because not too much fitting was needed.)

Step 4: Select the 21 situations for which the "rows-TIMES-columns" fit was largest and form the 21 x 8 table of fits against percent. Make a "rows-TIMES-columns" fit to this.

Step 5: Look at the residuals in the 21 x 8 table, comparing their size with the fit for each of the 21 situations separately. Set aside the 6 out of 21 for which residuals are largest and repeat step 4 for the remaining 15. The 8-vector for percents here is the result of the first phase of our analysis.

Step 6: Go back to the 32 [37] 20 x 8 tables, and regress each row of 8 values on the 8-vector thus obtained. (The regression should be chosen, iteratively, that

$$\text{median}\,(y-a-bx \mid 0.1\%,\ 0.5\%,\ 1\%) = 0$$

$$\text{median}\,(y-a-bx \mid 10\%,\ 25\%,\ 50\%) = 0$$

thus avoiding (i) the estimated variance, which can behave anomalously, and (ii) overemphasizing on any irregular value.)

Step 7: Repeat steps 2 and 3 for the "residuals from regression" just formed. Hopefully the resulting fits will be negligibly small. (If they are not, we carry out the subsequent steps in parallel, once on residuals from first step 6 and once on residuals formed by repeating steps 4, 5 and 6 after repeating steps 2 and 3. No further note of this possible branching will be taken in this outline.)

Step 8: Go back to the "residuals-from-regression" and repeat step 2 twice selecting 10 and 5 estimates, respectively, instead of 20.

Step 9: Reconstruct an "adjusted optimum estimate" 8-vector for each situation as follows:

(fitted regression for single best estimate)

PLUS

(fitted percent effect in step 8, for 5 or 10).

Step 10: Repeat the whole process, taking logarithms before step 1 and antilogarithms after step 9.

Step 11: We now have 4 8-vectors of adjusted optimum pseudovariances for each situation. Hopefully, they will agree quite well. Take the (percent by percent) median of the 4 as the standard against which all other estimates are to be compared. Give careful personal attention to each table of 4 8-vectors and the 8-vector of medians.

Step 12: Form the deficiencies for all variances and pseudovariances by the formula

$$\text{deficiency} = 1 - \text{efficiency} = 1 - \frac{\text{adjusted optimum}}{\text{actual}}$$

expressing the results in units of .001.

. . . . .

(Comment)

While careful examination of the actual details may suggest further modifications, the analysis just prescribed should be very considerably better than that used by Andrews, Bickel, Hampel, Huber, Rogers and Tukey, where the single "best" estimate was used in place of an "adjusted optimum."

. . . . .

Fitting "rows-PLUS-columns" by MEDIANS

Data: Y(I, J) I = 1 to R, J = 1 to C

Expanded data: I = 0 to R, J = 0 to C

Initialize Y(0, J) and Y(I, 0) to zero

Iterative loop

For each I, transfer med {Y(I, 1), Y(I, 2)... Y(I, C)} to Y(I, 0), namely increase Y(I, 0) by this amount and decrease each of Y(I, 1), Y(I, 2)... Y(I, C) by the same amount. (Note: I = 0 is included; J = 0 is special.) For each J, transfer med { Y(1, J), Y(2, J), ... Y(R, J) } to Y(0, J). (Note: J = 0 is included, I = 0 is special.)

Precision, stopping

All calculations in fixed point integers, input data multiplied by 10^k, k at least 1 (and usually no more), after expression as integers. Iterative loop traversed 6 times or until largest transfer does not exceed 2 in absolute value.

(end of quote)

STEPS 1, 2 AND 3 - WHAT WAS AND WHAT MIGHT HAVE BEEN DONE

Exhibit 2 shows the steps 1, 2 and 3 analyses for situations 2, 16 and 18. Data in the Data Set Description have been scaled by 1000 and rounded to integers in preparation for step 1 of the analyses. The results of fitting rows-PLUS-columns are shown in Panels A, E and I of exhibit 2. Note that computer printout displays of either rows-PLUS-columns or rows-TIMES-columns analysis of an R x C table is indexed identically to the iterative loop definition of fitting rows-PLUS-columns by medians given in the section on hasty specification. Thus the (0, 0) entry is the grand fit. The remaining C entries in the zero row are column

EXHIBIT 2. STEPS 1, 2, AND 3 ANALYSES

A) SITUATION 2 STEP 1 ROWS-PLUS-COLUMNS ANALYSIS

3779	-47	-727	-707	-495	48	630	1144	2762
2102201	7894042	-2070184	-1888021	-1865052	-1183187	1183188	6234820	7891222
16961	3291	-11232	-10952	-9544	-3290	8833	27689	105000
3415	178	-1303	-1302	-1166	-177	1256	3644	10268
813	22	-72	-69	-104	-21	46	210	1125
-631	9	254	273	191	-8	-279	-517	-1286
-846	-7	253	285	218	8	-297	-586	-1743
-656	-7	214	247	190	8	-284	-568	-1551
197	-7	14	52	16	8	-52	-129	-430
-206	1	18	50	0	-58	-92	-42	170
803	12040	-447	-380	-300	-61	62	183	1290
0	0	-55	-14	-18	20	3	16	0
-430	8	141	170	112	-7	-172	-275	-694
-983	5	187	223	174	-4	-280	-510	-1197
5505	108	-1711	-1702	-1464	-107	1457	3257	12614
3109	-34	-762	-761	-682	35	752	1512	4018
1964	-31	-367	-386	-383	32	467	829	1600
645	-2	-104	-86	-109	3	75	169	531
-38	8	77	107	57	-7	-128	-216	-349
-685	1	238	261	189	0	-275	-528	-1409
664412	51088	-662430	-650952	-649400	-51087	1500753	2198714	3677271
2371	7	-771	-738	-639	-6	720	1561	4589
1260	4	-458	-424	-374	-3	434	933	2404
179	18	-96	-59	-75	-17	21	103	723
-430	11	162	183	126	-10	-223	-400	-767
1757	-16	-176	-192	-228	17	230	436	992
736	-10	1	2	-52	-10	0	46	39
754	-3	1	0	-51	-2	11	54	82
-779	-7	176	234	206	8	-316	-612	-1530
20	-8	131	154	116	9	-171	-336	-1043
-363	-23	60	139	146	24	-222	-450	-1360
-374	-6	170	208	170	7	-239	-457	-1336
-719	-2	243	273	216	3	-302	-577	-1590
-1047	0	326	349	268	0	-367	-707	-1917
-856	-10	222	255	204	11	-284	-566	-1664
-251	-9	45	113	117	10	-180	-340	-983
-1138	5	330	348	257	-4	-350	-662	-1786
-1269	0	326	352	263	0	-353	-682	-1906
4702	1610	-3653	-3491	-3052	-1609	1927	15015	45498
-1418	11	401	416	301	-10	-395	-749	-2025
1285	-68	-556	-462	-329	69	415	713	837
-850	3	320	336	252	-2	-351	-665	-1758
-654	-30	106	154	142	31	-215	-470	-1553
376	-76	-647	-538	-381	77	407	655	1391
-395	0	-124	-44	1	78	81	75	-500
-436	0	-88	-10	29	92	62	0	-773
-747	-29	55	109	95	30	-147	-340	-1285
-776	-27	45	102	93	28	-139	-317	-1235
-874	-2	276	303	236	3	-334	-641	-1728
-208	0	0	27	0	-51	-122	-82	396
-867	-16	175	213	156	17	-218	-459	-1513
-940	21	221	247	182	-20	-298	-516	-1044
672	-27	-805	-702	-537	28	571	1058	3043
-225	86	-95	-47	-36	-6	8	204	1870
800	-42	-499	-355	-199	43	160	305	858
204	28	-290	-162	-27	165	170	109	-196
-828	-15	127	176	149	16	-221	-442	-1375
452	-11	-133	-86	-86	13	12	49	335
2145	-115	-1123	-1025	-833	116	1089	1907	3211
14795	2837	-11040	-10489	-8253	-2836	6054	29351	141061
49995	13619	-42487	-41049	-36215	-13618	24426	110212	299932
895625	4107261	-873489	-802145	-792055	-484337	484338	2529514	9097796
3833	182	-1622	-1570	-1346	-181	1463	4281	11373
702	-21	-193	-152	-149	22	110	185	522
-1442	16	418	428	309	-15	-416	-779	-1993
351	23	-117	-22	42	54	-104	-206	235

EXHIBIT 2, CONTINUED

B) SITUATION 2 BEST 20 DATA FOR STEP 2

2304	2027	2057	2150	2369	2549	2701	3104
2323	2033	2068	2165	2397	2594	2755	3096
2461	2107	2153	2276	2557	2785	2971	3364
2598	2243	2281	2402	2683	2919	3122	3615
2684	2330	2372	2504	2779	2993	3168	3575
2752	2254	2310	2473	2838	3144	3429	4359
2812	2332	2377	2525	2865	3169	3466	4555
2853	2453	2499	2645	2956	3199	3407	3937
2846	2358	2416	2571	2976	3322	3596	4159
2864	2417	2470	2631	2981	3267	3500	4019
2883	2520	2556	2684	2973	3206	3407	3931
2876	2457	2509	2654	2988	3264	3490	3950
2886	2349	2418	2603	3014	3358	3652	4336
2943	2447	2525	2709	3055	3312	3531	4230
2926	2319	2396	2599	3078	3492	3829	4528
2953	2358	2432	2630	3109	3513	3835	4507
3008	2574	2624	2779	3110	3386	3625	4230
3046	2603	2646	2786	3140	3447	3709	4445
3066	2608	2661	2816	3178	3467	3698	4332
3045	2502	2570	2770	3203	3538	3797	4332

C) SITUATION 2 STEP 2 ROWS-PLUS-COLUMNS ANALYSIS OF BEST 20

2923	-50	-498	-450	-283	51	332	571	1190
-586	19	190	172	98	-18	-119	-206	-422
-562	14	172	159	89	-13	-97	-175	-453
-413	3	97	95	51	-2	-55	-108	-335
-282	9	102	92	46	-8	-53	-89	-215
-191	4	98	92	57	-3	-70	-134	-346
-127	8	-41	-33	-37	-7	17	63	373
-88	29	-2	-5	-24	-19	3	61	531
-18	0	48	46	25	0	-37	-68	-157
-11	-13	-53	-43	-55	14	79	114	57
-4	-3	-1	2	-3	11	16	10	-89
5	6	91	79	40	-5	-53	-91	-186
11	-6	22	26	4	3	-2	-14	-174
27	-12	-101	-80	-62	13	76	131	195
63	8	-39	-9	7	17	-6	-25	53
79	-24	-183	-154	-118	25	158	256	336
108	-26	-173	-147	-116	27	150	233	286
136	0	14	16	4	0	-4	-4	-19
176	-2	2	-1	-29	-10	15	38	155
199	-4	-14	-9	-21	5	13	5	20
195	-22	-116	-96	-64	33	87	107	23

D) SITUATION 2 STEP 3 ROWS-TIMES-COLUMNS ANALYSIS OF BEST 20

1441	20	281	250	169	-21	-195	-350	-816
385	8	34	33	4	-6	-10	-11	30
386	3	16	20	-4	-1	11	19	0
266	-3	-9	0	-13	5	19	26	-21
193	3	24	22	-1	-2	1	8	11
281	-3	-15	-8	-11	5	9	8	-15
-246	15	59	56	22	-15	-52	-61	82
-319	38	127	110	53	-29	-87	-100	154
132	-3	-4	-1	-7	4	0	-1	-2
-95	-10	-14	-8	-32	11	51	65	-55
46	-4	-20	-13	-14	12	29	33	-35
173	1	21	17	-1	0	-5	-4	16
112	-9	-22	-13	-22	6	29	42	-42
-194	-6	-22	-9	-15	7	20	33	-33
-31	9	-26	2	14	16	-15	-41	16
-350	-14	-40	-27	-32	14	58	78	-77
-312	-17	-46	-34	-40	17	61	75	-82
19	0	7	9	0	0	0	5	3
-101	0	44	34	-4	-13	-13	-13	34
-20	-3	-5	-1	-16	4	7	-5	-4
-98	-19	-76	-61	-40	30	59	57	-93

EXHIBIT 2, CONTINUED

E) SITUATION 16 STEP 1 ROWS-PLUS-COLUMNS ANALYSIS

2549	17	-417	-390	-288	-15	350	702	2063
2804553	7192863	-2775004	-1842943	-1740207	-1516319	1516320	7192179	7190816
25222	7263	-20659	-19984	-16814	-7262	24813	59031	133727
4489	1240	-3340	-3185	-2592	-888	889	4241	18699
950	17	-652	-631	-506	-17	632	1447	3976
-161	-21	172	157	116	22	-138	-268	-966
-7	-32	232	218	169	33	-205	-441	-1499
-164	-16	193	170	126	17	-159	-299	-923
366	8	-219	-226	-191	-7	302	768	1527
-70	-4	117	100	69	5	-97	-124	-319
1448	485	-966	-953	-846	-445	446	3772	16432
1566421	8430996	-1543335	-875681	-801847	-657647	657648	6262058	8428949
-125	-17	149	132	97	17	-119	-211	-726
-486	5	273	240	172	-5	-256	-465	-694
6288	183	-4004	-3777	-2877	-182	2725	6229	28355
3314	49	-2090	-2013	-1615	-48	1725	3793	12357
1841	69	-1119	-1078	-863	-68	1110	2403	7620
518	5	-245	-243	-194	-4	249	629	1445
108	-7	0	-10	-16	0	21	149	188
-202	-30	206	190	146	30	-172	-364	-1306
857406	-49453	-855922	-840161	-838191	49453	1779537	2333160	2981885
1966	172	-925	-890	-714	-172	944	2991	7314
1010	109	-511	-503	-421	-109	516	1600	4292
241	57	-105	-113	-113	-56	65	394	1658
-51	0	78	66	35	-20	-104	-104	0
1411	-81	-452	-430	-285	82	443	782	592
497	14	-67	-67	-13	132	219	300	-205
508	17	-73	-73	-16	135	230	323	-143
-356	-17	334	301	222	17	-272	-543	-1617
-73	-43	19	24	40	62	-18	-114	-881
-419	-35	163	154	125	35	-151	-331	-1229
-335	-38	123	117	98	38	-114	-263	-1115
-520	-28	213	200	154	28	-188	-393	-1323
-609	-25	276	256	195	25	-239	-491	-1553
-456	-23	197	184	141	24	-172	-342	-1137
-302	-33	104	102	85	34	-99	-215	-929
-417	-27	277	257	196	28	-235	-493	-1604
-534	-21	343	314	234	22	-284	-581	-1777
8566	3260	-7670	-7473	-6760	-3021	3022	12333	51908
-475	-20	376	351	261	20	-316	-644	-1939
827	50	-395	-395	-335	-49	541	1360	2514
-461	-23	316	286	219	23	-275	-564	-1691
-315	0	71	48	1	-106	-249	-267	1306
-2548	-16	418	391	289	16	-349	-701	-2062
-2548	-16	418	391	289	16	-349	-701	-2062
-307	0	42	33	0	-80	-210	-257	271
-374	0	188	173	130	0	-218	-404	-748
-507	7	206	191	138	-6	-228	-408	-651
1533	364	-590	-619	-583	-363	478	2892	9741
-70	-33	0	0	3	29	0	0	-351
-294	-29	237	222	172	30	-210	-442	-1474
-502	-18	287	266	199	18	-251	-502	-1474
574	32	-227	-214	-186	-31	266	863	1931
0	-23	81	68	51	24	-51	-83	-551
275	64	-257	-253	-216	-63	142	500	2252
-89	-14	-42	-46	-39	15	77	231	192
-215	-15	315	280	205	15	-258	-509	-1475
587	54	-184	-189	-162	-53	113	503	2185
1589	272	-900	-891	-791	-271	1049	2815	13079
21489	16431	-19150	-18610	-15478	-3492	3493	12661	120556
70345	26022	-64205	-61416	-53291	-26021	97181	236016	778030
1272981	8724433	-1251956	-878451	-836225	-608768	608769	5934445	8722386
4783	1366	-3460	-3313	-2718	-978	978	5049	21154
662	53	-373	-364	-309	-53	368	980	3754
-670	-22	368	334	252	23	-306	-628	-1911
1839	44	839	731	454	-44	-522	-943	-2432

EXHIBIT 2, CONTINUED

F) SITUATION 16 BEST 20 DATA FOR STEP 2

1872	1828	1821	1841	1885	1921	1951	2029
1930	1797	1804	1845	1948	2049	2149	2448
2009	1939	1937	1959	2020	2079	2134	2299
2016	1823	1837	1893	2040	2189	2336	2767
2065	1829	1841	1890	2018	2162	2334	3452
2045	1916	1922	1957	2049	2145	2246	2635
2085	1918	1912	1946	2041	2156	2299	3431
2070	2032	2034	2046	2078	2107	2131	2197
2081	1986	1983	2018	2095	2162	2225	2459
2085	1871	1885	1944	2100	2269	2451	3017
2110	1874	1892	1965	2148	2327	2499	2962
2120	1990	1997	2038	2143	2245	2339	2589
2190	1944	1956	2015	2158	2305	2471	3488
2192	2109	2103	2126	2194	2270	2351	2638
2192	1919	1940	2023	2236	2449	2652	3161
2249	1886	1890	1945	2110	2333	2667	5602
2258	1865	1883	1952	2144	2380	2685	4575
2229	1932	1957	2042	2264	2496	2732	3379
2241	2073	2085	2137	2268	2393	2513	2842
2335	2231	2223	2250	2333	2425	2526	2921

G) SITUATION 16 STEP 2 ROWS-PLUS-COLUMNS ANALYSIS OF BEST 20

2111	-5	-164	-161	-118	6	126	246	665
-231	0	114	104	81	0	-82	-172	-513
-171	-2	23	27	25	3	-15	-35	-155
-95	0	89	84	63	0	-60	-125	-379
-82	-5	-39	-28	-15	6	35	62	74
-86	47	-30	-21	-15	-11	12	64	763
-63	4	34	37	29	-3	-27	-46	-76
-41	22	14	5	-3	-33	-38	-15	697
-36	2	123	122	91	-1	-92	-188	-541
-22	0	63	57	49	1	-51	-108	-293
-17	-1	-56	-45	-29	2	51	113	260
18	-12	-89	-74	-44	13	72	124	168
21	-5	23	27	25	6	-11	-37	-206
55	30	-56	-47	-31	-13	14	59	657
82	5	81	72	52	-4	-48	-87	-219
103	-15	-129	-111	-71	16	109	192	282
45	99	-104	-103	-91	-51	52	265	2781
85	68	-166	-151	-125	-57	58	243	1714
136	-11	-149	-127	-85	12	124	240	468
140	-3	-13	-4	4	11	16	16	-73
223	7	62	51	35	-6	-34	-53	-77

H) SITUATION 16 STEP 3 ROWS-TIMES-COLUMNS ANALYSIS OF BEST 20

3735	-31	73	68	55	20	-37	-130	-982
148	17	74	66	51	-9	-61	-100	28
44	2	11	15	16	0	-8	-13	5
109	14	59	56	41	-7	-45	-72	21
-23	-8	-33	-22	-11	8	32	51	-11
-204	22	26	31	27	3	-16	-36	10
24	7	27	31	24	-5	-23	-34	11
-182	0	64	51	33	-20	-64	-104	26
156	21	80	82	58	-12	-69	-111	32
85	9	39	35	31	-4	-39	-66	17
-74	-10	-35	-26	-13	7	40	76	-16
-52	-18	-75	-61	-34	17	65	98	-26
57	1	7	12	13	1	-3	-9	4
-177	9	-7	-2	5	0	-11	-27	3
65	13	63	55	38	-9	-38	-55	19
-87	-26	-105	-89	-53	22	97	149	-39
-748	9	100	87	62	3	-54	-100	32
-468	11	-37	-31	-28	-23	-8	13	-8
-138	-27	-111	-91	-56	22	104	171	-43
19	-1	-18	-8	1	9	19	25	-5
25	10	55	44	29	-8	-30	-41	15

EXHIBIT 2, CONTINUED

I) SITUATION 18 STEP 1 ROWS-PLUS-COLUMNS ANALYSIS

1706	-9	-66	-63	-36	1C	41	64	117
1411	30	-538	-507	-378	-29	4C4	922	3018
190	-3	-24	-19	-7	7	9	9	4
-13	1	3	5	3	C	-2	-4	-6
-93	2	8	11	5	-1	-5	-7	-7
-107	3	19	16	8	-2	-8	-12	-18
120	0	7	1	C	-1	-1	0	13
-71	3	20	17	1C	-2	-10	-17	-31
-92	0	7	7	2	-4	-8	-7	0
-80	0	0	0	-2	0	2	6	18
0	0	-13	-8	-6	1	6	12	31
-43	0	-13	-6	-5	1	4	9	24
-45	0	-1	-3	-5	0	5	11	31
17	0	-3	-7	-6	1	7	12	27
222	-5	-37	-30	-12	8	13	13	6
97	-2	-17	-12	-4	4	5	5	3
16	0	-6	-1	0	3	1	0	-1
-72	1	0	5	1	0	-2	-2	0
-103	2	5	7	3	-1	-4	-4	-2
-99	3	21	18	8	-2	-9	-13	-19
49	278	-40	-35	-32	-15	16	48	614
133	-4	-25	-21	-10	5	10	13	13
3	0	-8	-3	-2	1	2	3	3
-90	0	2	5	0	-1	-3	-2	4
-111	3	19	17	8	-2	-9	-13	-17
94	0	-18	-12	-3	9	10	9	0
-22	0	-8	-2	-2	1	2	3	5
-15	0	-9	-3	-3	1	3	4	7
-39	0	-11	-4	-1	3	3	3	1
-26	1	-1	4	2	2	0	-2	-11
-33	0	1	6	3	0	-3	-7	-16
-72	1	4	8	5	0	-4	-8	-17
-113	0	5	9	6	C	-6	-11	-22
-112	1	15	14	8	0	-9	-15	-27
-38	0	2	4	3	0	-3	-8	-23
-54	1	6	9	4	0	-6	-10	-22
-48	2	20	13	7	-1	-8	-11	-15
19	0	4	3	4	1	-5	-10	-24
-67	1	6	8	3	-2	-6	-6	0
145	4	42	20	7	-3	-7	-7	-3
114	-5	-41	-31	-16	6	16	22	29
227	0	0	-3	0	2	0	0	7
-52	0	5	5	3	0	-3	-4	-4
438	-16	-134	-109	-69	17	83	134	223
42	-1	-12	-7	-4	3	5	6	2
4	0	2	4	1	C	-1	-1	-4
-17	2	16	13	7	-1	-8	-13	-30
-21	0	12	11	6	0	-6	-11	-27
57	0	-8	-6	-4	1	3	6	16
21	-1	-10	-6	-5	2	7	11	15
127	-1	-11	-13	-10	2	12	21	38
-26	0	8	3	C	0	-2	-4	-10
226	-1	5	6	8	2	-12	-26	-67
75	0	10	4	0	0	0	-2	-11
197	-3	-58	-46	-30	4	34	62	150
162	0	-19	-17	-13	0	16	36	106
48	0	1	-3	C	4	2	0	-1
-38	1	4	7	4	C	-5	-8	-17
54	0	-7	-5	-4	0	5	11	25
344	-14	-89	-76	-43	15	47	67	102
529	-15	-115	-106	-62	16	72	117	219
940	-13	-258	-245	-164	14	179	344	932
36	1	-2	0	1	2	C	0	-3
-66	2	7	10	4	-1	-4	-4	-2
-82	3	31	16	9	-2	-9	-14	-21
2663	-52	25	-90	-105	-20	21	44	86

EXHIBIT 2, CONTINUED

J) SITUATION 18 BEST 20 DATA FOR STEP 2

1583	1530	1537	1561	1601	1626	1644	1686
1585	1541	1543	1564	1601	1624	1641	1682
1588	1547	1547	1566	1601	1626	1645	1693
1591	1550	1550	1569	1605	1630	1649	1696
1594	1540	1545	1568	1610	1638	1661	1716
1600	1561	1560	1578	1613	1638	1657	1703
1605	1554	1559	1581	1620	1648	1669	1721
1603	1553	1556	1578	1617	1645	1669	1730
1606	1551	1556	1579	1623	1653	1677	1736
1616	1587	1575	1595	1630	1654	1672	1718
1616	1558	1561	1586	1635	1669	1696	1760
1624	1570	1577	1601	1642	1669	1688	1731
1625	1567	1574	1598	1642	1672	1695	1750
1627	1587	1587	1607	1641	1664	1680	1719
1630	1578	1583	1605	1645	1673	1696	1755
1632	1580	1585	1607	1647	1676	1699	1753
1643	1591	1596	1619	1660	1686	1705	1745
1644	1591	1594	1619	1662	1690	1712	1765
1650	1611	1606	1628	1665	1690	1710	1758
1651	1592	1592	1618	1670	1707	1736	1808

K) SITUATION 18 STEP 2 ROWS-PLUS-COLUMNS ANALYSIS OF BEST 20

1624	-7	-59	-54	-31	8	35	56	104
-33	0	0	1	2	3	0	-2	-8
-30	0	8	5	3	0	-3	-7	-14
-29	1	12	7	3	0	-3	-5	-5
-25	1	12	7	3	0	-2	-4	-5
-21	0	-1	-1	-1	0	1	3	10
-17	1	14	8	3	0	-3	-5	-7
-10	0	1	1	0	0	0	0	4
-13	0	2	0	0	0	0	2	15
-9	0	-3	-3	-3	1	3	6	17
0	1	24	7	4	0	-3	-6	-8
1	-1	-7	-9	-7	2	8	14	30
8	0	-1	0	1	2	2	0	-4
9	0	-6	-4	-3	1	3	5	12
10	1	13	8	5	0	-4	-9	-18
13	0	0	0	0	0	0	2	13
15	0	0	0	0	0	1	3	9
26	0	0	0	0	2	0	-1	-9
29	0	-1	-3	-1	1	2	3	8
33	0	13	3	2	0	-2	-3	-3
36	-1	-8	-13	-10	2	11	19	43

L) SITUATION 18 STEP 3 ROWS-TIMES-COLUMNS ANALYSIS OF BEST 20

88	-35	-342	-272	-192	27	197	360	775
-94	0	-3	0	0	3	2	0	-2
-228	0	1	0	0	1	0	0	0
-160	1	7	3	1	0	0	0	5
-157	0	7	3	0	0	0	0	4
119	0	1	1	0	0	0	0	2
-189	1	9	4	0	0	0	0	5
27	0	2	2	0	0	0	0	2
131	0	6	3	1	-1	-2	-1	6
219	0	3	2	0	0	0	0	2
-242	0	16	1	0	0	0	1	7
421	0	5	0	0	1	1	1	1
-33	0	-2	0	0	3	3	1	-2
186	0	0	0	0	0	0	0	0
-307	0	3	0	0	0	0	0	2
124	0	4	3	1	0	-1	-1	4
95	0	3	2	1	0	0	0	2
-95	0	-2	-1	-1	2	2	1	-3
111	0	1	-1	0	1	0	0	0
-121	0	10	0	0	0	0	0	4
586	0	9	0	0	0	1	1	3

effects. The remaining R entries in the zero column are the row effects. The (I, J) entries for I = 1 to R and J = 1 to C are the residuals.

The use of means would have been considerably less satisfactory for step 1. In situation 2, for example, two entries in the first row were "+ ∞" represented by 9999992, yet the corresponding median-fitted terms were

common	row (of 65)	col (of 8)	residual
3779	2102201	-47	7894042
3779	2102201	2762	7891222

The 65 residuals in one of these columns (Panel A, Exhibit 2, column 2), – the one for the variance – can be described simply as follows:

one each at 7894042 and 4107261

one each at 51088 13619 and 12040

one each at 3291 2837 and 1610

one each at 182 178 108 and 86

one each at -115 -76 and -68

18 at 1 to 28

6 at 0

26 at -1 to -42

The sum of the residuals is about the sum of the first 8 above, namely 12,085,791, corresponding to a mean of about 185,000. With 50 of the 65 residuals between -42 and +28, a shift of 185,000 would be quite out of place. Clearly the median fitting has served us well – well enough for step 2 to pick out 20 well-performing estimates.

Notice that it does not matter to us whether we get the best 20 estimates. Almost any good 20 would do.

In step 2 we repeat the rows-PLUS-columns fit again, this time intending to take the results more seriously. In exhibit 2, Panels B, F and J are the data for the 20 good estimates, and Panels C, G and K the results of the rows-PLUS-columns analyses.

.....

(a side path)

In situation 2 the two fits (Panels A and C) agree extremely well, as shown in exhibit 3, only 5 of the 20 differing by more than ± 1 (these are ± 4, ± 4, ± 5, ± 7, ± 14).

If we had wanted only a relatively good fit by rows, we would not have done too badly with the median fit to the 8 x 65, where

14 of the 520 residuals exceed 1,000,000 in magnitude

12 more exceed 100,000 in magnitude

20 more exceed 10,000

68 more exceed 1,000.

To get the quality of row fits we have, in the presence of such large perturbations, would be wholly infeasible by relying on means.

.....

(back to the rationale)

Insofar as dependence on columns goes, the main results of step 2 are

- the fit by columns, and
- the residuals (ready for further analysis).

Exhibit 3. Comparison of fit to 20 selected rows (20 selected estimates) in the 8 x 65 and 8 x 20 of steps 1 and 2 for situation 2.

Estimate Number	Step 1 (65 x 8)			Step 2 (20 x 8)			Diff. Sums
	Common	Row	Sum	Common	Row	Sum	
64	3779	-1442	2337	2923	-586	2337	0
39	"	-1418	2361	"	-562	2361	0
37	"	-1269	2510	"	-413	2510	0
36	"	-1138	2641	"	-282	2641	0
33	"	-1047	2732	"	-191	2732	0
13	"	-983	2796	"	-127	2796	0
51	"	-940	2839	"	-88	2835	4
48	"	-874	2905	"	-18	2905	0
50	"	-867	2912	"	-11	2912	0
34	"	-856	2923	"	-4	2919	4
41	"	-850	2929	"	5	2928	1
6	"	-846	2933	"	11	2934	1
56	"	-828	2951	"	27	2950	1
28	"	-779	3000	"	63	2986	14
47	"	-776	3003	"	79	3002	1
46	"	-747	3032	"	108	3031	1
32	"	-719	3060	"	136	3059	1
19	"	-685	3094	"	176	3099	5
7	"	-656	3123	"	199	3122	1
42	"	-654	3125	"	195	3118	7

We feel that these residuals may well conceal varying multiples of a single pattern of dependence upon columns. Our next task is to pull this out, too.

There are a variety of ways to do this. We have already fitted rows-PLUS-columns, the least that we might do is to fit a single constant times the product, finding

common + row + col + (const)(row)(col)

a fit that is close to one degree of freedom for nonadditivity (e.g., Tukey, 1949). This is not doing very much, in our present situation, since it asks only "How far can we go by seeking no new information about possible dependence on columns?"

A step further, but still within the same restricted scope would be to fit

common + row + col + (col*)(row)

where common, row, and col all come from a rows-PLUS-cols fit and col* is at our disposal.

One way to go still further is to fit

common + row + col + (row*)(col*)

where the first three terms come from a rows-PLUS-cols fit and the last term is, in matrix language, an arbitrary term of rank 1. We now have col AND col* to look at. If they were the same, we could condense the fit to the term

common + row + (row*)(col*)

which might be a very interesting form to fit.

Or we could begin with the latter fit.

It does not seem quite clear whether, when doing any of these extended fits, we should use the common, row and col terms from an initial rows-PLUS-cols fit or whether we should do otherwise, perhaps iterating the two parts of the fit back and forth. In the present context we only notice these interesting possibilities, deciding to stick to the initial rows-PLUS-columns effects and common.

In the case at hand, the specification calls for the next-to-the-last fit, involving both col and col*, with the latter to be fitted to the residuals from the initial rows-PLUS-columns fit. As the specification indicates, in the absence of thinking through, we allowed this fit to be a classical least squares fit. The conventional calculation, used in this case, involves finding (directly or indirectly) the largest latent value of the data matrix and the corresponding latent vectors: one 8 wide, the other 20 high. (One approach is to multiply the data matrix by its transpose, in both orders, and to solve the two eigenvalue-eigenvector problems, whose eigenvalues duplicate one another and are the squares of the latent values and whose eigenvectors are the latent vectors.

Panels D, H and L of exhibit 2 show the results of step 3 rows-TIMES-columns fits to the step 2 residuals. The grand fit is the largest latent value. The remaining entries in the zero row and zero column are the multiplicative row* and col* effects, the latent vectors described above.

.....

(a comment)

At this point, had we been behaving sensibly, we would have looked to see how col and col* interrelate. Exhibit 4 shows the actual situation for situations 16 and 2, plus 8, 12 and 15. The ratio lines show decent constancy with the greatest tendency to deviate associated either (a) with small values – what else could we expect – or (b) with the 0.1% level – well known to be the most subject to variability of all the Monte Carlo estimates. (Without sophisticated Monte Carlo, estimating a 0.1% point after calculating only 600 values would clearly have been fruitless.)

Exhibit 4. Comparison of "col" and "col*" for five situations.

Situation and Vector	Pseudovariance at							
	50%	25%	10%	(**)	2.5	1%	0.5%	0.1%
16								
col	-164	-161	-118	-5	6	126	246	665
col*	73	68	55	20	-31	-37	-130	-982
(ratio)	(.44)	(.42)	(.46)	(4.0)	(5.2)	(.29)	(.53)	(1.5)
2								
col	-498	-450	-283	-50	53	332	571	1190
col*	281	250	169	20	-21	-195	-350	-816
(ratio)	(.57)	(.56)	(.60)	(.40)	(.40)	(.59)	(.61)	(.63)
8								
col	-113	-95	-61	-11	12	71	118	235
col*	378	304	201	42	-41	-238	-397	-710
(ratio)	(.30)	(.31)	(.30)	(.26)	(.29)	(.29)	(.30)	(.33)
12								
col	-19	-20	-14	-1	2	16	29	68
col*	187	182	131	6	-5	-121	-259	-911
(ratio)	(.10)	(.11)	(.11)	(.17)	(.40)	(.13)	(.11)	(.07)
15								
col	-28	-26	-17	-2	3	20	35	73
col*	240	202	143	5	-4	-149	-302	-814
(ratio)	(.12)	(.13)	(.12)	(.40)	(.75)	(.13)	(.12)	(.08)

** The variance.

Note that sign of "col*" is an accident of the computation, so that the obvious choice of ± col* has to be made in each line of ratios.

We ought also to ask something about how well, relatively, col and col* are determined. Looking at situation 16, where everything is at its largest, we find the difference

(row*-max) · (col*) - (row*-min) · (col*)

is roughly twice as large as

col

where row*-max and row*-min are the most positive and the most negative values of row*. This indicates that col and col* are determined with something like the same accuracy (or would be, if both were found resistantly).

.....

(back to the rationale)

Having done the above separately for all 32 (really 18) situations, we now start putting things together. At this point, the hasty specification missed the boat – fortunately for the overall study, that boat was not running very far the day it was missed.

In less allegorical language, the situation was this:

- we had extracted both col and col* from the data for each situation;
- we wanted to combine these patterns of dependence on column to get a good single one for analysis;
- we chose to forget all the col vectors, and to keep, and summarize selectively, the apparently most strongly determined col* vectors.

Had it been – and it seems unlikely that we could exclude this possibility a priori – that col and col* were meaningfully different, and that they each took out a meaningful share of the available variability, this choice (taking only one of them onward) would have been a mistake — because, although we needed to keep at least two different kinds of 8-vectors, we had kept only one.

In the actual situation, this difficulty was not violent; the direction of col and col*, as 8-vectors, was quite nearly the same. We still want to consider, however, the differences in behavior at 0.1% as to their cause, their importance, and their avoidance. What might we expect from a resistant fitting of (row*)(col*) in

common + row + col + (row*)(col*) ?

A Resistant Fitting of TIMES

To this end, the table of residuals after resistant fitting of

common + row + col

to the 20 x 8 for situation 16 was analyzed resistantly. A resistant analysis had to be invented for the purpose. The one chosen appears to work quite well enough in this instance, although its general utility is still unclear. The procedure was as follows:

- change signs of certain rows and certain columns with the aim of making nonnegative as many entries as possible (the question of a good algorithm for this purpose deserves further study, as does a crisper and more useful formulation of the objective);
- take logarithms of the results, making the logs of zeros and negative numbers "L" for "very low";
- apply a very resistant row-PLUS-column to the table of logarithms (iterative medians were used, skipping unmanageable columns);
- find antilogs of the column and row fits (after recombining common with one or the other) thus found;

- restore the changed signs; and

- take the results as col* and row* and look at the residuals thus defined.

Exhibit 5 shows the original entries and the results of changing signs. Two of the eight columns (the two where the residuals were smallest) come out barely manageable in the sense that 9 out of the 20 entries are - or 0. For these 9 the logs will be L (corresponding to $-\infty$) so the medians for these columns will be found from the two most negative numerical entries. (This corresponds to treating negative entries as "like 0, although more so.")

If we had ten or more L's in a column, or four or more L's in a row, we would have to regard that column, or row, as unmanageable. This would mean:

- setting it aside, as far as the fitting of row-PLUS-cols to the logs goes; and

- treating the corresponding fitted value as "L" and setting its antilog as zero.

Exhibit 6 shows the upper part (upper 4 lines) of the actual iterative fitting by medians. The resulting fits, when we throw the common back with rows are:

Rows (for logs, in 0.01) 96, 29, 86, 52, 25, 50, 8, 100, 72, 71, 86, 36, 54, 82, 106, 99, 114, 108, 0, 36

Cols (for logs, in 0.01) -170, 8, 4, -4, -220, -2, 23, 42

Rows* (as factor) 91, 20, 72, 33, 18, 33, 12, 100, 52, 51, 72, 23, 35, 66, 115, 98, 130, 120, 10, 23

Cols* (as factor) 0.02, -1.20, -1.12, -.87, .02, .91, 1.74, 2.51.

Exhibit 5. Residuals from rows-PLUS-cols fit to the 20 x 8 from situation 16, as was and the signs after changes.

Original Values								Signs after Change	
0	114	104	81	0	-82	-172	-513	0+++0+++	-
-2	23	27	25	3	-15	-35	-155	-+++-+++	-
0	89	84	83	0	-60	-125	-379	0+++0+++	-
-5	-39	-28	-15	6	35	62	74	++++++++	+
47	-30	-21	-15	-11	12	64	763	-+++-+++	+
4	34	37	29	-3	-27	-46	-76	++++++++	-
22	14	5	-3	-33	-38	-15	697	+++-+++-	-
2	123	122	71	-1	-92	68	-541	++++++++	-
0	63	57	49	1	-51	-108	-293	0+++-+++	-
-1	-56	-45	-29	2	51	113	260	++++++++	+
-12	-89	-74	-44	13	72	124	168	++++++++	+
-5	23	27	25	6	-11	-37	-206	-+++-+++	-
30	-56	-17	-31	-13	14	59	657	-+++-+++	+
5	81	72	52	-4	-48	-87	-219	++++++++	-
-15	-129	-111	-71	16	109	192	282	++++++++	+
99	-104	-103	-81	-51	52	265	2781	-+++-+++	+
68	-166	-151	-125	-57	58	243	1714	-+++-+++	+
-11	-149	-127	-85	12	124	240	468	++++++++	+
-3	-13	-4	4	11	16	16	-73	++++++++	+
7	62	51	35	-6	-34	-53	-77	++++++++	-
							(Changes)	----++++	
							Number of zeros	30002000	
							Number of minuses	60107002	

Exhibit 6. Parts of the stages of analysis of the logs (logs are of residuals/10 and in units of .01). Top four lines and summaries only.

A) ORIGINAL LOGS

				Values					Medians
L	106	102	91	L	91	123	71		91
L	36	43	40	L	18	54	119		38
L	95	92	80	L	78	110	158		86
-40	59	45	18	-22	54	79	87		50
			(16 more rows)						

B) After FIRST ADJUSTMENT

				Residuals				Rows
L	15	11	0	L	0	32	-20	91
L	-2	5	2	L	-20	16	81	38
L	9	6	-6	L	-8	24	74	86
-90	9	-5	-32	-72	4	29	37	50
			(16 more rows)					
Medians								
-164	8	5	-6	-164	-4	24	40	68
Cols								
0	0	0	0	0	0	0	0	0

C) After SECOND ADJUSTMENT

				Second Residuals				Medians	Rows
L	7	6	6	L	4	8	-60	5	23
L	-10	0	8	L	-16	-8	41	-9	-30
L	-1	1	0	L	-4	0	34	0	18
74	-1	-10	-26	92	8	5	-3	2	18
			(16 more rows)						
Cols									
-164	8	5	-6	-164	-4	24	40	0	68

D) After THIRD ADJUSTMENT

L	2	1	1	L	-1	3	-65	28
L	-1	9	17	L	-7	1	50	-39
L	-1	1	0	L	-4	0	32	18
72	-3	-12	-28	80	6	3	-5	20
			(16 more rows)					
Medians								
-6	0	-1	2	-56	2	-1	-2	4
Cols								
-164	8	5	-6	-164	-4	24	40	68

E) After FOURTH ADJUSTMENT

								Medians	
L	2	2	-1	L	-3	4	-63	0	24
L	-1	10	15	L	-9	2	52	0	-43
L	-1	2	-2	L	-6	1	34	0	14
78	-3	-11	-30	136	4	4	-3	0	16
			(16 more rows)						
Cols									
170	8	4	-4	-220	-2	23	42	1	72

Note: Only 3 row medians are nonzero at the last step above. They are -1, -1 and -2 and their introduction leaves all column medians zero. (To the accuracy kept) the iteration is complete. (Making the -1, -1 and -2 changes is almost certainly unimportant.)

We can now find the residuals after

common + rows + col + row* · col*

When we do, we find that further fitting of "+row** + col**" is possible, with row** from -4 to 13 and col** from -2 to 4. Since row** can be included in row and col** in col, we may as well do this fitting. (We could start from the new "common + row · col" and fit "row* · col*" again, if we wished.)

We need now to compare the result of using this resistant fit with the result of using the nonresistant eigenvalue fit. A glance at the two tables of residuals shows a clear difference in how the columns are treated. Exhibit 7 brings this out relatively clearly:

- columns 1 and 8 were better fitted by the eigenfit, with spreads shrunk by factors of 2 and 20 respectively;
- column 5 was fitted as well by one as the other;
- columns 2, 3, 4, 6 and 7 were better fitted by the resistant fit, by factors of 5 to 13 in spread.

The first diagnosis is clear:

- the eigenfit took column 8 seriously, the resistant fit did not;
- column 1 seems to behave like a junior version of column 8; and
- the other columns can be much better fitted (in this form) if column 8 can be forgotten about.

Notice that the total, for all columns except 8, of the sums of 4 in exhibit 7 are:

Exhibit 7. Comparison of residuals after two fits of the form common + row + col + row* · col*.

A) RESIDUALS after EIGENFIT, by column (abstracted from Panel E, Exhibit 3).

	Column							
	1	2	3	4	5	6	7	8
highest	22	100	87	62	22	104	171	32
3rd highest	17	74	66	57	17	65	98	28
5th highest	13	64	57	38	8	32	51	21
5th lowest	-8	-33	-31	-11	-9	-45	-72	-11
3rd lowest	-18	-75	-61	-34	-12	-61	-100	-26
lowest	-27	-111	-91	-56	-23	-69	-111	-43
Sum of 4*	56	246	209	134	46	203	221	86

B) RESIDUALS after RESISTANT FIT, by columns

	1	2	3	4	5	6	7	8
highest	96	15	16	25	14	20	94	2531
3rd highest	47	6	5	8	5	8	25	727
5th highest	26	2	3	6	2	4	10	567
5th lowest	-11	-5	-3	-4	-11	-8	-10	-150
3rd lowest	-20	-10	-5	-6	-29	-23	-13	-202
lowest	-27	-15	-6	-8	-54	-61	-17	-290
Sum of 4*	104	23	16	24	47	43	58	1646

C) COMPARISON of ABOVE SUMS of 4

	1	2	3	4	5	6	7	8
Smaller	Eig	Res	Res	Res	~	Res	Res	Eig
by factor of	2	11	13	6		5	6	19

* Sum of 3rd and 5th highest and of absolute values of 3rd and 5th lowest.

Eigenfit: 1115

Resistant fit: 315

Clearly all but column 8 is better fitted by the resistant fit.

What of a deeper diagnosis? What is there in common between columns 8 and 1? How can understanding the difference indicate to us which analysis we ought to want to use?

Column 8 is the pseudovariance based on an estimated 0.1% point. Column 1 is the ordinary variance. These are the two of the eight most sensitive to the appearance of a single unusual shape among the 600 to 1000 on which the Monte Carlo is based. We have to believe that

- columns 8 and 1 are the most sensitive to accidents of sampling, and
- when we must choose, we ought to prefer the fit associated with the other columns.

In this analysis, then, we have understandable reasons for choosing the resistant row* · col* fit. (We also like the fact that it can – in extremis, as for the example above – be done with pencil, paper and slide rule or few-place log table.)

For situation 16, then, we have the following 8-vectors to compare:

	1	2	3	4	5	6	7	8
col	-5	-164	-161	-118	6	126	246	665
col*(eigen)	-31	73	68	55	20	-37	-130	-982
col*(resist)	.02	-1.20	-1.12	-.87	.02	.91	1.74	2.51

These can be more easily compared after some simple rescaling with the following results (sum of absolute values for columns 2, 3, 4, 6 and 7 brought to 725, signs adjusted):

	1	2	3	4	5	6	7	8
col	-5	-164	-161	-118	6	126	246	665
col*(eigen)	62	-146	-136	-110	-40	74	260	1964
col*(resist)	-2	-151	-140	-108	-2	113	216	312

We can take considerable satisfaction in the degree of agreement of these three vectors, particularly in the agreement of the first (whose value we have for all situations) with the third (whose value we have only for situation 16).

We recall where we stand:

- we have col and col* (the latter calculated nonresistively) for each situation;
- we have evidence, though we planned to concentrate on col*, that we might do better to use col (as a substitute for a resistant analysis); and
- col and col* are determined with crudely comparable accuracy.

STEPS 4 AND 5 – WHAT WAS AND WHAT MIGHT HAVE BEEN DONE

Step 4 called for looking at the 21 situations (21 picked relative to 32) for which the rows-TIMES-columns fit was largest (by dropping the 11 little ones, we hoped to avoid unnecessary noise) and then fit rows-TIMES-columns, which we will identify as

row** · col**

to the 21 x 8 table. (Again the fitting was classical rather than resistant.)

In fact, such fits were made: for all 32 (really 18), for the "best 21" (really 10 sets of two duplicates and one single), and for the "best 15" of these 21. Exhibit 8 shows the results of these classical rows-TIMES-columns fits.

With the sums of absolute values for columns 2, 3, 4, 6 and 7 adjusted to 725 (for convenient comparison) by multiplication with a constant factor, these results become

(all 32)	-18,	-155,	-161,	-96,	18,	122,	193,	347
(sel 21)	-7,	-171,	-149,	-99,	8,	106,	206,	565
(sel 15)	-5,	-163,	-142,	-100,	5,	106,	213,	591

As an alternative procedure, we could take the col 8-vectors for each of the 18 truly different situations, select the 8 largest and fit, resistantly, row**-TIMES-col** to the resulting 8 x 8. Exhibit 9 shows some steps in the process. (Panel C shows that we could have introduced the sum of the middle six 8-vectors as a ninth 8-vector without too much distortion.) The fitted values of col**, when reduced in the same way as above, are not as different from the results of the 21 x 8 on col* as we might have feared.

We have

(8 x 8 on col)	-17,	-173,	-151,	-97,	19,	115,	190,	398
(15 x 8 on col*)	-5,	-163,	-142,	-100,	5,	106,	213,	591
(21 x 8 on col*)	-7,	-171,	-149,	-99,	8,	106,	206,	565
(32 x 8 on col*)	-18,	-155,	-161,	-96,	18,	122,	193,	347

where we have included the 32 x 8 on col* for further comparison.

EXHIBIT 8. CLASSICAL ROWS-TIMES-COLUMNS FIT TO COL^{X} (EIGEN) EFFECTS

A) FOR ALL 32 SITUATIONS

									SITUATION
17	36	311	322	191	-37	-243	-386	-742	1
1441	20	281	250	169	-21	-195	-350	-816	2
213	-9	-230	-203	-148	9	164	326	865	3
24	70	385	359	204	-68	-221	-361	-700	4
52	-46	-491	-305	-179	47	217	352	676	5
38	-27	-433	-278	-171	26	192	355	735	6
54	40	284	278	188	-28	-205	-367	-791	7
707	42	378	304	201	-41	-238	-397	-710	8
974	18	302	238	159	-14	-174	-332	-826	9
91	45	354	334	202	-44	-234	-374	-722	10
79	37	292	329	191	-33	-229	-383	-752	11
430	-5	187	182	131	6	-121	-259	-911	12
163	-20	136	143	109	21	-92	-226	-941	13
2798	5	186	174	129	1	-125	-308	-897	14
210	5	240	202	143	-4	-149	-302	-875	15
3735	-31	73	68	55	20	-37	-130	-982	16
589	64	558	414	184	-61	-217	-341	-557	17
88	-35	-342	-272	-192	27	197	360	775	18
92	-46	-357	-340	-204	47	229	373	720	19
974	18	303	239	159	-14	-174	-332	-826	20
2798	-4	-185	-172	-128	0	126	309	898	21
210	-4	-237	-201	-142	5	152	304	876	22
3735	33	-72	-67	-54	-19	38	131	983	23
706	41	378	303	200	-41	-239	-397	-710	24
55	38	297	290	191	-23	-199	-352	-789	25
51	42	494	306	186	-41	-211	-348	-675	26
430	7	-185	-179	-128	-6	124	262	912	27
79	46	300	334	204	-41	-225	-374	-748	28
163	19	-137	-140	-110	-18	92	228	942	29
24	-52	-375	-349	-209	53	229	372	704	30
203	-12	-237	-213	-152	11	172	336	855	31
38	31	451	272	174	-31	-199	-340	-729	32

B) FOR BEST 21 SITUATIONS

									SITUATION
4485	-11	-259	-225	-150	12	161	311	857	
-207	-43	-168	-141	-85	31	112	159	-183	16
208	44	169	142	86	-30	-110	-158	184	23
-221	-7	-71	-50	-20	13	34	0	-45	14
222	8	73	51	21	-12	-33	0	46	21
-221	7	22	25	18	-9	-35	-40	38	2
-222	6	43	13	8	-2	-13	-21	29	9
-222	5	43	14	8	-2	-14	-21	29	20
-215	30	126	85	54	-29	-82	-95	121	8
-215	29	125	84	53	-29	-83	-95	121	24
-197	53	327	213	50	-50	-74	-64	203	17
221	19	72	44	21	-18	-35	-46	61	27
-220	-17	-70	-41	-19	18	37	49	-60	12
223	2	29	21	2	-2	4	15	9	3
-222	-6	-19	-22	-7	7	11	8	-18	15
223	7	22	23	8	-6	-7	-6	19	22
223	0	22	12	-1	0	12	25	-1	31
-217	-32	-117	-77	-38	33	64	77	-103	13
218	31	116	80	37	-30	-64	-75	104	29
217	-34	-104	-120	-56	35	73	70	-113	19
-216	33	101	115	55	-32	-77	-70	112	10
220	-23	-84	-49	-43	15	38	52	-72	18

C) FOR BEST 15 SITUATIONS

									SITUATION
3843	-7	-239	-208	-147	7	155	312	869	
-258	-3	-52	-34	-18	7	28	1	-31	14
259	4	53	35	18	-6	-27	0	32	21
-258	11	43	42	22	-15	-41	-40	48	2
-258	10	64	30	12	-8	-20	-22	39	9
-258	10	64	30	12	-8	-20	-22	39	20
-250	33	147	101	57	-34	-89	-97	126	24
259	15	52	28	18	-12	-29	-47	47	27
-258	-13	-51	-25	-16	13	32	50	-46	12
260	-1	9	5	0	3	10	15	-3	3
-259	-2	0	-6	-4	1	5	9	-6	15
260	3	2	7	5	0	-1	-7	6	22
260	-4	2	-3	-4	5	18	24	-12	31
-255	-28	-99	-62	-36	27	59	79	-87	13
256	27	98	65	35	-25	-59	-77	88	29
256	-27	-106	-66	-46	21	45	53	-80	18

Exhibit 9. Resistant row**-TIMES-col** fit to "col" 8-vectors for selected situations.

A) The 18 8-VECTORS, one per situation

Situation	1	2	3	4	5	6	7	8
16	-5	-164	-161	-118	6	126	246	665
14	-36	-752	-601	-405	37	518	904	2369
2	-50	-498	-450	-283	51	332	571	1190
9	-29	-193	-181	-111	30	130	195	355
8	-11	-113	-95	-61	12	71	118	235
17	-24	-149	-119	-83	25	87	118	174
3	-5	-53	-52	-34	6	40	68	140
18	-7	-59	-54	-31	8	35	56	124
12	-1	-19	-20	-14	2	16	29	68
15	-2	-28	-26	-17	3	20	35	73
13	-1	-14	-12	-8	2	9	16	32
10	0	1	0	-1	0	0	1	3
11	0	-2	-2	-2	1	3	4	8
7	0	-7	-2	-3	0	4	8	18
5	0	3	2	0	0	0	0	-1
6	0	3	0	-1	0	0	1	3
4	0	3	3	2	0	-1	-1	-3
1	0	0	1	0	0	0	0	1
Sum of 6*	-4	-69	-62	-45	8	52	93	202

B) The RESISTANT row**-TIMES-col** FIT – to the 8 x 8

col**	-16	-158	-138	-89	17	105	174	363
row**								
1.17	14	21	1	-14	-14	3	42	239
4.57	37	-27	31	3	-41	-46	-31	709
3.17	1	3	-12	-1	3	-2	18	40
1.23	-9	-8	-11	-1	9	1	-19	91
.68	0	-2	0	0	0	-2	-2	-16
.89	-16	-8	4	-4	16	-15	-37	-150
.38	1	7	0	0	0	0	2	2
.37	-1	0	-3	2	2	-4	-8	-30

C) FIT of MULTIPLE of col* to the sum of 6

Sum of 6:	-4	-69	-62	-45	8	52	93	202
col*:	-9	-76	-68	-43	8	51	84	175

* The 6 in the middle, from #12 to #7.

We would probably have done better, successively, to use 8-vectors obtained further down in the following list:

15 x 8 on col* (classical, not resistant)

21 x 8 on col* (classical, not resistant)

32 x 8 on col* (resistant)

21 x 8 on col* (resistant)

8 x 8 on col (resistant)

8 x 8 on the result of fitting "row+row*-TIMES-col" resistantly (resistant)

(The top choice was prescribed, and used.)

STEPS 6 AND 7 – WHAT WAS AND WHAT MIGHT HAVE BEEN DONE

From step 5 we were to have a relatively good 8-vector, hopefully reflecting, up to a constant multiple for each estimate-situation combination, the general pattern of variation across columns for those estimates that performed relatively well in the situation considered. Sufficiently resistant regression on this 8-vector ought to eliminate this general pattern from the 8-vectors (rows) for the good estimates, leaving over residuals that deserve attention – at least to see what they may be trying to say.

The 8-vector that was available – that from the 21 x 8 on col* (exhibit 8, Panel B, first row) – is not quite as good as we might wish, but it is far from being bad. The regression rule prescribed in step 6 is presumably not as good as we would now choose – we would drop the 0.1% point from the first median – but, again, medians are pretty resistant, and can usually be handled rather incautiously without serious ill effects (something we cannot say about means).

Thus step 6, though not optimum, should be fairly effective. These median-fitted regressions for situations 2, 16 and 18 are shown in exhibit 10. The data analyzed appear in the 20 x 8 arrays of Panels B, F and I of exhibit 2.

Step 7 called for looking at the residuals from regression in terms of steps 2 and 3. This could have been interpreted in various ways, since it was incautiously worded. It is not clear that what was actually done –

- keep the 20's selected in step 1,
- do the regressions for these 20,
- fit first row-PLUS-col and then row*-TIMES-col* to the residuals from regression –

was less than the best that could have been done. The real question at issue now is "Was enough consistent behavior taken out by the row*-TIMES-col* fit to concern us?" The step 7 analyses for situations 2, 16 and 18 are shown in exhibit 11.

Exhibit 12 sets out the latent value involved in the fits for all 18 situations, both in step 3 and step 7. Notice that we are not looking at the first and second latent values for each table. In step 3 we looked at the first eigenvector for each table – and then we adopted, and regressed out, a compromise vector. In step 7 we looked at what was left, which will, on the average, be larger than the second latent value would be, since compromise regression will not remove all of the first latent value.

The ratio has a median of about 2.4, with a few situations appreciably smaller. These few include the three furthest from Gaussian, the double exponential (with its anomalous sharp break in the center) and t-on-3 (about which there can be various theories).

Before coming to a conclusion as to what we might have done about a second vector, we need to look at the 8-vectors that

EXHIBIT 10. INTERCEPT AND RESIDUALS FROM REGRESSION

A) SITUATION 2

2320	-5	0	-6	11	41	40	0	-277
2347	-12	0	-4	12	41	44	0	-388
2483	-8	0	-1	26	64	60	0	-479
2625	-13	0	-10	13	47	47	0	-395
2694	2	0	-3	35	75	63	0	-439
2765	5	0	-8	24	59	49	0	-257
2825	4	0	-17	5	26	26	0	-56
2868	0	0	-6	34	77	63	0	-433
2896	-31	0	-10	7	65	78	0	-688
2888	-7	0	-6	34	80	75	0	-575
2906	-8	0	-12	17	57	51	0	-372
2906	-14	0	-4	25	70	68	0	-583
2916	-9	0	-2	38	83	77	0	-632
2929	30	-18	0	66	113	84	0	-377
2976	-26	0	-6	29	84	93	0	-827
3000	-24	0	-7	27	91	98	0	-820
3031	-6	0	-7	30	66	60	0	-456
3084	-20	1	-17	0	43	53	0	-382
3082	0	0	-6	27	83	79	0	-467
3065	0	0	-3	53	122	109	0	-773

B) SITUATION 16

1876	-1	9	-4	0	7	8	0	-54
1943	-7	13	0	0	1	4	0	-69
2016	-3	14	0	0	1	4	0	-45
2037	-13	15	0	-1	-3	3	0	-102
2039	34	17	0	-8	-26	-23	0	590
2052	-1	13	0	-2	-6	-2	0	42
2067	24	29	0	-10	-30	-25	0	718
2073	-1	3	0	0	3	5	0	-36
2085	0	9	-7	0	7	8	0	-16
2106	-11	18	0	-4	-12	0	17	-4
2137	-17	15	-1	0	3	10	0	-183
2135	-9	12	0	0	4	9	0	-114
2163	35	18	0	0	-10	-10	0	466
2203	-6	21	0	-5	-11	-5	0	22
2226	-22	21	0	0	1	11	0	-252
2202	59	42	0	-35	-101	-99	0	2103
2205	66	29	0	-24	-70	-63	0	1032
2267	-25	20	0	-4	-12	0	4	-174
2258	-9	11	-1	0	4	8	0	-126
2345	-4	26	0	-7	-15	-9	0	71

C) SITUATION 18

1580	5	0	0	11	20	14	0	-72
1584	2	0	-3	6	15	11	0	-59
1590	0	0	-4	3	10	9	0	-50
1593	0	0	-4	3	11	9	0	-52
1593	3	0	-1	8	16	11	0	-66
1603	0	0	-5	1	9	8	0	-50
1604	3	0	0	8	15	12	0	-63
1603	1	0	-2	6	12	9	0	-55
1606	2	0	-1	7	16	12	0	-67
1620	-1	7	-9	0	9	8	0	-46
1618	0	0	-4	5	15	12	0	-75
1622	4	0	0	11	19	14	0	-75
1623	4	0	0	10	18	13	0	-73
1627	1	0	-4	5	12	10	0	-54
1629	2	0	-1	7	14	11	0	-59
1632	2	0	-1	7	14	11	0	-65
1641	4	0	0	9	18	13	0	-74
1644	2	0	-3	8	17	12	0	-68
1654	-2	0	-9	1	10	8	0	-51
1655	0	0	-7	2	14	12	0	-73

EXHIBIT 11. STEP 7 ANALYSES

A) SITUATION 2

879	38	4	-4	-24	-76	-91	5	992
200	-4	0	1	-10	-13	-3	0	-1
77	-4	2	6	-9	-18	-6	2	-1
-31	1	1	7	0	-5	-1	1	0
70	-3	4	2	-6	-11	-2	4	0
7	5	-4	0	4	2	0	-4	0
221	5	0	0	3	6	7	0	1
458	1	3	-2	-6	-6	7	3	0
13	2	-4	-3	3	4	0	-4	0
-266	-11	4	0	-20	-18	0	5	-1
-143	4	0	0	4	1	3	0	0
94	0	3	0	-3	-1	2	3	0
-151	-1	1	2	-3	-9	-4	1	0
-210	2	-1	1	4	-2	-2	0	0
48	5	-49	-21	10	17	-2	-30	1
-430	-5	1	0	-6	-13	-2	1	0
-423	-2	2	-1	-8	-6	3	2	0
-6	1	0	0	4	-2	0	0	0
88	-7	9	0	-15	-10	8	8	0
-22	8	0	0	0	12	16	0	2
-379	14	-3	-2	13	22	13	-3	3

B) SITUATION 16

2653	41	11	0	-15	-45	-48	0	997
-10	2	-7	-5	1	10	2	0	1
-14	0	-2	0	2	4	-1	0	0
-6	1	-2	0	0	5	0	0	0
-27	-5	0	1	0	-1	-3	0	0
234	13	-5	0	3	6	3	0	0
27	0	-3	1	1	2	-2	0	0
282	-1	4	0	2	8	7	0	1
-3	2	-13	0	1	7	0	0	0
3	1	-9	-9	0	10	1	-2	1
9	-6	3	2	0	-4	0	18	0
-58	-5	1	0	0	2	0	0	0
-31	0	-2	0	1	5	1	0	0
187	19	-3	0	8	16	9	0	1
20	-2	5	1	-1	-2	-5	1	0
-84	-9	6	0	-2	-4	-3	0	0
806	-22	1	0	0	1	1	0	1
402	27	1	0	-5	-17	-14	0	-2
-53	-11	9	4	-1	-10	-6	7	0
-36	0	-2	0	2	6	1	1	0
38	-3	9	0	-4	-5	-7	0	0

C) SITUATION 18

53	-5	215	-52	-99	-112	-12	146	952
-206	0	0	0	1	1	0	0	1
97	0	0	0	0	1	0	0	1
322	0	-1	1	0	0	0	0	0
288	0	0	1	0	0	0	0	0
-50	1	0	1	0	0	0	0	0
348	0	0	1	0	0	0	0	0
8	0	0	1	1	0	0	0	0
183	0	-1	1	1	0	-1	0	0
-60	0	0	1	0	0	0	0	0
463	0	6	-1	0	1	1	0	0
-162	0	3	0	-1	0	1	2	0
-259	0	0	0	0	0	0	0	0
-220	0	0	0	0	0	0	0	0
219	0	-1	0	0	0	0	0	0
87	0	0	2	1	0	0	0	1
-13	0	0	2	1	0	0	0	0
-226	0	1	0	0	0	0	0	0
-77	0	1	0	1	2	1	0	0
348	0	0	-1	0	1	1	1	0
-85	0	3	-2	-2	1	3	3	0

Exhibit 12. Latent values corresponding to row*-TIMES-col* fits in steps 3 and 7.

Situation	Latent Value		Ratio of Latent Values	(Notes)
	Step 3	Step 7		
16	3735	2653	1.4	(3S/4)
14	2798	2262	1.2	(S)
2	1441	879	1.6	(Cauchy)
9	974	343	2.8	
8	707	331	2.1	
17	589	418	1.4	(dex)
12	430	126	3.4	
3	213	57	3.7	
15	210	48	4.4	
13	163	65	2.5	
10	91	40	2.3	
18	88	53	1.7	(t_3)
11	79	30	2.6	
7	54	25	2.2	
5	52	23	2.3	
6	38	15	2.5	
4	24	11	2.2	
1	17	5	3.4	

arise in step 7. Exhibit 13 shows them arranged – and summarized by medians – first in three groups and then all together. Clearly there is a quite appreciable consistency among these 8-vectors, over and above the fact that 90-odd % of the sum of squares is usually associated with column 8.

For reasons that become clear below, it is natural to ask if the row* vectors obtained in steps 3 and 7 show any apparent relationship. Exhibit 14 shows a comparison for situations 16, 14 and 2. There appears to be a relationship for situation 16 but not for the other two.

At this point, then

- there seems to be consistent indication of a second 8-vector of probable importance;
- the corresponding 20-vectors do not seem obviously related to those of step 3; and
- we might have gone on taking out a second vector.

The actual decision, made at the scene of the computations, was NOT to take out a second vector.

STEPS 8 AND 9 – WHAT WAS AND WHAT MIGHT HAVE BEEN DONE

At this point then, the residuals-from-regression found in step 6 were treated as in step 2, and both the best-fitted ten estimates and the best-fitted five were selected and fitted with row***-PLUS-col***. Exhibit 15 shows the results of this step 8 analysis for situations 2, 16 and 18.

Step 9 was modified, at the computing scene, so instead of trying both the best-5 and best-10 col*** vectors, the mean of these two was taken, and the result added back to the regression fitted in step 6. This was done for the one estimate for each situation which appeared to perform best, as shown by the intercept of the

Exhibit 13. The col* values found in step 7.

Situation	1	2	3	4	5	6	7	8
16	41	11	0	-15	-45	-48	0	997
14	22	38	6	-27	-72	-71	0	992
2	38	4	-4	-24	-76	-91	5	992
9	-3	85	86	-70	-199	-135	36	959
8	10	67	2	-53	-138	-124	3	978
17	-34	258	37	-155	-200	-93	71	923
18	-5	215	-52	-99	-112	-12	146	952
12	73	96	13	-19	-78	-64	-1	986
3	15	127	24	-50	-131	-108	33	975
15	13	93	18	-71	-147	-112	7	974
13	55	38	7	-36	-107	-85	2	987
10	2	18	47	-71	-158	-130	32	973
11	37	-63	125	-27	-138	-144	-2	968
7	35	-220	145	1	-179	-140	-4	936
5	4	309	9	-95	-157	-133	33	923
6	3	511	0	-76	-88	-46	62	846
4	14	57	53	-121	-210	-89	37	961
1	78	105	152	-63	-156	-179	17	947
COLUMN MEDIANS – for the three groups above								
	10	67	2	-55	-112	-91	36	978
	26	86	21	-42	-139	-110	4	974
	14	105	53	-76	-156	-133	33	936
Median	14	67	21	-55	-139	-110	33	974

Exhibit 14. The row* values obtained in steps 3 and 7 for three far-from-Gaussian situations.

Situation 16			Situation 14			Situation 2		
Est.	Step 3	Step 7	Est.	Step 3	Step 7	Est.	Step 3	Step 7
[8]	156	-3	[1]	192	318	[2]	386	77
[1]	148	-10	[10]	191	9	[1]	385	200
[3]	109	-6	[18]	150	-102	[5]	281	7
[9]	85	3	[7]	132	-13	[3]	266	-31
[14]	68	20	[6]	99	-75	[4]	193	70
[12]	57	-31	[9]	95	-20	[11]	173	94
[2]	44	-14	[3]	89	210	[8]	132	13
[20]	25	38	[8]	36	231	[12]	112	-151
[6]	24	27	[5]	28	-17	[10]	46	-143
[19]	19	-36	[4]	12	330	[17]	19	-6
[4]	-23	-27	[17]	-1	-312	[19]	-20	-22
[11]	-52	-58	[12]	-26	-17	[14]	-31	48
[10]	-74	9	[11]	-33	382	[9]	-95	-266
[15]	-87	-84	[2]	-47	428	[20]	-98	-379
[18]	-138	-53	[13]	-83	-256	[18]	-101	88
[13]	-177	187	[20]	-172	-126	[13]	-194	-210
[7]	-182	282	[16]	-263	106	[6]	-246	221
[5]	-264	234	[15]	-282	55	[16]	-312	-423
[16]	-468	806	[14]	-369	-99	[7]	-319	458
[17]	-748	402	[19]	-727	373	[15]	-350	-430

EXHIBIT 15. STEP 8 ANALYSES

A) SITUATION 2 BEST-FITTED TEN ESTIMATES

0	-6	0	-7	21	61	54	0	-414
0	1	0	1	-9	-19	-14	0	136
-2	-2	3	6	-5	-15	-6	3	29
2	-2	-1	5	3	2	4	-1	-66
-4	-1	5	2	-2	-8	-2	5	24
7	3	-6	-2	8	7	3	-6	-31
0	12	0	0	3	-1	-4	0	157
-4	16	5	-4	-10	-29	-23	5	362
4	3	-3	-2	9	12	5	-3	-23
1	0	0	1	13	19	20	0	-161
0	-22	1	0	-11	5	25	1	-272

ROWS-PLUS-COLUMNS ANALYSIS OF THE TEN ROWS OF RESIDUALS IN PANEL A OF EXHIBIT 11. THE ROWS SELECTED WERE ASSOCIATED WITH THE TEN SMALLEST INTERCEPTS IN PANEL A.

B) SITUATION 2 BEST-FITTED FIVE ESTIMATES

0	-12	0	-6	13	47	47	0	-395
0	7	0	0	-1	-5	-6	0	117
0	0	0	2	0	-5	-2	0	7
5	0	-4	1	8	12	8	-4	-89
0	0	0	-2	0	0	0	0	0
9	6	-8	-5	13	18	7	-8	-52

C) SITUATION 16 BEST-FITTED TEN ESTIMATES

0	-1	14	0	0	0	5	1	-27
0	0	-3	-4	2	8	4	0	-26
0	-4	0	0	2	2	0	0	-40
0	-1	0	0	1	2	0	0	-17
0	-10	3	1	0	-1	0	1	-73
0	36	4	0	-6	-25	-27	0	618
0	0	0	0	0	-5	-6	0	70
0	26	15	0	-8	-29	-30	0	746
0	0	-9	0	2	4	1	0	-8
2	0	-5	-8	0	6	2	-1	9
-1	-7	7	2	-1	-9	-1	18	24

D) SITUATION 16 BEST-FITTED FIVE ESTIMATES

0	-3	14	0	0	1	4	0	-55
1	0	-5	-5	0	5	2	0	0
0	-2	0	0	2	0	0	1	-11
0	0	0	0	0	0	0	0	11
0	-7	3	1	0	-3	0	1	-44
0	38	3	0	-7	-28	-27	0	647

E) SITUATION 18 BEST-FITTED TEN ESTIMATES

1	1	0	-3	5	14	10	0	-62
3	0	-3	0	2	2	0	-2	-12
0	0	0	0	0	0	0	0	2
-1	0	2	0	0	-2	0	2	13
1	1	0	1	1	1	0	0	-5
-1	0	2	0	0	-1	0	2	11
-2	0	3	0	-1	-2	0	3	14
0	0	0	0	0	-1	-1	0	6
0	1	0	2	2	0	0	0	-1
1	0	0	1	1	0	0	0	-5
0	0	0	0	0	1	2	0	-12

F) SITUATION 18 BEST-FITTED FIVE ESTIMATES

1	1	0	-3	6	14	10	0	-60
3	0	-3	0	2	1	0	-2	-14
0	0	0	0	0	0	0	0	0
-1	0	2	0	-1	-2	0	2	11
0	1	0	1	1	0	0	0	-7
-1	0	2	0	-1	-2	0	2	9

regression fit (this choice is reasonable, since the 8-vector regressed on has 4 negative and 4 positive components). Exhibit 16 panels show several stages of this calculation. At this point, in Panel D we have (for each situation) an adjusted value for each of the 8 columns for the estimate which seems to perform best in that situation. It has been adjusted by a lengthy process, whose main constituents are

- finding a (hopefully) desirable 8-vector (common to situations);
- for each situation, regressing the 8-vectors for first 20 and then 10 and 5 well-behaved estimates on this common 8-vector;
- finding any consistent pattern of residuals from regression for the 10 and 5 selected well-behaved estimates; and
- using this to modify the regression fitted to that one estimate for each situation which appears to perform best.

The kinds of questions that concern us as "might have beens" are:

- Should we have picked a different common vector?
- Should we have used two or more common vectors?
- Did we make an analysis in the right terms (with the quantities to be analyzed expressed in a way most useful to the analysis)?
- Was the whole approach sound?

We have discussed the first two questions as we went along. If we had been able to do the analysis as data analyses need to be done, a step at a time, with opportunities for exploration of

EXHIBIT 16. STEP 9 ANALYSES

A) FITTED REGRESSION OF THE BEST PERFORMING ESTIMATE FOR EACH SITUATION

								SITUATION
1000	1000	1000	1000	1000	1000	1000	1000	1
2310	2027	2064	2139	2328	2509	2701	3382	2
1533	1487	1493	1505	1536	1565	1596	1706	3
1156	1159	1159	1158	1156	1154	1152	1145	4
1308	1311	1311	1310	1308	1306	1304	1297	5
1435	1434	1434	1434	1436	1437	1438	1442	6
1779	1779	1779	1779	1779	1780	1780	1781	7
3121	3050	3059	3078	3125	3171	3219	3390	8
6052	5840	5868	5924	6066	6202	6346	6858	9
1127	1127	1127	1127	1127	1127	1127	1127	10
1245	1242	1242	1243	1245	1247	1249	1256	11
1660	1642	1644	1649	1661	1672	1684	1726	12
1189	1180	1181	1183	1189	1195	1201	1222	13
5761	5317	5376	5493	5790	6074	6375	7444	14
1337	1311	1314	1321	1338	1355	1372	1434	15
1874	1819	1826	1841	1878	1913	1951	2084	16
1272	1119	1139	1179	1282	1379	1483	1851	17
1578	1531	1537	1550	1581	1612	1644	1759	18
1127	1127	1127	1127	1127	1127	1127	1127	19
6052	5840	5868	5924	6066	6202	6346	6857	20
5761	5317	5376	5493	5790	6074	6374	7443	21
1337	1311	1314	1321	1338	1355	1372	1434	22
1874	1819	1826	1841	1878	1913	1951	2084	23
3121	3050	3059	3078	3125	3171	3219	3391	24
1779	1779	1779	1779	1779	1780	1780	1782	25
1308	1312	1311	1310	1308	1306	1304	1297	26
1660	1642	1645	1649	1661	1672	1684	1727	27
1245	1242	1242	1243	1245	1247	1249	1257	28
1189	1180	1181	1184	1189	1195	1201	1222	29
1156	1159	1158	1158	1156	1154	1151	1144	30
1533	1487	1493	1505	1535	1565	1596	1706	31
1435	1434	1434	1434	1435	1436	1438	1441	32

B) BEST-10 COL*** VECTORS FOR EACH SITUATION

									SITUATION
0	0	1	0	0	0	0	0	1
-6	0	-7	21	61	54	0	-414	2
-1	3	-2	0	4	5	0	-33	3
0	0	1	0	0	0	0	4	4
0	0	1	0	0	0	0	4	5
0	4	0	0	0	0	0	1	6
1	-2	3	0	0	0	0	0	7
0	-5	0	5	15	14	0	-99	8
1	0	-16	14	52	42	0	-258	9
0	3	0	0	0	1	0	0	10
0	0	0	0	0	1	0	-4	11
0	4	0	0	2	2	0	-13	12
0	-1	0	0	2	2	0	-12	13
6	-32	0	0	1	10	0	-127	14
0	0	0	0	4	4	0	-24	15
-1	14	0	0	0	5	1	-27	16
7	-11	-9	24	44	37	-6	-200	17
1	0	-3	5	14	10	0	-62	18
0	3	0	0	0	0	0	0	19
1	0	-17	14	51	42	0	-258	20
6	-32	0	0	1	9	0	-126	21
0	0	0	1	4	4	0	-25	22
-1	14	1	0	0	4	1	-27	23
-1	-5	0	5	15	14	0	-99	24
0	-1	3	0	0	0	0	0	25
0	0	0	0	0	0	0	3	26
0	4	0	0	2	2	0	-14	27
0	1	0	0	0	1	0	-3	28
0	-1	0	0	2	2	0	-12	29
0	0	1	0	0	0	0	4	30
-1	2	-2	0	4	5	0	-33	31
0	4	0	0	0	0	0	2	32

EXHIBIT 16, CONTINUED

C) BEST-5 COL*** VECTORS FOR EACH SITUATION

								SITUATION
0	0	1	0	0	0	0	0	1
-12	0	-6	13	47	47	0	-395	2
-1	2	-1	0	4	5	0	-35	3
0	0	1	0	0	0	0	4	4
0	0	1	0	0	0	0	1	5
0	4	0	0	0	0	0	1	6
1	-1	3	0	0	0	0	0	7
-1	-1	0	3	8	10	0	-79	8
1	0	-13	14	46	38	0	-227	9
0	2	0	0	0	0	0	0	10
0	1	0	0	0	1	0	-1	11
0	4	0	0	2	2	0	-9	12
0	-2	0	0	2	2	0	-11	13
22	-19	0	1	-4	-2	0	361	14
0	0	0	0	3	3	0	-25	15
-3	14	0	0	1	4	0	-55	16
8	-33	-7	26	48	42	-7	-211	17
1	0	-3	6	14	10	0	-60	18
0	2	0	0	0	0	0	0	19
1	0	-14	14	45	38	0	-227	20
22	-19	0	1	-4	-2	0	362	21
0	0	0	0	3	3	0	-26	22
-3	14	0	0	1	4	0	-55	23
-1	-1	0	2	9	11	0	-79	24
0	-1	3	0	0	0	0	1	25
0	0	1	0	0	0	0	1	26
0	4	0	0	1	2	0	-10	27
0	0	0	0	0	0	0	-2	28
0	-1	0	0	2	2	0	-10	29
0	0	1	0	0	0	0	5	30
-1	2	-2	0	4	5	0	-35	31
0	4	0	0	0	0	0	2	32

D) FINAL ADJUSTED COLUMN VECTORS* FOR EACH SITUATION

								SITUATION
1000	1000	1001	1000	1000	1000	1000	1000	1
2300	2027	2057	2156	2382	2560	2701	2977	2
1531	1489	1491	1505	1540	1570	1596	1671	3
1156	1159	1160	1157	1155	1153	1152	1149	4
1308	1311	1311	1309	1307	1306	1304	1299	5
1435	1438	1434	1434	1435	1436	1438	1443	6
1780	1776	1782	1779	1779	1780	1780	1781	7
3119	3046	3059	3082	3137	3183	3219	3300	8
6053	5839	5852	5938	6115	6242	6345	6614	9
1127	1130	1127	1126	1127	1127	1127	1126	10
1245	1242	1242	1242	1245	1248	1249	1253	11
1659	1646	1644	1649	1662	1674	1684	1714	12
1189	1178	1181	1184	1191	1197	1201	1210	13
5775	5291	5376	5494	5788	6077	6375	7560	14
1336	1311	1313	1321	1342	1358	1372	1408	15
1872	1833	1827	1840	1878	1918	1951	2042	16
1279	1096	1130	1204	1327	1419	1476	1645	17
1579	1530	1533	1555	1595	1622	1643	1696	18
1127	1130	1127	1126	1127	1127	1127	1127	19
6053	5839	5852	5938	6114	6242	6345	6614	20
5775	5291	5376	5494	5788	6077	6374	7561	21
1336	1311	1313	1321	1342	1358	1372	1408	22
1872	1833	1827	1840	1878	1918	1951	2043	23
3119	3045	3059	3081	3137	3183	3219	3301	24
1780	1776	1781	1779	1779	1780	1780	1782	25
1308	1312	1312	1309	1308	1306	1304	1299	26
1660	1646	1645	1649	1663	1674	1684	1714	27
1245	1242	1242	1242	1245	1248	1249	1253	28
1189	1178	1181	1184	1191	1197	1201	1210	29
1156	1159	1160	1157	1154	1153	1152	1148	30
1531	1489	1490	1505	1540	1570	1596	1671	31
1435	1438	1434	1433	1435	1436	1438	1443	32

* PANEL A + (PANEL B + PANEL C)/2

results, thought, reconsideration, and going back to try again, we would probably have done better. (Given that the steps had to be written down hastily and without guidance, we probably have not done too badly, in comparison with a reasonable standard.)

STEPS 10 AND 11 – WHAT MIGHT HAVE BEEN DONE

No doubt in view of the heavy computing already carried out, those at the scene failed to do step 10, which would have repeated the analyses on log variances and log pseudovariances instead of raw variances and pseudovariances. This is more regrettable since it came to light after the workshop was held that

$$\text{log pseudovariance} \doteq \text{constant} + 2h \cdot (\text{deviate})^2$$

where "deviate" is the unit Gaussian deviate for the tail area involved in the pseudovariance, is not too bad an approximation for many of the estimate distributions. Having no analysis of logs, however, we cannot check either

- the general quality of this approximation, or
- whether the fit in terms of logs is better behaved than that in terms of raw pseudovariances.

The only thing we can do is to look at the 8-vectors we found in terms of $(\text{deviate})^2$ and see if this analysis, carefully examined, would have hinted at this relationship. Exhibit 17 shows the behavior of two "first" 8-vectors when plotted against $(\text{deviate})^2$. (The variance does not necessarily correspond to any one value of $(\text{deviate})^2$, but the leading terms of the Fisher-Cornish expansion suggest plotting it near $(\text{deviate})^2 = 3$.)

The linearity of "8 x 8 on col" is really quite good, and suggests doing regression on

$$(\text{deviate})^2 - 3.5$$

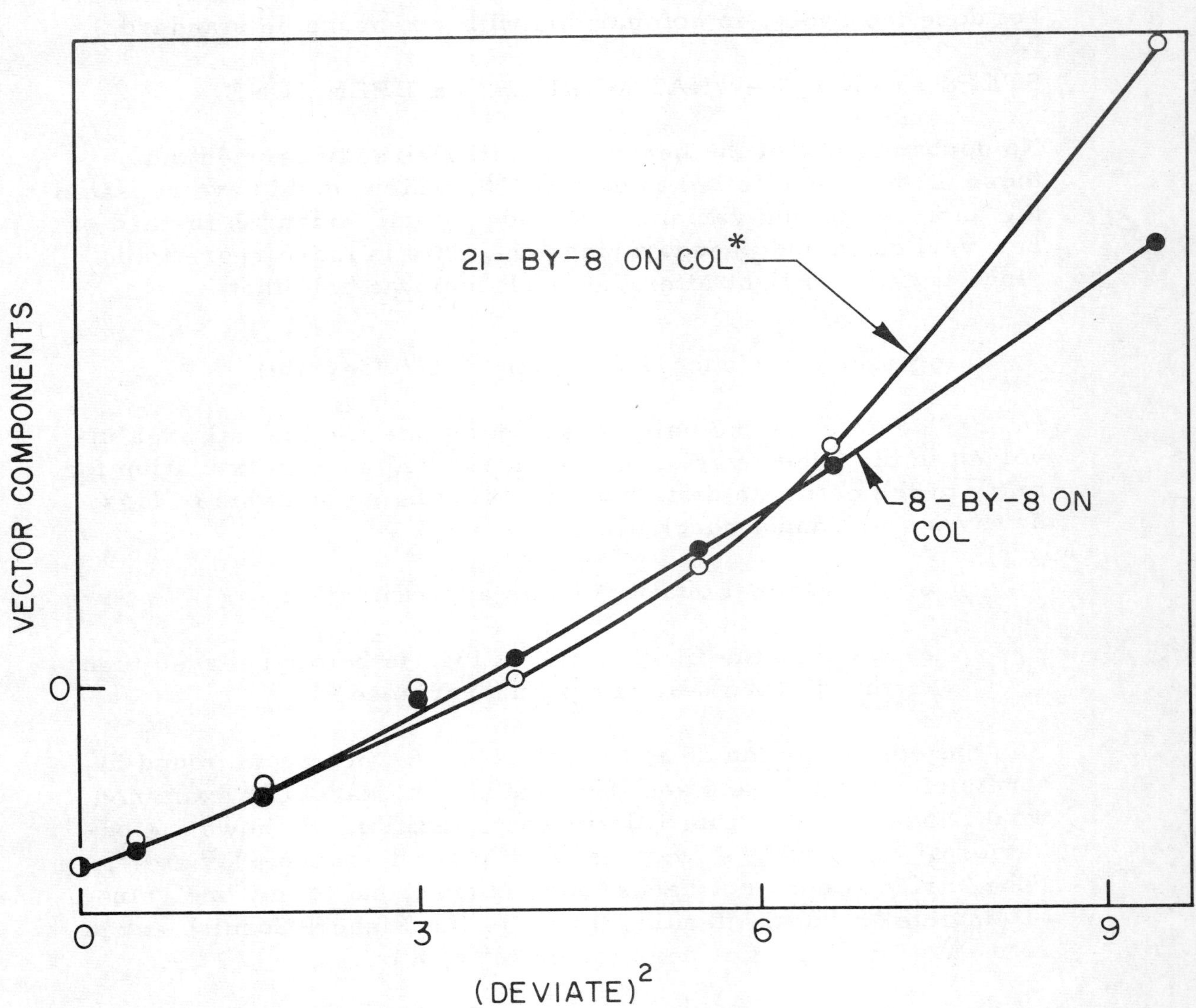

Exhibit 17. Two 8-vectors plotted against (deviate)^2.

as a substitute for what was done, since we had come to believe that "8 x 8 on col" was probably more appropriate than "15 x 8 on col*".

Doing regression (still by medians) on the best estimate for situation 16 yields residuals

(on $(\text{deviate})^2$): -2, 15, 0, -4, -8, -5, 0, 116

(on 15 x 8 on col*): -1, 9, -4, 0, 7, 8, 0, -54

corresponding to the following several series of pseudovariances:

(original data):
1872, 1828, 1821, 1941, 1885, 1921, 1951, 2029
(regression on 15 x 8):
1873, 1819, 1825, 1841, 1878, 1913, 1951, 2084
(same adjusted further):
1871, 1833, 1821, 1841, 1885, 1917, 1951, 2042
(regression on $(\text{deviate})^2$):
1874, 1813, 1821, 1837, 1893, 1926, 1951, 1913
(same adjusted further):
?, ?, ?, ?, ?, ?, ?, ?

(Since we have not applied regression on $(\text{deviate})^2$ to all 5 or 10 best choices, we are not prepared to adjust the regression on $(\text{deviate})^2$ further, and are leaving it not fully comparable with the main adjustment.) We leave the comparison of these answers to the reader.

Proceeding further with looking at the 8-vectors, exhibit 18 plots two others against $(\text{deviate})^2$, namely,

- the second vector found in exhibit 12, and
- 10 times the difference between

(15 x 8 on col*) and (8 x 8 on col).

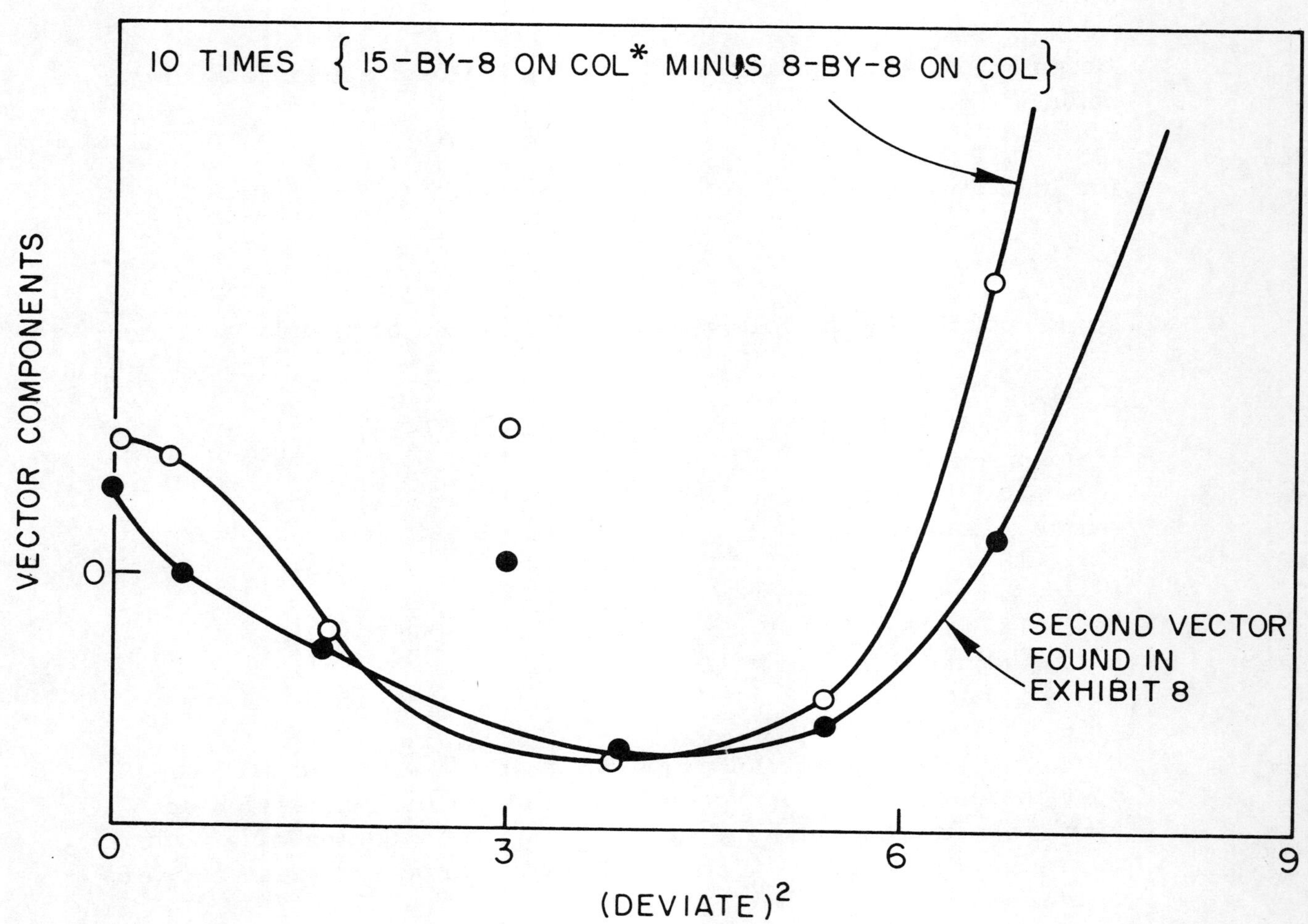

Exhibit 18. Two more 8-vectors plotted against (deviate)^2

While the detailed match is not overclose, the general run of the curves is quite similar.

This leaves us wondering how we would have done had we regressed out either "8 x 8 on col" or a resistantly obtained vector instead of "15 x 8 on col*". What would have been left?

Since the best 5 and best 10 were arbitrarily combined, and since step 10 was not taken, there was no opportunity to take step 11.

STEP 12 – WHAT WAS AND WHAT MIGHT HAVE BEEN DONE

The whole purpose of our analysis was to try to do a good job of step 12 – to try to obtain more consistent and also more useful comparisons of the performance of the various estimators with the best available for the situation at hand. Exhibit 19 shows the final calculation of estimate deficiencies for situations 2, 16 and 18.

Let us look first at part of the result for situation 22 (the same as situation 15, except that values for estimates 43 and 44 are present, instead of being indicated as zero). Exhibit 20 sets out the results for some selected estimators. First we show all of the 65 estimators that perform within 5% of the best of the 65. These tend to have reasonably constant (pseudo) discrepancies, although #53 moves upward and #61 downward as we move into the tails. The next three (including the median) show quite constant discrepancies near 26%. The remainder show steady tendencies, either up or down.

The purposes of our adjustments were to make this table more uniform and interpretable. In this case, they surely have not ruined matters.

A hasty glance suggests that the adjustment has perhaps been most satisfactory in situation 18. Panel C of exhibit 19 shows the complete analysis for all 65 estimates of situation 18.

EXHIBIT 19. FINAL CALCULATION OF ESTIMATE DEFICIENCIES

A) SITUATION 2

ID	ESTIMATE	VARIANCE	ONVAR	25.0%	10.0%	2.5%	1.0%	0.5%	0.1%
1	MEAN (AVERAGE)	1000	942	991	991	997	999	1000	1000
2	5% SYM TRIM MEAN	904	769	773	799	864	915	946	977
3	10% SYM TRIM MEAN	686	607	603	610	663	718	775	853
4	15% SYM TRIM MEAN	496	465	461	460	484	514	546	649
5	MIDMEAN	260	242	242	242	253	268	284	356
6	MEDIAN	200	175	180	188	203	216	226	246
7	GASTWIRTH	250	223	227	234	250	262	270	313
8	TRIMEAN	413	379	381	384	409	438	459	528
9	JAECKEL ADAPT TRIM	348	292	294	299	331	377	422	542
10	BICKEL ADAPT TRIM	861	405	411	431	479	515	543	655
11	SYM JAECKEL AD TRI	384	323	327	340	381	420	453	545
12	2-CHOICE ADAPT TRI	305	266	268	273	297	327	359	450
13	ADAPT TRIM LINCOM	164	101	110	128	161	186	212	317
14	HUBER PROP 2, K=2.	754	704	701	706	742	775	803	879
15	HUBER PROP 2, K=1.	662	624	621	622	658	690	717	782
16	HUBER PROP 2, K=1.	594	564	558	557	591	626	650	705
17	HUBER PROP 2, K=1.	474	436	433	436	468	501	529	614
18	HUBER PROP 2, K=1.	379	344	345	347	370	396	421	516
19	HUBER PROP 2, K=0.	245	221	223	226	241	257	272	330
20	1ST-HU 1.5*IQS MEA	997	597	876	882	996	999	999	999
21	1ST-HU 2.0*IQS MED	624	564	563	570	615	659	695	779
22	1ST-HU 1.5*IQS MED	540	474	474	483	531	581	620	708
23	1ST-HU 1.0*IQS MED	414	353	355	363	403	445	481	600
24	1ST-HU 0.7*IQS MED	305	271	272	276	296	318	340	443
25	HUBER 2.0*ADS	580	562	556	552	575	600	620	680
26	HUBER 1.5*ADS	484	465	460	456	477	502	527	593
27	1S-HU 1.5*ADS MED	487	467	462	459	480	505	529	596
28	M (2.0,2.0,5.5)*AD	219	172	186	204	220	227	235	296
29	M (2.5,4.5,9.5)*AD	386	367	366	370	382	399	414	460
30	M (2.2,3.7,5.9)*AD	312	262	277	297	317	330	343	382
31	M (2.1,4.0,8.2)*AD	313	288	292	300	311	325	340	383
32	M (1.7,3.4,8.5)*AD	235	213	216	224	234	244	255	296
33	M (1.2,3.5,8.0)*AD	143	130	133	139	143	145	147	167
34	M (ADA,4.5,8.0)*AD	197	161	167	181	201	216	228	259
35	M-TYPE (SINE)	337	288	299	315	336	356	376	439
36	"OLSHEN'S"	115	96	98	102	112	123	135	176
37	T-SKIP "OLSHEN'S"	65	38	45	53	68	81	91	115
38	MEAN LIKELIHOOD	771	506	520	563	656	768	890	948
39	CAUCHY MAX LIKELIH	10	3	6	4	6	13	20	38
40	LEAST FAVORABLE	535	464	472	491	540	581	610	656
41	ITER S-SKIP TRIMEA	202	196	195	197	199	202	207	243
42	ITER C-SKIP TRIMEA	245	190	200	222	256	276	289	313
43	MX(3T,2)-SKIP MEXT	429	271	293	342	443	507	546	642
44	3T-SKIP MEAN	310	199	218	254	321	375	413	473
45	MX(3T,2)-SKIP MEAN	302	198	216	250	316	365	398	441
46	MX(5T,2)-SKIP MEAN	221	140	154	180	234	271	296	339
47	MN(5T,.6)-SKIP MEA	214	126	142	171	226	267	295	342
48	C&T&S-SKIP TRIMEAN	194	174	177	185	194	200	207	244
49	TAKEUCHI ADAPTIVE	347	287	288	299	332	372	417	557
50	MAXIMUM EST LIKE	192	140	149	161	200	229	249	284
51	JOHNS ADAPTIVE	182	131	135	146	169	192	221	346
52	HOGG 67 ON KURTOSI	474	305	324	369	474	547	594	710
53	HOGG 69 ON KURTOSI	360	257	265	286	337	389	449	636
54	T-SKIP HOGG 67	488	395	415	445	490	523	552	637
55	T-SKIP HOGG 69	420	316	339	377	432	465	484	545
56	ADAPT FROM SKIPPIN	203	137	150	172	210	238	260	313
57	HODGES-LEHMANN	449	399	402	409	445	475	502	594
58	BICKEL-HODGES	601	502	509	531	609	665	699	750
59	2 TIMES FOLDED MED	892	702	721	781	849	899	945	982
60	3 TIMES FOLDED MED	966	808	829	874	941	968	984	992
61	4 TIMES FOLDED MED	1000	920	979	980	994	998	999	1000
62	3-FOLDED 1-TRIM ME	703	615	614	626	681	736	793	863
63	3-FOLDED 2-TRIM ME	479	431	432	438	477	510	535	617
64	CAUCHY PITMAN (LOC	2	0	0	-2	-5	-3	0	41
65	SHORTH	440	383	395	414	437	450	467	582

EXHIBIT 19, CONTINUED

B) SITUATION 16

ID	ESTIMATE	VARIANCE	ONVAR	25.0%	10.0%	2.5%	1.0%	0.5%	0.1%
1	MEAN (AVERAGE)	1000	942	998	998	999	1000	1000	1000
2	5% SYM TRIM MEAN	947	726	753	828	908	964	978	988
3	10% SYM TRIM MEAN	774	441	472	557	694	768	837	927
4	15% SYM TRIM MEAN	470	245	262	319	458	572	655	786
5	MIDMEAN	214	144	152	169	215	262	308	414
6	MEDIAN	259	222	229	240	266	286	303	342
7	GASTWIRTH	215	151	156	172	212	255	300	420
8	TRIMEAN	363	195	205	244	350	462	555	686
9	JAECKEL ADAPT TRIM	248	158	165	185	239	297	361	516
10	BICKEL ADAPT TRIM	584	298	311	357	469	600	770	909
11	SYM JAECKEL AD TRI	1000	927	997	998	998	999	1000	1000
12	2-CHOICE ADAPT TRI	228	150	156	176	225	277	330	457
13	ADAPT TRIM LINCOM	102	44	45	55	80	110	151	405
14	HUBER PROP 2, K=2.	793	585	609	676	783	839	876	948
15	HUBER PROP 2, K=1.	684	454	472	535	676	758	812	899
16	HUBER PROP 2, K=1.	582	357	374	432	564	672	740	855
17	HUBER PROP 2, K=1.	394	237	249	288	383	477	556	689
18	HUBER PROP 2, K=1.	298	181	190	217	289	367	444	584
19	HUBER PROP 2, K=0.	198	142	149	165	204	240	273	342
20	1ST-HU 1.5*IQS MEA	998	493	906	914	998	999	999	999
21	1ST-HU 2.0*IQS MED	602	422	435	476	566	670	762	853
22	1ST-HU 1.5*IQS MED	492	303	315	354	453	567	667	794
23	1ST-HU 1.0*IQS MED	346	191	201	229	309	402	498	686
24	1ST-HU 0.7*IQS MED	256	150	159	180	236	301	369	552
25	HUBER 2.0*ADS	519	407	418	456	533	596	642	691
26	HUBER 1.5*ADS	392	284	294	330	406	470	518	584
27	1S-HU 1.5*ADS MED	394	285	295	331	409	473	522	590
28	M (2.0,2.0,5.5)*AD	146	131	131	135	144	155	170	226
29	M (2.5,4.5,9.5)*AD	235	118	134	174	255	316	363	442
30	M (2.2,3.7,5.9)*AD	113	22	35	64	125	176	219	310
31	M (2.1,4.0,8.2)*AD	146	45	58	91	160	217	264	354
32	M (1.7,3.4,8.5)*AD	72	-4	6	28	79	124	165	262
33	M (1.2,3.5,8.0)*AD	30	-19	-12	3	36	64	92	166
34	M (ADA,4.5,8.0)*AD	102	20	31	54	105	155	204	323
35	M-TYPE (SINE)	160	51	67	99	170	232	286	396
36	"OLSHEN'S"	117	79	85	97	123	146	166	211
37	T-SKIP "OLSHEN'S"	68	55	57	61	70	77	86	112
38	MEAN LIKELIHOOD	870	394	438	547	767	868	919	969
39	CAUCHY MAX LIKELIH	96	98	102	101	96	90	84	70
40	LEAST FAVORABLE	456	285	294	331	432	551	641	743
41	ITER S-SKIP TRIMEA	101	77	79	88	103	113	123	169
42	ITER C-SKIP TRIMEA	168	28	34	54	110	178	268	635
43	MX(3T,2)-SKIP MEXT	0	0	0	0	0	0	0	0
44	3T-SKIP MEAN	0	0	0	0	0	0	0	0
45	MX(3T,2)-SKIP MEAN	171	17	30	57	124	194	273	554
46	MX(5T,2)-SKIP MEAN	145	57	66	87	130	168	210	414
47	MN(5T,.6)-SKIP MEA	94	-1	8	27	69	113	164	408
48	C&T&S-SKIP TRIMEAN	581	404	405	427	493	609	746	871
49	TAKEUCHI ADAPTIVE	239	110	125	161	246	322	386	512
50	MAXIMUM EST LIKE	165	116	124	139	172	199	224	281
51	JOHNS ADAPTIVE	85	43	50	60	83	106	131	225
52	HOGG 67 ON KURTOSI	410	260	274	305	389	487	584	713
53	HOGG 69 ON KURTOSI	264	171	179	204	265	326	384	497
54	T-SKIP HOGG 67	356	147	162	206	315	422	515	714
55	T-SKIP HOGG 69	239	83	96	136	236	335	425	567
56	ADAPT FROM SKIPPIN	198	179	178	182	195	209	228	301
57	HODGES-LEHMANN	416	276	285	315	387	467	550	723
58	BICKEL-HODGES	577	350	360	398	512	654	745	894
59	2 TIMES FOLDED MED	954	590	637	778	908	931	948	986
60	3 TIMES FOLDED MED	981	778	835	905	960	989	994	998
61	4 TIMES FOLDED MED	1000	921	995	996	997	999	1000	1000
62	3-FOLDED 1-TRIM ME	785	469	497	575	704	779	851	933
63	3-FOLDED 2-TRIM ME	430	243	256	296	402	512	601	774
64	CAUCHY PITMAN (LOC	0	-2	-2	1	3	2	0	-6
65	SHORTH	579	619	614	596	566	545	529	492

EXHIBIT 19, CONTINUED

C) SITUATION 18

ID	ESTIMATE	VARIANCE	ONVAR	25.0%	10.0%	2.5%	1.0%	0.5%	0.1%
1	MEAN (AVERAGE)	497	391	398	425	485	545	600	729
2	5% SYM TRIM MEAN	161	152	154	160	166	167	166	159
3	10% SYM TRIM MEAN	62	60	61	62	62	62	62	59
4	15% SYM TRIM MEAN	16	16	17	17	15	16	15	14
5	MIDMEAN	7	13	11	9	6	5	4	0
6	MEDIAN	131	134	131	131	130	131	131	133
7	GASTWIRTH	29	36	34	32	28	25	22	13
8	TRIMEAN	15	15	15	15	13	14	15	19
9	JAECKEL ADAPT TRIM	23	18	18	20	24	28	31	36
10	BICKEL ADAPT TRIM	69	59	61	65	71	75	78	85
11	SYM JAECKEL AD TRI	44	33	37	40	46	50	53	59
12	2-CHOICE ADAPT TRI	44	39	37	39	45	50	53	62
13	ADAPT TRIM LINCOM	78	74	72	74	80	85	87	91
14	HUBER PROP 2, K=2.	174	161	164	172	180	182	180	173
15	HUBER PROP 2, K=1.	118	110	112	117	122	123	122	117
16	HUBER PROP 2, K=1.	78	72	74	78	80	81	81	77
17	HUBER PROP 2, K=1.	28	24	26	27	28	30	31	31
18	HUBER PROP 2, K=1.	9	7	8	8	9	10	11	11
19	HUBER PROP 2, K=0.	13	20	17	15	11	10	8	4
20	1ST-HU 1.5*IQS MEA	220	72	74	78	88	105	120	318
21	1ST-HU 2.0*IQS MED	135	124	126	132	139	142	143	138
22	1ST-HU 1.5*IQS MED	70	63	66	68	72	74	75	73
23	1ST-HU 1.0*IQS MED	17	14	15	15	17	19	20	23
24	1ST-HU 0.7*IQS MED	6	11	9	7	3	2	1	-1
25	HUBER 2.0*ADS	118	108	110	117	123	124	123	115
26	HUBER 1.5*ADS	56	49	52	54	58	61	62	60
27	1S-HU 1.5*ADS MED	60	52	55	58	62	65	66	65
28	M (2.0,2.0,5.5)*AD	46	37	40	45	50	52	52	49
29	M (2.5,4.5,9.5)*AD	55	50	53	55	56	57	56	49
30	M (2.2,3.7,5.9)*AD	50	48	50	51	51	51	50	43
31	M (2.1,4.0,8.2)*AD	28	26	28	29	28	28	27	20
32	M (1.7,3.4,8.5)*AD	2	0	3	4	3	2	1	-5
33	M (1.2,3.5,8.0)*AD	4	7	7	6	3	1	0	-8
34	M (ADA,4.5,8.0)*AD	47	45	46	48	49	48	46	36
35	M-TYPE (SINE)	39	38	40	40	39	38	36	28
36	"OLSHEN'S"	43	50	46	45	42	40	39	35
37	T-SKIP "OLSHEN'S"	79	80	79	81	80	78	76	66
38	MEAN LIKELIHOOD	31	31	32	31	30	30	31	33
39	CAUCHY MAX LIKELIH	145	163	152	147	141	139	139	136
40	LEAST FAVORABLE	125	106	111	120	131	136	138	137
41	ITER S-SKIP TRIMEA	179	180	178	180	180	179	178	175
42	ITER C-SKIP TRIMEA	39	38	38	40	40	40	40	39
43	MX(3T,2)-SKIP MEXT	254	213	222	237	265	285	299	317
44	3T-SKIP MEAN	91	83	85	89	94	96	97	91
45	MX(3T,2)-SKIP MEAN	72	70	71	71	72	73	73	69
46	MX(5T,2)-SKIP MEAN	60	65	64	62	59	57	55	44
47	MN(5T,.6)-SKIP MEA	57	61	60	59	58	56	53	43
48	C&T&S-SKIP TRIMEAN	99	94	94	97	101	103	104	105
49	TAKEUCHI ADAPTIVE	80	72	74	77	82	86	88	87
50	MAXIMUM EST LIKE	133	128	127	129	135	140	144	147
51	JOHNS ADAPTIVE	54	56	53	54	55	56	55	50
52	HOGG 67 ON KURTOSI	178	182	182	183	179	173	166	144
53	HOGG 69 ON KURTOSI	109	113	109	109	109	110	108	101
54	T-SKIP HOGG 67	164	139	145	153	168	180	190	218
55	T-SKIP HOGG 69	151	141	142	144	150	157	165	189
56	ADAPT FROM SKIPPIN	95	94	91	95	98	98	97	92
57	HODGES-LEHMANN	48	46	48	48	48	47	46	39
58	BICKEL-HODGES	98	92	93	95	99	102	105	108
59	2 TIMES FOLDED MED	221	192	197	210	231	241	247	252
60	3 TIMES FOLDED MED	285	254	257	272	294	309	320	340
61	4 TIMES FOLDED MED	398	341	344	364	402	434	462	541
62	3-FOLDED 1-TRIM ME	89	86	87	89	90	91	90	86
63	3-FOLDED 2-TRIM ME	32	32	33	32	31	32	33	32
64	CAUCHY PITMAN (LOC	23	36	27	25	21	19	17	13
65	SHORTH	633	647	636	632	634	634	633	629

Exhibit 20. Adjusted discrepancies (in .001) for some selected estimators for situation 22.

		Pseudovariance						
Est.	(Var)	50%	25%	10%	2.5%	1%	0.5%	0.1%
54	(1)	0	1	1	0	0	0	0
55	(2)	2	2	2	1	1	1	3
46	(2)	2	1	1	3	4	5	8
56	(3)	8	8	6	1	-2	-5	-13
51	(11)	12	12	12	11	10	9	6
53	(26)	18	17	20	26	32	39	60
61	(23)	27	26	25	23	21	19	13
57	(32)	39	40	37	30	25	20	7
52	(38)	36	36	38	39	40	41	41
58	(38)	31	32	34	38	42	45	56
19	(105)	102	103	103	103	103	105	107
5*	(111)	109	110	109	109	110	111	123
26	(127)	124	126	126	127	127	129	137
6*	(260)	264	264	262	258	256	253	248
32	(138)	153	152	145	133	124	117	99
20	(143)	117	122	128	142	157	171	225
37	(394)	189	196	225	307	409	554	821

* Note that 5 is the midmean and 6 the median.

Exhibit 21 shows the pseudovariances for the 16 estimates that perform within 4% of the best one. The bumps that have been introduced into the discrepancies for the best estimate, #32, seem to have done a reasonable job of smoothing out the discrepancies for the other good ones.

WHAT HAVE WE LEARNED OF GENERAL APPLICATIONS?

We have again confirmed the effectiveness of (iterative) median analysis for row-PLUS-column fitting. Missing and "infinite" values hampered steps 1 and 2 inappreciably.

We have learned that a resistant fit of row-TIMES-column can be quite different from an eigenfit, and that sometimes resistant fits can be quite clearly preferable.

We have learned one way to try to make a resistant row-TIMES-column fit – namely, chasing out minus signs, taking logs, fitting row-PLUS-column, and taking antilogs.

We have learned that a few values "beyond infinity" – like logs of small negative numbers, for example – need not hinder us (as long as we use median fitting).

We have learned that, by these devices, getting a "first important component" out of a rectangular table need not require more computing power than a pen and pencil.

We are prepared to think of taking out successive "important components" by a repetition of this process.

We have learned to stop and think about such hybrid fits as

$$\text{common} + \text{row} + \text{row}^* \cdot \text{col}^*$$

before going all the way to

$$\text{common} + \text{row} + \text{col} + \text{row}^* \cdot \text{col}^*$$

Exhibit 21. Adjusted discrepancies (in .001) for selected estimates in situation 18.

		Pseudovariances						
Est.	(Var)	50%	25%	10%	2.5%	1%	0.5%	0.1%
32	(2)	0	3	4	3	2	1	-5
33	(4)	7	7	6	3	1	0	-8
24	(6)	11	9	7	3	2	1	-1
5	(7)	13	11	9	6	5	4	0
18	(9)	7	8	8	9	10	11	11
19	(13)	20	17	15	11	10	8	4
4	(16)	16	17	17	15	14	15	14
23	(17)	14	15	15	17	19	20	23
64	(23)	36	27	25	21	19	17	13
9	(23)	18	18	20	24	28	31	36
7	(29)	36	34	32	28	25	22	13
17	(28)	24	26	27	28	30	31	31
31	(28)	26	28	29	28	28	27	20
38	(31)	31	32	31	30	30	31	33
63	(32)	32	33	32	31	32	33	32
42	(39)	38	38	40	40	40	40	39

with the danger that our concentration on col* will tend to make us forget col.

The freedom to use row-TIMES-column resistantly should not be underestimated. Doing row-PLUS-column is very important, but being able to go further adds very considerably to our tool kit.

HOW WOULD WE TRY THIS ANALYSIS IF WE WERE TO START AGAIN?

Surely just what we would plan to do would depend on the day of the week and the phase of the moon, but not arbitrarily so.

We have learned that there is consistency from situation to situation, even as far as the second 8-vector goes. It would thus seem natural to pool much more of our analysis across situations.

Were we to go back to this same data, we would have 18 x 65 = 1170 rows to deal with. Let us suppose that we feel, based on other intervening analyses, that log pseudovariances should receive our attention. We might consider beginning as follows:

1. Set up the 1170 8-vectors of log pseudovariances.

2. Regress each on "$(\text{deviate})^2 - 3.5$" and sort out by regression coefficient.

3. Select 4 or 5 groups, each of 50 to 100 8-vectors, beginning with the largest coefficients. (Elimination of some strays may be needed.)

4. Process each group according to, first

   common + row + row* · col*

   then

   common + row + row* · col* + row** · col**

and so on, with some exploration of iterative refitting of earlier terms once later ones are fitted. (Doing all fitting resistantly, of course.)

5. Decide how many product terms are needed for each group.

6. Look at the corresponding low dimensional linear spaces (any fit in terms of col* and col** is also a fit in terms of 3col*-2col** and a-col*-b· col**) of 8-vectors, one for each group, and see if these are reasonably compatible.

7. Given a small set of 8-vectors, try adjusting all 1170 8-vectors in their terms.

8. Look hard at the resulting residuals to see if they are apparently random, except for explained effects.

9. THINK HARD, and then proceed.

DATA SET DESCRIPTION

DATA - FIRST PHASE OF THE PRINCETON ROBUSTNESS STUDY (J. W. TUKEY)

CASES AND VARIABLES

18 CASES (SITUATIONS)
10 VARIABLES - 8 MEASUREMENT (1 VARIANCE AND 7 PSEUDOVARIANCES), 2 CATEGORICAL (SITUATION IDENTIFICATION AND ESTIMATE IDENTIFICATION)

BRIEF DESCRIPTION

THESE DATA COME FROM THE FIRST PHASE OF THE PRINCETON ROBUSTNESS STUDY - AN EXAMINATION OF POINT ESTIMATES OF LOCATION FOR SAMPLES FROM SYMMETRIC DISTRIBUTIONS OR POLYSAMPLES FROM SYMMETRIC SITUATIONS (BOTH CALLED SITUATIONS). FOR A NUMBER OF SITUATIONS THE SAMPLING DISTRIBUTIONS OF 65 ESTIMATES WERE CHARACTERIZED WITH MONTE CARLO GENERATED VARIANCE AND PERCENTILES. THE DATA CONSIST OF 18 TABLES (ONE PER SITUATION) OF SIZE 65 BY 8 (ESTIMATES BY VARIANCE AND PSEUDOVARIANCES).

PURPOSE OF STUDY

THE PURPOSE OF THE STUDY WAS TO EVALUATE THE PERFORMANCE OF POINT ESTIMATES OF LOCATION FOR EACH OF A BROAD CLASS OF SYMMETRIC SITUATIONS. THE IDEA WAS TO MEASURE INDIVIDUAL ESTIMATE PERFORMANCE RELATIVE TO ''BEST ESTIMATE PERFORMANCE'' FOR THE SPECIFIC SITUATION.

REFERENCE

ANDREWS, D. F., P. J. BICKEL, F. R. HAMPEL, P. J. HUBER, W. H. ROGERS, AND J. W. TUKEY. ROBUST ESTIMATES OF LOCATION: SURVEY AND ADVANCES. PRINCETON UNIVERSITY PRESS, PRINCETON, NEW JERSEY, 1972.

DETAILED DESCRIPTION

VAR	COL	FORMAT	DESCRIPTION
***	***	******	***********
1	1-2	XX.	SITUATION 1 = N(0,1) 2 = CAUCHY 3 = 75% OF N(0,1) AND 25% OF N(0,1)/U(0,1) 4 = 19 FROM N(0,1) AND 1 FROM N(0,9) 5 = 18 FROM N(0,1) AND 2 FROM N(0,9) 6 = 17 FROM N(0,1) AND 3 FROM N(0,9) 7 = 15 FROM N(0,1) AND 5 FROM N(0,9) 8 = 10 FROM N(0,1) AND 10 FROM N(0,9) 9 = 5 FROM N(0,1) AND 15 FROM N(0,9) 10 = 19 FROM N(0,1) AND 1 FROM N(0,100) 11 = 18 FROM N(0,1) AND 2 FROM N(0,100) 12 = 15 FROM N(0,1) AND 5 FROM N(0,100) 13 = 90% OF N(0,1) AND 10% OF N(0,1)/U(0,1) 14 = N(0,1)/U(0,1) 15 = 90% OF N(0,1) AND 10% OF N(0,1)/U(0,1/3) 16 = 75% OF N(0,1) AND 25% OF N(0,1)/U(0,1/3) 17 = DOUBLE EXPONENTIAL 18 = STUDENT'S T WITH 3 DEGREES OF FREEDOM

```
IN THE PRECEDING CCDES
  (1) N(MU,SIGMA SQUARED) IS A GAUSSIAN
      DISTIBUTION WITH MEAN MU AND VARIANCE
      SIGMA SQUARED
  (2) U(A,B) IS A (UNIFORM) RECTANGULAR
      DISTRIBUTION BETWEEN A AND B
  (3) A ''/'' INDICATES A RATIO OF INDEPEN-
      DENT DEVIATES
  (4) ''20-K FROM ... AND K FROM ...''
      IMPLIES DRAWING OF FIXED NUMBERS FROM
      EACH DISTRIBUTION
  (5) ''(100-P)% CF ... AND P% CF ...''
      IMPLIES 20 FROM A MIXED DISTRIBUTION.

2    4-5      XX.      ESTIMATE
                          1 = MEAN (AVERAGE)
                          2 = 5% SYM TRIM MEAN
                          3 = 10% SYM TRIM MEAN
                          4 = 15% SYM TRIM MEAN
                          5 = MIDMEAN
                          6 = MEDIAN
                          7 = GASTWIRTH
                          8 = TRIMEAN
                          9 = JAECKEL ADAPT TRIM
                         10 = BICKEL ADAPT TRIM
                         11 = SYM JAECKEL AD TRIM
                         12 = 2-CHOICE ADAPT TRIM
                         13 = ADAPT TRIM LINCOM
                         14 = HUBER PROP 2, K=2.0
                         15 = HUBER PROP 2, K=1.7
                         16 = HUBER PROP 2, K=1.5
                         17 = HUBER PROP 2, K=1.2
                         18 = HUBER PROP 2, K=1.0
                         19 = HUBER PROP 2, K=0.7
                         20 = 1ST-HU 1.5*IQS MEAN
                         21 = 1ST-HU 2.0*IQS MED
                         22 = 1ST-HU 1.5*IQS MED
                         23 = 1ST-HU 1.0*IQS MED
                         24 = 1ST-HU 0.7*IQS MED
                         25 = HUBER 2.0*ADS
                         26 = HUBER 1.5*ADS
                         27 = 1S-HU 1.5*ADS MED
                         28 = M (2.0,2.0,5.5)*AD
                         29 = M (2.5,4.5,9.5)*AD
                         30 = M (2.2,3.7,5.9)*AD
                         31 = M (2.1,4.0,8.2)*AD
                         32 = M (1.7,3.4,8.5)*AD
                         33 = M (1.2,3.5,8.0)*AD
                         34 = M (ADA,4.5,8.0)*AD
                         35 = M-TYPE (SINE)
                         36 = ''OLSHEN'S''
                         37 = T-SKIP ''OLSHEN'S''
                         38 = MEAN LIKELIHOOD
                         39 = CAUCHY MAX LIKELIHOOD
                         40 = LEAST FAVORABLE
                         41 = ITER S-SKIP TRIMEAN
                         42 = ITER C-SKIP TRIMEAN
                         43 = MX(3T,2)-SKIP MEXTR
                         44 = 3T-SKIP MEAN
                         45 = MX(3T,2)-SKIP MEAN
                         46 = MX(5T,2)-SKIP MEAN
                         47 = MN(5T,.6)-SKIP MEAN
                         48 = C&T&S-SKIP TRIMEAN
                         49 = TAKEUCHI ADAPTIVE
                         50 = MAXIMUM EST LIKE
                         51 = JOHNS ADAPTIVE
```

```
                          52 = HOGG 67 ON KURTOSIS
                          53 = HOGG 69 ON KURTOSIS
                          54 = T-SKIP HOGG 67
                          55 = T-SKIP HOGG 69
                          56 = ADAPT FROM SKIPPING
                          57 = HODGES-LEHMANN
                          58 = BICKEL-HODGES
                          59 = 2 TIMES FOLDED MED
                          60 = 3 TIMES FOLDED MED
                          61 = 4 TIMES FOLDED MED
                          62 = 3-FOLDED 1-TRIM MED
                          63 = 3-FOLDED 2-TRIM MED
                          64 = CAUCHY PITMAN (LOC)
                          65 = SHORTH

   3    7-14  XXXX.XXX  VARIANCE
   4   16-23  XXXX.XXX  PSEUDOVARIANCE AT 50%
   5   25-32  XXXX.XXX  PSEUDOVARIANCE AT 25%
   6   34-41  XXXX.XXX  PSEUDOVARIANCE AT 10%
   7   43-50  XXXX.XXX  PSEUDOVARIANCE AT 2.5%
   8   52-59  XXXX.XXX  PSEUDOVARIANCE AT 1.0%
   9   61-68  XXXX.XXX  PSEUDOVARIANCE AT 0.5%
  10   70-77  XXXX.XXX  PSEUDOVARIANCE AT 0.1%

                        THE DEFINITION OF THE ESTIMATES IS CONTAINED IN
                        THE ABOVE REFERENCE FOR THE DATA.

   FORMAT IS (F2.0,1X,F2.0,8(1X,F8.3))       N = 18


 LOCATION OF DATA

   CARD IMAGE - FS.CC73.CTUKE1 (TABLE 2)
   CARD IMAGE - FS.CC73.CTUKE2 (TABLE 16)
   CARD IMAGE - FS.CC73.CTUKE3 (TABLE 18)


 LISTING OF TABLE 2 (THE FIRST TWO LINES ARE COLUMN GUIDES)

   0        1         2         3         4         5         6         7
   1234567890123456789012345678901234567890123456789012345678901234567890123456 7

    2  1 9999.992    35.067   217.250   240.431   922.841  3289.801  8341.964  9999.992
    2  2   23.984     8.780     9.079    10.700    17.497    30.203    49.574   128.502
    2  3    7.324     5.162     5.183     5.531     7.064     9.080    11.983    20.224
    2  4    4.566     3.792     3.815     3.992     4.619     5.268     5.947     8.479
    2  5    3.108     2.674     2.712     2.843     3.187     3.497     3.774     4.622
    2  6    2.876     2.457     2.509     2.654     2.988     3.264     3.490     3.950
    2  7    3.066     2.608     2.661     2.816     3.178     3.467     3.698     4.332
    2  8    3.921     3.263     3.321     3.497     4.032     4.553     4.991     6.307
    2  9    3.525     2.862     2.914     3.075     3.561     4.109     4.673     6.504
    2 10   16.575     3.407     3.494     3.786     4.569     5.275     5.910     8.634
    2 11    3.731     2.996     3.056     3.265     3.847     4.412     4.940     6.541
    2 12    3.309     2.762     2.810     2.965     3.388     3.805     4.217     5.415
    2 13    2.752     2.254     2.310     2.473     2.838     3.144     3.429     4.359
    2 14    9.344     6.845     6.873     7.324     9.225    11.371    13.686    24.660
    2 15    6.806     5.398     5.419     5.710     6.971     8.270     9.545    13.668
    2 16    5.663     4.647     4.648     4.863     5.823     6.840     7.717    10.105
    2 17    4.374     3.592     3.630     3.819     4.475     5.129     5.738     7.717
    2 18    3.701     3.090     3.139     3.302     3.780     4.241     4.668     6.152
    2 19    3.046     2.603     2.646     2.786     3.140     3.447     3.709     4.445
    2 20  719.232     5.032    16.530    18.294   617.152  2169.575  2868.051  4348.226
    2 21    6.110     4.651     4.704     5.015     6.191     7.500     8.856    13.501
    2 22    4.995     3.852     3.906     4.168     5.083     6.103     7.117    10.205
    2 23    3.928     3.133     3.190     3.386     3.988     4.609     5.206     7.443
    2 24    3.311     2.782     2.823     2.978     3.385     3.754     4.092     5.342
    2 25    5.471     4.631     4.635     4.811     5.601     6.396     7.117     9.290
```

2	26	4.456	3.788	3.809	3.966	4.552	5.145	5.706	7.316
2	27	4.481	3.806	3.825	3.985	4.578	5.174	5.732	7.377
2	28	2.943	2.447	2.525	2.709	3.055	3.312	3.531	4.230
2	29	3.743	3.203	3.246	3.420	3.856	4.257	4.607	5.517
2	30	3.343	2.747	2.846	3.065	3.487	3.822	4.109	4.816
2	31	3.349	2.846	2.904	3.078	3.459	3.794	4.091	4.829
2	32	3.008	2.574	2.624	2.779	3.110	3.386	3.625	4.230
2	33	2.684	2.330	2.372	2.504	2.779	2.993	3.168	3.575
2	34	2.864	2.417	2.470	2.631	2.981	3.267	3.500	4.019
2	35	3.469	2.845	2.933	3.149	3.586	3.976	4.331	5.305
2	36	2.598	2.243	2.281	2.402	2.683	2.919	3.122	3.615
2	37	2.461	2.107	2.153	2.276	2.557	2.785	2.971	3.364
2	38	10.044	4.100	4.281	4.933	6.919	11.038	24.641	56.741
2	39	2.323	2.033	2.068	2.165	2.397	2.594	2.755	3.096
2	40	4.947	3.779	3.893	4.238	5.181	6.109	6.922	8.663
2	41	2.883	2.520	2.556	2.684	2.973	3.206	3.407	3.931
2	42	3.045	2.502	2.570	2.770	3.203	3.538	3.797	4.332
2	43	4.031	2.780	2.908	3.278	4.280	5.192	5.955	8.308
2	44	3.334	2.530	2.630	2.888	3.509	4.094	4.602	5.644
2	45	3.293	2.526	2.623	2.876	3.482	4.034	4.487	5.330
2	46	2.953	2.358	2.432	2.630	3.109	3.513	3.835	4.507
2	47	2.926	2.319	2.396	2.599	3.078	3.492	3.829	4.528
2	48	2.853	2.453	2.499	2.645	2.956	3.199	3.407	3.937
2	49	3.523	2.843	2.890	3.075	3.567	4.078	4.632	6.728
2	50	2.846	2.358	2.416	2.571	2.976	3.322	3.596	4.159
2	51	2.812	2.332	2.377	2.525	2.865	3.169	3.466	4.555
2	52	4.376	2.918	3.041	3.418	4.527	5.652	6.654	10.256
2	53	3.591	2.729	2.797	3.020	3.594	4.191	4.902	8.185
2	54	4.489	3.352	3.515	3.884	4.670	5.369	6.029	8.199
2	55	3.963	2.965	3.112	3.460	4.196	4.783	5.237	6.548
2	56	2.886	2.349	2.418	2.603	3.014	3.358	3.652	4.336
2	57	4.172	3.370	3.437	3.649	4.293	4.874	5.425	7.328
2	58	5.760	4.072	4.190	4.594	6.088	7.643	8.976	11.897
2	59	21.363	6.805	7.376	9.824	15.785	25.258	49.070	162.397
2	60	67.345	10.558	12.016	17.062	40.203	78.830	165.131	356.469
2	61	5006.636	25.186	96.550	106.853	415.115	1384.375	3430.065	9999.992
2	62	7.747	5.262	5.333	5.770	7.478	9.705	13.038	21.747
2	63	4.412	3.560	3.620	3.836	4.551	5.221	5.811	7.765
2	64	2.304	2.027	2.057	2.150	2.369	2.549	2.701	3.104
2	65	4.106	3.285	3.400	3.677	4.233	4.655	5.068	7.127

LISTING OF TABLE 16 (THE FIRST TWO LINES ARE COLUMN GUIDES)

```
0         1         2         3         4         5         6         7
1234567890123456789012345678901234567890123456789012345678901234567890123456
```

16	1	9999.992	31.679	963.767	1066.604	1290.765	4323.773	9999.992	9999.992
16	2	35.052	6.694	7.396	10.668	20.492	52.935	87.505	163.562
16	3	8.295	3.279	3.461	4.156	6.133	8.277	11.981	27.800
16	4	3.533	2.428	2.476	2.703	3.465	4.481	5.648	9.538
16	5	2.382	2.141	2.153	2.214	2.393	2.598	2.820	3.483
16	6	2.525	2.355	2.368	2.421	2.558	2.685	2.801	3.104
16	7	2.384	2.159	2.163	2.221	2.385	2.574	2.786	3.523
16	8	2.940	2.277	2.297	2.434	2.891	3.567	4.385	6.505
16	9	2.490	2.177	2.187	2.258	2.467	2.730	3.055	4.221
16	10	4.499	2.612	2.652	2.861	3.535	4.793	8.471	22.492
16	11	9999.992	25.216	692.898	766.834	911.307	2226.968	7831.738	9999.992
16	12	2.423	2.155	2.165	2.232	2.425	2.654	2.914	3.760
16	13	2.085	1.918	1.912	1.946	2.041	2.156	2.299	3.431
16	14	9.037	4.414	4.668	5.670	8.638	11.912	15.768	39.255
16	15	5.929	3.354	3.458	3.958	5.798	7.938	10.358	20.283
16	16	4.476	2.852	2.920	3.237	4.305	5.850	7.495	14.073
16	17	3.089	2.403	2.432	2.583	3.046	3.666	4.398	6.575
16	18	2.666	2.239	2.255	2.351	2.641	3.028	3.508	4.908
16	19	2.333	2.135	2.146	2.204	2.361	2.524	2.684	3.103
16	20	810.520	3.615	19.403	21.474	909.394	2639.843	3193.818	3843.905

16	21	4.705	3.172	3.234	3.512	4.327	5.810	8.209	13.893
16	22	3.686	2.630	2.665	2.849	3.434	4.426	5.862	9.915
16	23	2.864	2.266	2.285	2.387	2.717	3.205	3.886	6.511
16	24	2.514	2.157	2.172	2.243	2.460	2.742	3.094	4.560
16	25	3.895	3.089	3.138	3.385	4.026	4.753	5.444	6.615
16	26	3.077	2.561	2.588	2.744	3.162	3.616	4.049	4.904
16	27	3.091	2.565	2.592	2.751	3.176	3.637	4.082	4.976
16	28	2.192	2.109	2.103	2.126	2.194	2.270	2.351	2.638
16	29	2.448	2.077	2.109	2.226	2.521	2.806	3.063	3.657
16	30	2.110	1.874	1.892	1.965	2.148	2.327	2.499	2.962
16	31	2.192	1.919	1.940	2.023	2.236	2.449	2.652	3.161
16	32	2.016	1.823	1.837	1.893	2.040	2.189	2.336	2.767
16	33	1.930	1.797	1.804	1.845	1.948	2.049	2.149	2.448
16	34	2.085	1.871	1.885	1.944	2.100	2.269	2.451	3.017
16	35	2.229	1.932	1.957	2.042	2.264	2.496	2.732	3.379
16	36	2.120	1.990	1.997	2.038	2.143	2.245	2.339	2.589
16	37	2.009	1.939	1.937	1.959	2.020	2.079	2.134	2.299
16	38	14.392	3.026	3.250	4.065	8.077	14.487	24.150	65.086
16	39	2.070	2.032	2.034	2.046	2.078	2.107	2.131	2.197
16	40	3.443	2.562	2.589	2.751	3.310	4.267	5.438	7.953
16	41	2.081	1.986	1.983	2.018	2.095	2.162	2.225	2.459
16	42	2.249	1.886	1.890	1.945	2.110	2.333	2.667	5.602
16	43	0.0	0.0	0.0	0.0	0.0	0.0	0.0	0.0
16	44	0.0	0.0	0.0	0.0	0.0	0.0	0.0	0.0
16	45	2.258	1.865	1.883	1.952	2.144	2.380	2.685	4.575
16	46	2.190	1.944	1.956	2.015	2.158	2.305	2.471	3.488
16	47	2.065	1.829	1.841	1.890	2.018	2.162	2.334	3.452
16	48	4.463	3.073	3.071	3.209	3.702	4.910	7.676	15.886
16	49	2.461	2.060	2.087	2.192	2.491	2.828	3.180	4.189
16	50	2.241	2.073	2.085	2.137	2.268	2.393	2.513	2.842
16	51	2.045	1.916	1.922	1.957	2.049	2.145	2.246	2.635
16	52	3.172	2.477	2.517	2.647	3.075	3.739	4.688	7.117
16	53	2.542	2.212	2.226	2.311	2.557	2.847	3.167	4.060
16	54	2.905	2.148	2.179	2.318	2.744	3.316	4.026	7.139
16	55	2.461	1.998	2.021	2.130	2.458	2.886	3.392	4.714
16	56	2.335	2.231	2.223	2.250	2.333	2.425	2.526	2.921
16	57	3.207	2.533	2.555	2.684	3.066	3.599	4.341	7.384
16	58	4.427	2.819	2.855	3.057	3.850	5.537	7.655	19.280
16	59	40.486	4.469	5.036	8.270	20.529	27.881	37.401	146.657
16	60	98.934	8.271	11.087	19.314	46.856	170.426	309.613	852.989
16	61	9999.992	23.155	396.688	439.016	666.746	1884.649	7210.687	9999.992
16	62	8.716	3.454	3.628	4.325	6.338	8.661	13.084	30.550
16	63	3.282	2.420	2.456	2.613	3.142	3.930	4.894	9.029
16	64	1.872	1.828	1.821	1.841	1.885	1.921	1.951	2.029
16	65	4.450	4.810	4.729	4.554	4.328	4.216	4.147	4.019

LISTING OF TABLE 18 (THE FIRST TWO LINES ARE COLUMN GUIDES)

```
0         1         2         3         4         5         6         7
1234567890123456789012345678901234567890123456789012345678901234567890123456
```

18	1	3.138	2.512	2.546	2.702	3.098	3.563	4.104	6.253
18	2	1.883	1.805	1.812	1.852	1.913	1.947	1.970	2.017
18	3	1.683	1.628	1.633	1.658	1.701	1.730	1.751	1.802
18	4	1.605	1.554	1.559	1.581	1.620	1.648	1.669	1.721
18	5	1.591	1.550	1.550	1.569	1.605	1.630	1.649	1.696
18	6	1.817	1.767	1.763	1.789	1.834	1.866	1.891	1.956
18	7	1.627	1.587	1.587	1.607	1.641	1.664	1.680	1.719
18	8	1.603	1.553	1.556	1.578	1.617	1.645	1.669	1.730
18	9	1.616	1.558	1.561	1.586	1.635	1.669	1.696	1.760
18	10	1.696	1.626	1.633	1.663	1.717	1.754	1.783	1.854
18	11	1.652	1.582	1.591	1.620	1.673	1.708	1.736	1.803
18	12	1.651	1.592	1.592	1.618	1.670	1.707	1.736	1.808
18	13	1.713	1.653	1.651	1.680	1.734	1.772	1.800	1.867
18	14	1.912	1.823	1.833	1.878	1.946	1.982	2.005	2.051
18	15	1.791	1.718	1.726	1.761	1.817	1.849	1.872	1.922

18	16	1.713	1.649	1.656	1.686	1.735	1.765	1.787	1.837
18	17	1.625	1.567	1.574	1.598	1.642	1.672	1.695	1.750
18	18	1.594	1.540	1.545	1.568	1.610	1.638	1.661	1.716
18	19	1.600	1.561	1.560	1.578	1.613	1.638	1.657	1.703
18	20	2.024	1.648	1.655	1.686	1.749	1.812	1.867	2.486
18	21	1.825	1.747	1.753	1.792	1.854	1.891	1.917	1.969
18	22	1.698	1.633	1.641	1.669	1.720	1.752	1.776	1.829
18	23	1.606	1.551	1.556	1.579	1.623	1.653	1.677	1.736
18	24	1.588	1.547	1.547	1.566	1.601	1.626	1.645	1.693
18	25	1.791	1.715	1.723	1.760	1.819	1.852	1.874	1.917
18	26	1.673	1.608	1.616	1.644	1.694	1.727	1.751	1.805
18	27	1.680	1.614	1.622	1.650	1.701	1.735	1.759	1.814
18	28	1.656	1.588	1.597	1.628	1.679	1.711	1.734	1.784
18	29	1.671	1.611	1.619	1.645	1.691	1.720	1.741	1.784
18	30	1.663	1.607	1.614	1.639	1.682	1.710	1.729	1.772
18	31	1.624	1.570	1.577	1.601	1.642	1.669	1.688	1.731
18	32	1.583	1.530	1.537	1.561	1.601	1.626	1.644	1.686
18	33	1.585	1.541	1.543	1.564	1.601	1.624	1.641	1.682
18	34	1.657	1.602	1.607	1.633	1.677	1.704	1.722	1.759
18	35	1.643	1.591	1.596	1.619	1.660	1.686	1.705	1.745
18	36	1.650	1.611	1.606	1.628	1.665	1.690	1.710	1.758
18	37	1.715	1.662	1.664	1.692	1.735	1.760	1.778	1.817
18	38	1.630	1.578	1.583	1.605	1.645	1.673	1.696	1.755
18	39	1.846	1.827	1.807	1.822	1.857	1.885	1.908	1.964
18	40	1.804	1.711	1.724	1.766	1.836	1.877	1.906	1.966
18	41	1.924	1.866	1.865	1.897	1.945	1.975	1.998	2.057
18	42	1.644	1.591	1.594	1.619	1.662	1.690	1.712	1.765
18	43	2.118	1.943	1.970	2.038	2.171	2.269	2.343	2.484
18	44	1.737	1.669	1.676	1.707	1.761	1.795	1.819	1.867
18	45	1.701	1.645	1.650	1.674	1.719	1.749	1.772	1.822
18	46	1.680	1.637	1.637	1.658	1.696	1.720	1.738	1.774
18	47	1.675	1.629	1.631	1.653	1.693	1.718	1.736	1.773
18	48	1.753	1.688	1.692	1.722	1.774	1.808	1.834	1.896
18	49	1.716	1.649	1.656	1.684	1.738	1.775	1.802	1.859
18	50	1.822	1.755	1.755	1.786	1.845	1.887	1.919	1.988
18	51	1.670	1.621	1.618	1.643	1.689	1.718	1.739	1.785
18	52	1.921	1.871	1.874	1.904	1.944	1.961	1.970	1.981
18	53	1.772	1.725	1.721	1.745	1.791	1.822	1.843	1.886
18	54	1.889	1.777	1.792	1.835	1.917	1.978	2.029	2.170
18	55	1.859	1.781	1.786	1.817	1.877	1.925	1.968	2.091
18	56	1.745	1.689	1.686	1.718	1.768	1.798	1.819	1.869
18	57	1.658	1.604	1.610	1.634	1.676	1.702	1.722	1.766
18	58	1.750	1.685	1.690	1.718	1.770	1.806	1.835	1.902
18	59	2.026	1.893	1.908	1.969	2.074	2.138	2.181	2.269
18	60	2.210	2.052	2.064	2.135	2.260	2.348	2.416	2.570
18	61	2.622	2.320	2.336	2.444	2.670	2.866	3.054	3.695
18	62	1.734	1.673	1.678	1.707	1.754	1.784	1.806	1.855
18	63	1.632	1.580	1.585	1.607	1.647	1.676	1.699	1.753
18	64	1.616	1.587	1.575	1.595	1.630	1.654	1.672	1.718
18	65	4.307	4.328	4.214	4.227	4.358	4.432	4.478	4.572

REFERENCES

Andrews, D.F., P.J. Bickel, F.R. Hampel, P.J. Huber, W.H. Rogers and J.W. Tukey (1972). Robust Estimates of Location: Survey and Advances, Princeton, New Jersey, Princeton University Press.

Bahadur, R.R. (1960). On the asymptotic efficiency of tests and estimates. Sankhya 22, 229-252.

Tukey, J.W. (1949). One degree of freedom for non-additivity. Biometrics, 5, No. 3.

DISCUSSION OF THE TUKEY CHAPTER

J. Hartigan

It seems to me that Dr. Tukey's analysis was a sophisticated, second-stage one, and it might be appropriate to try a naive first-look analysis. The differences between various methods of measuring error seem less interesting than the differences between estimators, differences between situations, and interactions between classes of similar estimators and classes of similar situations.

A two-way clustering on the data was made, using the 65 estimators, the 18 situations for a "middle" error estimate, and the 10% error estimate. (This is proportional to the difference between the 95% point and 5% point of the sample, after some smoothing.) A quick look at the data shows entirely different scales of error for different situations (or parent populations), a small number of missing values, and quite a large number of very large values. These large values dominate the first clustering, since the algorithm spends all its time isolating them in clusters of size 1 (see table 1).

```
TABLE 1.  TWO-WAY CLUSTERING OF DATA USING 65 ESTIMATORS, 18 SITUATIONS
AND THE 10% ERROR ESTIMATE


KEY

** =   VALUES GREATER THAN 99
-  =   VALUES MISSING

COLUMNS
   1 = N(0,1)
   2 = CAUCHY
   3 = 75% OF N(0,1) AND 25% OF N(0,1)/U(0,1)
   4 = 19 FROM N(0,1) AND 1 FROM N(0,9)
   5 = 18 FROM N(0,1) AND 2 FROM N(0,9)
   6 = 17 FROM N(0,1) AND 3 FROM N(0,9)
   7 = 15 FROM N(0,1) AND 5 FROM N(0,9)
   8 = 10 FROM N(0,1) AND 10 FROM N(0,9)
   9 = 5 FROM N(0,1) AND 15 FROM N(0,9)
  10 = 19 FROM N(0,1) AND 1 FROM N(0,100)
  11 = 18 FROM N(0,1) AND 2 FROM N(0,100)
  12 = 15 FROM N(0,1) AND 5 FROM N(0,100)
  13 = 90% OF N(0,1) AND 10% OF N(0,1)/U(0,1)
  14 = N(0,1)/U(0,1)
  15 = 90% OF N(0,1) AND 10% OF N(0,1)/U(0,1/3)
  16 = 75% OF N(0,1) AND 25% OF N(0,1)/U(0,1/3)
  17 = DOUBLE EXPONENTIAL
  18 = STUDENT'S T WITH 3 DEGREES OF FREEDOM


                                       00000000000000000000000000000000000000000000000000000
                                       00000000000000000000000000000000000000000 00000000000
                                       00000000000000000000000000 00000000000000 00000000000
                                       00000000000000000000000000 00000 00000000 00000000000
                                       00000000000000000000000000 00000 00000000 00000000 00
                                       00000000000000 00000000000 00000 00000000 00000000 00
                                       00000000000000 00000000000 00000 00000000 00000 00 00
                                       00000000 00000 00000000000 00000 00000000 00000 00 00
                                       00000000 00000 00000000000 00 00 00000000 00000 00 00
                                       00000000 00000 00000000000 00 00 00000 00 00000 00 00
                                       00000000 00000 00000000000 00 00 00000 00 00 00 00 00
                                       00000000 00000 00000000 00 00 00 00000 00 00 00 00 00
                                       00000000 00000 00000000 00 00 00 00 00 00 00 00 00 00
                                       00 00000 00000 00000000 00 00 00 00 00 00 00 00 00 00
                                       00 00000 00000 00 00000 00 00 00 00 00 00 00 00 00 00
                                       00 00000 00 00 00 00000 00 00 00 00 00 00 00 00 00 00
                                      CO 9  CO 7  CO 8  CO 5  CO 1  CO10  CO11  CO14  CO16
                                         CO18  CO17  CO 4  CO 6  CO13  CC 3  CO15  CO 2  CO12
                                        1  2  3  4  5  6  7  8  9 10 11 12 13 14 15 16 17 18
                                      ======================================================
M (ADA,4.5,8.0)*AC          34         1H 4H 2  3H 7  6H 1  2  2H 4H 1  1  1  2  1H 6 10  2 5H
                                       H  H     H     H        H  H              H           H
M (2.2,3.7,5.9)*AC          30         2H 7H 2  2H10 12H 1  C  0H 4H 1  0  0  0  0H11 19  3 3H
                                       H  H     H     H        H  H              H           H
MX(3T,2)-SKIP MEAN          45         3H10H 3  3H13 17H 1  1  1H 5H 3  1  0  1  0H10 15  2 4H
                                       H  H     H     H        H  H              H           H
MN(5T,.6)-SKIP MEAN         47         4H 9H 2  2H11 14H 1  1  1H 5H 4  2  0  3  1H 6  9  1 2H
                                       H  H     H     H        H  H              H           H
MX(5T,2)-SKIP MEAN          46         5H 9H 3  3H12 15H 1  1  1H 5H 4  2  1  3  1H 7 10  4 9H
                                       H  H     H     H        H  H              H           H
M (1.7,3.4,8.5)*AC          32         6H 4H 0  0H 6  5H 1  C  0H 6H 2  1  0  0  0H 7 13  1 2H
                                       H  H     H     H        H  H              H           H
JOHNS ADAPTIVE              51         7H 6H 2  3H 6  5H 4  4  3H 6H 4  8  3  8  4H 3  8  3 2H
                                       H  H     H     H        H  H              H           H
ADAPT TRIM LINCOM           13         8H 7H 3  3H 8  6H 4  5  4H 7H 5  8  4  8  4H 5  7  2 4H
                                       H  H     H     H        H  H              H           H
ITER C-SKIP TRIMEAN         42         9H 5H 1  2H 7  8H 4  2  2H 8H 6  5  2  5  3H 7 13  2 3H
                                       H  H     H     H        H  H              H           H
M (1.2,3.5,8.0)*AC          33        10H 2H 0  0H 3  1H 4  3  1H 9H 4  4  1  2  1H 3  7  0 1H
                                       H  H     H     H        H  H              H           H
M (2.0,2.0,5.5)*AC          28        11H 9H 2  0H 8  7H 5  3  1H11H16  4  1  2 12H 6 12  7 0H
```

TABLE 1, CONTINUEC

```
                                 H  H     H     H        H  H              H           H
CAUCHY PITMAN (LOC)       64   12H 4H 1  1H 1  0H 6  5  3H12H 7  7  4  6  4H 0  0  0  3H
                                 H  H     H     H        H  H              H           H
MAXIMUM EST LIKE          50   13H 3H 6  9H 4  7H 8 12  8H14H11 10 10 12 11H 6  9  7 11H
                                 H  H     H     H        H==H              H           H
''OLSHEN'S''              36   14H 0H 2  3H C  0H 9  7  4H16H10 10  5 10  7H 1  5  5  8H
                                 H  H     H     H        H  H              H           H
T-SKIP ''OLSHEN'S''       37   15H 3H 4  4H 2  4H10  9  6H18H14 12  7 10  8H 2  2  3  6H
                                 H  H     H     H        H  H              H           H
ITER S-SKIP TRIMEAN       41   16H16H10  8H13 10H12 1C 1CH20H18 14 10 13 10H 7 11  4  8H
                                 H  H     H     H        H  H              H           H
CAUCHY MAX LIKELIHOOD     39   17H 3H 8  9H 2  1H19 17 13H31H21 2C 13 17 14H C  0  5  7H
                                 H  H     H=====H        H==H              H===========H
HUBER PROP 2, K=0.7       19   18H 0H 0  0H 2  CH 4  3  2H 9H 5  6  3  8  6H 6 13  9 14H
                                 H  H     H     H        H  H              H           H
MIDMEAN                    5   19H 0H 0  0H 2  CH 4  3  2H 9H 5  6  3  8  5H 7 14  9 14H
                                 H  H     H     H        H  H              H           H
GASTWIRTH                  7   20H 1H 1  1H 3  0H 5  5  2H11H 7  8  4 10  7H 6 14  9 15H
                                 H  H     H     H        H  H              H           H
1ST-HU 0.7*IQS MEC        24   21H 1H 0  0H 3  1H 4  7  1H 8H 4  5  3  8  5H 8 17 10 15H
                                 H  H     H     H        H  H              H           H
MEDIAN                     6   22H 2H 7  9H 2  3H14 13 10H22H17 17 12 20 16H 6 11 14 22H
                                 H  H     H     H        H  H              H           H
HUBER PROP 2, K=1.0       18   23H 2H 0  0H 6  4H 2  1  0H 5H 3  4  2  7  4H13 23 13 19H
                                 H  H     H     H        H  H              H           H
TRIMEAN                    8   24H 3H 0  1H 5  5H 3  2  1H 7H 0  6  3  9  6H16 27 14 21H
                                 H  H     H     H        H  H              H           H
2-CHOICE ADAPT TRIM       12   25H 4H 1  1H 6  5H 2  2  2H 5H 3  5  3  8  5H 9 17 10 15H
                                 H  H     H     H        H  H              H           H
JAECKEL ADAPT TRIM         9   26H 4H 0  0H 7  5H 2  2  1H 5H 3  4  3  8  4H10 19 10 16H
                                 H  H     H     H        H  H              H           H
1ST-HU 1.0*IQS MEC        23   27H 3H 0  0H 7  5H 2  1  0H 5H 2  4  3  7  4H15 25 13 20H
                                 H  H     H     H        H  H              H           H
ADAPT FROM SKIPPING       56   28H 6H 4  7H 6 10H 6  6  6H 7H11 15  9 16 10H 7 10 10 14H
                                 H  H     H     H        H  H              H           H
M (2.1,4.0,8.2)*AC        31   29H 5H 1  1H 9  9H 0  C  0H 3H 0  0  0  0  0H11 19  4  5H
                                 H  H     H     H        H  H              H           H
M-TYPE (SINE)             35   30H 5H 1  2H10 11H 0  C  1H 3H 0  0  0  0  0H13 21  5  5H
                                 H  H     H     H        H  H              H           H
TAKEUCHI ADAPTIVE         49   31H 6H 3  5H10 12H 1  2  3H 2H 1  1  1  2  2H12 19  9 19H
                                 H  H     H     H        H  H              H           H
C&T&S-SKIP TRIMEAN        48   32H13H 5  5H11 11H11  7  5H18H11 10  7 10 12H 6 10 32  6H
                                 H  H     H     H        H  H              H           H
M (2.5,4.5,9.5)*AC        29   33H 6H 2  4H11 12H 0  C  1H 2H 0  0  1  0  0H16 26  9 10H
                                 H  H     H     H        H  H              H           H
HOGG 69 ON KURTOSIS       53   34H 7H 5  7H12 14H 2  5  5H 3H 4  7  4 10  6H10 18 12 16H
                                 H  H     H=====H        H  H              H===========H
3T-SKIP MEAN              44   35H10H 4  4H14H19H 1  2  2H 5H C  2  1  1  0H11H15H 0H 4H
                                 H  H     H  H  H        H  H              H  H  H  H  H
T-SKIP HOGG 69            55   36H14H 7  7H18H23H 3  4  5H 6H 4  3  4  2  2H18H26H 7H 7H
                                 H  H     H  H  H        H  H              H  H  H  H  H
T-SKIP HOGG 67            54   37H16H 8  9H21H25H 4  4  5H 7H 4  3  5  3  3H23H34H12H11H
                                 H  H     H  H  H        H  H              H  H  H  H  H
MX(3T,2)-SKIP MEXTR       43   38H15H14 17H20H27H11 15 12H14H 0  9  9  6  0H20H23H 0H 9H
                                 H  H     H=====H===========H==============H===========H
HOGG 67 ON KURTOSIS       52   39H11H10 15H19 23H 5  9 11  2H 8 14H10 20 13H16H26H19H35H
                                 H  H     H     H           H     H        H  H  H  H  H
HODGES-LEHMANN            57   40H 2H 2  2H 7  7H 0  1  2  3H 2  4H 5 10  7H19H30H20H29H
                                 H  H     H     H           H     H        H  H  H  H  H
BICKEL ADAPT TRIM         10   41H 2H 3  3H 6  6H 2  3  2  4H 5 14H 7 22 13H21H32H24H40H
                                 H  H     H     H           H     H        H  H  H  H  H
HUBER PROP 2, K=1.2       17   42H 4H 1  1H 8  8H 0  C  0  3H 2  3H 3  7  5H20H33H18H25H
                                 H  H     H     H           H     H        H  H  H  H  H
3-FOLDED 2-TRIM MED       63   43H 4H 1  1H 9  8H 1  C  0  4H 2  4H 3  7  5H21H33H19H27H
                                 H  H     H     H           H     H        H  H  H  H  H
HUBER 1.5*ADS             26   44H 4H 2  3H10 10H C  C  1  2H 1  4H 4  9  7H22H35H22H31H
                                 H  H     H     H           H     H        H  H  H  H  H
1S-HU 1.5*ADS MED         27   45H 5H 2  3H10 10H 0  C  1  2H 1  4H 4  9  7H22H36H22H31H
                                 H  H     H     H           H     H        H  H  H  H  H
```

TABLE 1, CONTINUEC

```
15% SYM TRIM MEAN         4 46H 3H 0  0H 8  7H 1  C  0  4H 2  3H 3  7  5H23H36H21H33H
                              H  H     H     H           H     H        H  H  H  H  H
1ST-HU 1.5*IQS MEC       22 47H 5H 3  4H11 11H C  C  1  2H 2  4H 5 10  7H25H39H24H34H
                              H  H     H     H           H     H        H  H  H  H  H
LEAST FAVORABLE          40 48H 7H 6  7H14 17H C  2  4  1H 2  6H 7 12  8H24H40H22H28H
                              H  H     H     H           H     H        H==H==H==H==H
MEAN LIKELIHOOD          38 49H 6H 1  0H 4  2H11  5  2 23H11  3H11  2 13H33H51H48H27H
                              H  H     H     H           H     H        H  H  H=====H
BICKEL-HODGES            58 50H 6H 4  5H13 12H 1  3  4  2H 3  7H 8 13 10H30H46H28H39H
                              H  H     H     H           H     H        H  H  H=====H
HUBER PROP 2, K=1.5      16 51H 6H 3  4H13 14H C  C  1  1H 1  3H 5 10  8H34H50H32H46H
                              H  H     H     H           H     H        H  H  H  H  H
HUBER 2.0*ADS            25 52H 5H 6  8H14 15H C  2  4  0H 2  6H 9 16 11H33H49H35H49H
                              H  H     H     H           H     H        H  H  H  H  H
1ST-HU 2.0*IQS MEC       21 53H 6H 7  9H15 16H 1  3  5  CH 3  7H10 17 12H36H52H38H53H
                              H  H     H=====H           H==============H===========H
SYM JAECKEL AD TRIM      11 54H -H -  -H -  -H -  -  -  -H -  -H -  -  -H -  -  -  -H
                              H=========================================H===========H
SHORTH                   65 55H49H64 68H47 40H94H86 78H**H** **H77 89 87H37 30H57H56H
                              H=========================================H===========H
HUBER PROP 2, K=1.7      15 56H 7  6  7H15 17H 0  1  3  1H 2  4H 9 13 11H44 62 46H71H
                              H        H     H           H     H        H        H==H
10% SYM TRIM MEAN         3 57H 5  3  3H12 13H 0  C  1  2H 2  3H 6  8 10H42 60 50H92H
                              H        H     H           H     H        H        H  H
3-FOLDED 1-TRIM MED      62 58H 6  4  5H13 14H C  C  2  1H 2  4H 8 11 12H44 63 53H96H
                              H        H     H           H     H========H============
HUBER PROP 2, K=2.0      14 59H 7  9 13H19 21H C  3  7  0H 3  6H15 20 18H60 84 75H**H
                              H        H     H           H     H========H===========H
2 TIMES FOLDED MED       59 60H 8 12 16H21 24H 1  5  9  0H 5  9H17 23 22H78 ** ** **H
                              H        H     H           H     H========H           H
5% SYM TRIM MEAN          2 61H 7  8 12H18 20H C  2  6  1H 4  4H23 47 30H98 ** ** **H
                              H        H     H           H=====H========H===========H
3 TIMES FOLDED MEC       60 62H 8 16 23H24 25H 2 1C 15  0H10 12H38H**H48H** ** **H**H
                              H        H     H           H     H=====H==H        H==H
1ST-HU 1.5*IQS MEAN      20 63H 5  3  3H11 11H 0  C  1  2H 6 10H89H10H97H** ** **H38H
                              H=========================================H===========H
4 TIMES FOLDED MEC       61 64H 8H25 26H26 26H 7H14 20H 0H** 83 ** ** **H** ** ** **H
                              H  H     H     H  H     H  H              H           H
MEAN (AVERAGE)            1 65H 8H31 30H29 27H10H2C 25H 0H** ** ** ** **H** ** ** **H
                              H=====================================================H
```

The data were transformed as follows to bring the variables to the same scale and remove the effect of outliers:

- Each observation was divided by the minimum observation in its column. (Thus, each estimate for a given situation is measured by its error compared to the smallest observed error for that situation.)

- The ratios were square-rooted, because standard deviations are less skew in distribution.

- Any ratio exceeding 2 was reduced to 2 to suppress such bad estimates in the clustering.

- The final ratios were transformed to 100 (ratio -1) so that the numbers in the table represent percent excess over the minimum error.

The table can't be understood without reference to the row names and column names. Even these do not completely describe the estimators, although the situations (or populations) can be guessed at. Look at the top of the table for the good estimators, and at the bottom for the bad ones, like the mean. The densely hatched area at the bottom right-hand corner shows the program splitting off the estimators and situations that are really bad (over twice the standard deviation of the best estimator).

The basic model is that values within blocks are equal, with blocks outlined by H's and ='s. Look at big blocks to find interactions between sets of similar estimators and situations.

<u>H. L. Lucas</u>

Preparation of this manuscript was difficult. Dr. Tukey's paper presents analyses of data about data analysis; it was only after much wheel-spinning that I distinguished between questions his data per se are about and questions associated with his analysis of data. My confusion was compounded by the fact that he invoked (justifiably) an extended version of one of the procedures by which his data were gathered for comparison in order to perform an analysis to make the comparison.

At first I floundered among many questions and thoughts about the nature of the different estimators studied in the Princeton project, what might be expected of them under different conditions, when some might be better than others, and so on. Early drafts of this discussion revolved mainly around such matters. That thrust was abandoned, however, when I came across "Robust Estimates of Location" (Andrews et al, Princeton University Press, 1972) cited by Dr. Tukey. Practically everything that had occurred to me was well covered in that book.

Upon making a new start, I apparently took a broader view. Phraseologies like "robust attitudes about" and "robust approaches to" data analysis came to mind. These had strong appeal and my discussion is structured in that vein.

Robust approaches to data analysis involve a number of matters, the basic ones being

- obtaining good estimates of items of interest, and
- obtaining good indices of dispersion of those estimates.

I shall say a number of things about those two points, but feel it very important also to consider

- choosing good model forms.

It is convenient and, I think, sufficient for me to employ the framework of the regression model, $\underline{y} = \underline{\phi}(\underset{\sim}{x}) + \underline{\epsilon}$, where $\underline{y}$ (a column vector) and $\underset{\sim}{x}$ (a matrix) are the data at hand, $\underline{\epsilon}$ is the "error" part of $\underline{y}$, and $\underline{\phi}(\underset{\sim}{x})$ is a measure ($\underline{x}$-dependent) of location for $\underline{y}$. Under such formulation, the measure of location for $\underline{\epsilon}$ can always be set at $\underline{0}$.

I note in passing that Andrews et al chose this form, with $\underline{\phi}(\underset{\sim}{x}) \equiv \underline{0}$, in studying robust estimates. Dr. Tukey also chose this form for his paper. Dr. Tukey let $\underline{y}$ be the data on indices of dispersion of the robust estimates, and for $\underline{\phi}(\underset{\sim}{x})$, usually representing the median, he chose forms like those commonly used in the analysis of variance for cross classified data with and without regression features.

To ease presentation, I adopt some conventions:

- "Model" or "model form" refers to $\underline{\phi}(\underset{\sim}{x})$.
- "Error structure" refers to the distribution of $\underline{\epsilon}$ and will be symbolized by S_ϵ.
- Because, in general, interest lies in estimating (using the data to decide on) a form for $\underline{\phi}(\underset{\sim}{x})$ and/or estimating the parameters of $\underline{\phi}$, estimators will often by symbolized as $\hat{\underline{\phi}}$. If $\underline{\phi}(\underset{\sim}{x})$ is restricted to the form $\underset{\sim}{x}\,\underline{\theta}$, then $\hat{\underline{\theta}}$ will be used.
- Indices of dispersion for $\hat{\underline{\phi}}$ or $\hat{\underline{\theta}}$ will be designated as $\underset{\sim}{D}(.)$ and their estimators as $\hat{\underset{\sim}{D}}(.)$. If variances and covariances are adequate indices of dispersion, then $\underset{\sim}{V}$ will be used in place of $\underset{\sim}{D}$.

In Andrews et al, in Dr. Tukey's paper, and in the literature generally, robustness has to do with

- How good, over a class of S_ϵ, is a $\hat{\underline{\phi}}$ that is "best" for a subclass of S_ϵ (often with one member)? and/or
- How good is a $\hat{\underset{\sim}{D}}(\hat{\underline{\phi}})$, often in the context of testing hypotheses or making confidence statements about $\underline{\phi}$?

Although I am somewhat concerned about the first question, I am <u>very</u> concerned about the second. These days, automated procedures are leading to the accumulation of large masses of data with complicated S_ϵ, and high-speed computing is leading to volumes of statistical analyses carried out with "recipes" thought to be quite general by many people but which actually hold only for simple S_ϵ.

The Princeton project and its direct forebearers attend particularly to how good different $\hat{\phi}$'s are for one type of S_ϵ, i.e., mainly symmetric independently distributed cases with longer tails than the Gaussian (the outlier situations). Other work (e.g., by Box and coworkers, by Cochran and some others of the Fisher school, by Dr. Tukey and other time-series analyzers) has focused on situations in which S_ϵ takes a pattern which is qualitatively but not quantitatively identifiable. Perhaps unjustifiably, I am less concerned about the outlier matter than I am about the patterned S_ϵ, the latter being a prime reason that $\hat{\underset{\sim}{D}}(\hat{\underline{\phi}})$ is not good in many prevalent data analyses.

The reason I am not very worried about outliers is that the data I have met through the years are mostly well behaved. Some type of screening of the data per se or of first-fit residuals (often conveniently done by examining ranges, variance or something more detailed after dividing the data into subsets based on $\underset{\sim}{x}$ or on S_ϵ pattern) is always needed to trace out and correct misrecorded or mispunched data (which may be very numerous). Many apparent outliers fall into line if available ancillary information is structured into $\underline{\phi}(\underset{\sim}{x})$. The rare remaining data that are almost certain to be outliers I simply reject with a note. Perhaps I am really using a kind of robust estimator, but it is too informal to be graced with the name.

At any rate I have relied mainly on least squares estimators, introducing (when it seemed worthwhile) various features to handle asymmetry and/or heterogeneity and dependence patterns in S_ϵ. Occasionally, I have used estimators robust against outliers when $\underline{\phi}(\underset{\sim}{x})$ was very simple.

Failure to obtain good $\hat{\underset{\sim}{D}}(\hat{\phi})$ can come about for several reasons. If the form of $\underline{\phi}(\underset{\sim}{x})$ is $\underset{\sim}{x}\,\underline{\theta}$ and is fixed, so only its parameters, $\underline{\theta}$, are to be estimated, poor $\hat{\underset{\sim}{D}}(\hat{\underline{\theta}})$ can result from failing to make reasonable assumptions about S_ϵ. If the form of $\underline{\phi}(\underset{\sim}{x})$ is fixed but is not $\underset{\sim}{x}\,\underline{\theta}$, or if the form of $\underline{\phi}$ as well as its parameters are to be estimated, poor $\hat{\underset{\sim}{D}}(\hat{\phi})$ can result not only from bad assumptions about S_ϵ, but also by failing to realize that the distribution of $\hat{\phi}$ is in general complicated even with simple, well behaved S_ϵ.

It is possible to consider only a few selected examples of error structure and forms of $\underline{\phi}$. I hope they provide some view of the more general picture. For these examples, it seems sufficient to assume that S_ϵ is well behaved in the sense that it has all moments finite and is of a form adequately described by $\mathcal{E}(\underline{\epsilon}) = \underline{0}$ and moments no higher than second.

To address one matter, let us substitute $\sum_{i=1}^{p} \underset{\sim}{z}_i \underline{\delta}_i$ for $\underline{\epsilon}$ in the previous regression model. At the moment it will be useful also to let $\underline{\phi}(\underset{\sim}{x}) = \underset{\sim}{x}\underline{\theta}$ with all elements of $\underline{\theta}$ to be estimated.

That is, write the regression model as $\underline{y} = \underset{\sim}{x}\,\underline{\theta} + \sum_{i=1}^{p} \underset{\sim}{z}_i \underline{\delta}_i$, where $\mathcal{E}(\underline{\delta}_i) = \underline{0}_i$, $\mathcal{E}(\underline{\delta}_i \underline{\delta}_j') = \underset{\sim}{0}_{ij}$, and $\mathcal{E}(\underline{\delta}_i \underline{\delta}_i') = \underset{\sim}{I}_i \rho_i \sigma^2$, with $\rho_p = 1$. Thus, $\mathcal{E}(\underline{\epsilon}) = \underline{0}$ and $\underset{\sim}{V}(\underline{\epsilon}) = (\sum_{i=1}^{p} \underset{\sim}{z}_i \underset{\sim}{z}_i' \rho_i)\sigma^2$.

This is the "nested" or "components of variance" error structure and can serve well for many varieties of complicated data. In fact, it can serve as a way to view time series with non-stationary mean but stationary variance-covariance; also, negative intraclass correlation cases are encompassed by letting appropriate ρ_i be negative. The $\underset{\sim}{z}_i$ in this formulation depend on the randomization scheme employed or reasonably assumable.

Unfortunately, many kinds of data well suited by the foregoing model form and error structure are being analyzed routinely under the assumption $\underset{\sim}{V}(\underline{\epsilon}) = \underset{\sim}{I}\sigma^2$. There are cases in which the $\hat{\underline{\theta}}$ obtained is not bad. In fact, for certain neat data patterns (e.g., nested designs with equal numbers and split-plot designs, both with no covariables), $\hat{\underline{\theta}}$ is the BLUE. In all cases, however, if $\hat{\underset{\sim}{V}}(\hat{\underline{\theta}})$ is computed under $\underset{\sim}{I}\sigma^2$, the result is not good. The ρ_i enter different elements of $\underset{\sim}{V}(\hat{\underline{\theta}})$ in different proportions, but in $\hat{\underset{\sim}{V}}(\hat{\underline{\theta}})$ obtained under $\underset{\sim}{I}\sigma^2$ the ρ_i enter all elements in the same proportion.

It is not rare to see bad $\hat{\underset{\sim}{V}}(\hat{\underline{\theta}})$ computed in the neat cases (they are even found in some textbooks). Bad $\hat{\underset{\sim}{V}}(\hat{\underline{\theta}})$ is shockingly wide-spread, however, in the nonneat cases, shocking because statisticians too often are involved in the analyses. One reason, no doubt, is that better ways of analysing the data are not widely enough appreciated. Another reason, perhaps, is that better analyses are not easy to accomplish with available programs and statistical analysis systems.

Much work needs to be done on ways to analyze data of this sort and to develop more convenient computer programs. At the same time, approaches are available that can be handled with current programs by applying some ingenuity.

One approach begins in the way incomplete block designs (they have the nested error structure) are usually analyzed. Here the model is $\underset{\sim}{y} = \underset{\sim}{x}\underline{\theta} + \underset{\sim}{z}_1\underline{\delta}_1 + \underset{\sim}{z}_2\underline{\delta}_2$, with $\underline{\delta}_1$ the block effects, $\underline{\delta}_2$ the plot effects, $\underset{\sim}{z}_2 = \underset{\sim}{I}$ and $\underset{\sim}{V}(\underline{\varepsilon}) = (\underset{\sim}{z}_1\underset{\sim}{z}_1'\rho + I)\sigma^2$. Although algorithms may disguise it, the analysis is conceptually a two-step process; i.e., two estimates of $\underline{\theta}$ and $\underset{\sim}{V}(\underline{\theta})$ are obtained, namely, $\hat{\underline{\theta}}_1$ and $\hat{\underset{\sim}{V}}(\hat{\underline{\theta}}_1)$ by analyzing the individual plots as though $\underline{\delta}_1$ were not an error component, and $\hat{\underline{\theta}}^*$ and $\hat{\underset{\sim}{V}}(\hat{\underline{\theta}}^*)$ by analyzing the block means. A single estimate, $\hat{\underline{\theta}}_2$, is then computed as a mean of $\hat{\underline{\theta}}^*$ and $\hat{\underline{\theta}}_1$, with weights inverse to their estimated variances, and $\hat{\underset{\sim}{V}}(\hat{\underline{\theta}}_2)$ is computed as the reciprocal of the sum of reciprocals of their estimated variances.

Although neither $\hat{\underline{\theta}}_2$ nor $\hat{\underset{\sim}{V}}(\hat{\underline{\theta}}_2)$ is unbiased, and their distributions are such that the tests, etc., are not exact, this approach is generally a good one. The estimate, $\hat{\underline{\theta}}_1$, or at least inference made using it, can, however, sometimes be better. This is because $\underset{\sim}{V}(\hat{\underline{\theta}}^*)$ contains ρ, but $\underset{\sim}{V}(\underline{\theta}_1)$ does not. Hence if ρ is large and/or the number of degrees of freedom carried by $\hat{\underset{\sim}{V}}(\hat{\underline{\theta}}^*)$ is small, $\underline{\theta}_2$ could be noisier (bias plus variance) than $\underline{\theta}_1$ which is unbiased. Even if not, inferences could be noisier, because $\hat{\underset{\sim}{V}}(\hat{\underline{\theta}}_2)$ is noisier (bias plus variance) than $\hat{\underset{\sim}{V}}(\hat{\underline{\theta}}_1)$ which is unbiased.

Textbook incomplete block designs ordinarily have neatness features, which most practical data with nested error patterns do not share. One is that all elements of $\underline{\theta}$ are estimable at both the block and plot levels. Interestingly, for split plots (a special case of incomplete blocks) some elements are estimable only at one level and others only at the other level. The other feature is equal numbers in the nests. Nevertheless, good analyses of data with unequal numbers can be accomplished by striking means of some sort at the different nesting levels and, at each level, analyzing the means under the assumption $\underset{\sim}{I}\sigma^2$. The means at a given level do not, of course, have uniform variance, but variance heterogeneity in general is a much less worrisome matter than the correlation question we are focusing on here. I shall have a little more to say about that later.

It seems useful by way of illustration to outline the analysis performed on a set of data brought to me a few years ago. It required some manipulations not usually employed for incomplete blocks, which I will introduce by considering alternate ways of analyzing incomplete blocks.

An unbiased estimate, $\hat{\underline{\theta}}_3$, could be obtained straight out under the assumption $\underset{\sim}{I}\sigma^2$. Then by using the block contrasts and the within-block contrasts, both cleared of $\underline{\theta}$, unbiased estimates, $\hat{\rho}$ and $\hat{\sigma}^2$, could be obtained and linearly combined (a little algebra would show how) to obtain an unbiased, though somewhat shaky $\hat{\underset{\sim}{V}}(\hat{\underline{\theta}}_3)$. Another version would be to iterate; i.e., having obtained $\hat{\rho}$, compute $(\underline{z}\,\underline{z}'\hat{\rho} + \underset{\sim}{I})$ and using its inverse do weighted least squares to obtain $\hat{\underline{\theta}}_4$ and $\hat{\underset{\sim}{V}}(\hat{\underline{\theta}}_4)$. Still another approach would be to use past experience about ρ (call it ρ^*), compute $(\underset{\sim}{z}\,\underset{\sim}{z}'\rho^* + \underset{\sim}{I})$, and, using that, do weighted least squares to obtain $\hat{\underline{\theta}}_5$. An estimate of $\hat{\underset{\sim}{V}}(\hat{\underline{\theta}}_5)$ could then be obtained by combining ρ^* and information obtained about ρ from the data at hand. Perhaps iteration would again be appropriate.

It seems likely that there are fairly broad conditions for which it would be pointless to try to choose among the several estimators on grounds other than ease of getting the analysis done. The important point is that all give good $\hat{\underset{\sim}{V}}(\hat{\underline{\theta}})$ for whatever $\hat{\underline{\theta}}$ is computed.

In the data I met a few years ago, the number of plots per block (n_i) varied from 1 to 4, and the design pattern otherwise was very irregular. Some elements of $\underline{\theta}$, $\underline{\theta}_1$, were estimable only from main-plot means; some, $\underline{\theta}_2$, only from sub-plot contrasts; some, $\underline{\theta}_3$, at both levels; and some, $\underline{\theta}_4$, only if main-plot and sub-plot information were combined. A saving feature was that there were many blocks and most contained more than one plot. The model, $\underset{\sim}{y} = \underset{\sim}{x}\,\underline{\theta} + \underline{\epsilon}$ with $\mathcal{E}(\underline{\epsilon}\,\underline{\epsilon}') = (\underset{\sim}{z}\,\underset{\sim}{z}'\rho + \underset{\sim}{I})\,\sigma^2$ was employed. The following computations were done:

1. Unweighted analysis of block means to yield a residual mean square, s_b^2, with expectation, $(\rho + 1/h)\,\sigma^2$, where h is the harmonic mean of the n_i. This also yielded estimates of $\underline{\theta}_1$ and of $\underline{\theta}_3$.

2. Unweighted analysis of individual data under the assumption, block effects nonrandom, to yield s^2 with expectation, σ^2; this also yielded an estimate of $\underline{\theta}_2$ and a second estimate of $\underline{\theta}_3$.

3. Estimation of ρ as $\hat{\rho} = s_b^2/s^2 - 1/h$, and computing $\hat{W} = \underset{\sim}{x}\,\underset{\sim}{x}'\rho + \underset{\sim}{I}$.

4. Weighted analysis of the individual data, using $\hat{W}^{-1}$ and not fitting block effects, to yield $\hat{\underline{\theta}}$ and $\hat{V}(\hat{\underline{\theta}})$.

Since the members of degrees of freedom, f_b and f for s_b^2 and s^2, respectively, were large, I did not worry about using a "Satterthwaite type" adjustment. I simply used f_b when making inferences about anything involving $\underline{\theta}_1$, $\underline{\theta}_3$ and $\underline{\theta}_4$, and f if only $\underline{\theta}_2$ was involved.

I considered iteration; e.g., by fitting block effects in step (4), reestimating ρ, forming a new $\hat{W}$, etc., but this did not seem worthwhile.

Suppose there had not been the complication of $\underline{\theta}_4$. I think I would not have done step (4). I would simply have

- used $\hat{\underline{\theta}}_1$ from step (1) and estimated its variance in the standard way using s_b^2,

- used $\hat{\underline{\theta}}_2$ from step (2) with its variance estimated from s^2, and
- probably used the incomplete block approach, employing s_b^2 and s^2, to combine the two estimates of $\underline{\theta}_3$ and to estimate the variance of the combination.

Judging from what is going on these days, most people would have performed the analysis under $\underset{\sim}{I}\sigma^2$. Some might have regarded the block (animal) effects to be nonerror, fitted them as well as $\underline{\theta}$, and computed $\hat{\underset{\sim}{V}}(\hat{\underline{\theta}})$ purely from the plot-within-block mean square. This clearly would have been misleading for the elements of $\underline{\theta}_1$ because $\hat{\rho}$ was about 2, and the animals used were a sample of the kind of animals of interest, not the only animals of interest. Others might have fitted only $\underline{\theta}$ and computed $\hat{V}(\hat{\underline{\theta}})$ from the resulting residual mean square. This would have underestimated $\hat{\underset{\sim}{V}}(\hat{\underline{\theta}}_1)$ ($\underline{\theta}_1$ included effects of prime interest), overestimated $\hat{\underset{\sim}{V}}(\hat{\underline{\theta}}_2)$, but might not have done too badly for $\hat{V}(\underline{\theta}_3)$ and $\hat{V}(\underline{\theta}_4)$.

Now for a few remarks about variance heterogeneity. Interpretations about $\underline{\theta}$ seem to be rather robust against variance heterogeneity which is not related to $\underline{x}$; hence, if the relation appears to be weak, I am not concerned unless the ratio of largest to smallest estimated variance at different points encompassed by $\underset{\sim}{x}$ is rather large. In the data set just discussed, that ratio was only about 4/3, and I would not have been concerned if the ratio had been, say, 5, because there appeared to be little relation to $\underline{x}$. If there appears to be a relation to $\underline{x}$, or if there is experimental or theoretical reason to believe one exists, the matter should be examined and if necessary with weights inverse to the variance.

Sometimes cumulative experience provides the basis for choosing weights. Sometimes there is a theoretical basis for choosing a smoothing relation, say $\Gamma(\underline{x})$, as an expression for variances. In such an event, one can estimate variances for different points (or limited regions) of $\underline{x}$-space from the data at hand, estimate the parameters of Γ, and then weight by $\hat{\Gamma}$. Other times one must use an empirical smoothing formula to arrive at, say $\Gamma^*(\underline{x})$. Finding $\hat{\Gamma}$ or $\hat{\Gamma}^*$ are, as Dr. Tukey puts it, "borrowing strength." (Such, of course, is also what one is doing in fitting $\underline{\phi}$). Depending on how many observations there are at a given point (or limited region) of $\underline{x}$-space, one may wish to iterate on a joint fit of $\underline{\phi}$ and

Γ (or Γ^*). If the data set is small, however, iteration, or even trying to take heterogeneity into account at all, may introduce sufficient noise (bias plus variance) into $\hat{\underline{\phi}}$ and $\hat{\underset{\sim}{V}}(\hat{\underline{\phi}})$ to overcome its value.

The situation is analogous to that found in the Princeton work. Their estimators of location can all be regarded as weighted means of the observations, with weights depending to varying extents on the observations themselves. Generally speaking, the estimators with weights strongly dependent on the observations were better than those with weights weakly dependent on the observations when sample sizes were large enough, but were poorer when the sample sizes were small enough. The relative goodness was conditioned by the nature of the background populations.

I have just been dealing with how failure to take error structure, S_ϵ, properly into account when analyzing data can lead to poor, often very poor, estimates of the variance of estimates of $\underline{\phi}(\underset{\sim}{x})$. I now turn to the other reasons noted earlier.

If the form of $\underline{\phi}(\underset{\sim}{x})$ is fixed, then finding $\hat{\underline{\phi}}$ consists simply of estimating the parameters, $\underline{\theta}$. If $\underline{\phi}(\underset{\sim}{x}) \neq \underset{\sim}{x}\,\underline{\theta}$ (i.e., if the model is nonlinear), however, the distribution of $\hat{\underline{\theta}}$ is complicated and generally asymmetric, and arriving at good indices of dispersion, $\underset{\sim}{D}(\hat{\underline{\theta}})$ is a worrisome problem. Asymptotic indices may work well for large sets of data (which sometimes means very large indeed by usual standards). In data sets ordinarily met, however, $\hat{\underline{\theta}}$ is often severely biased, and sampling studies have shown that $\hat{\underset{\sim}{V}}(\hat{\underline{\theta}})$ can be 20-30 times larger than the asymptotic value.

One approach for arriving at a $\hat{\underset{\sim}{D}}(\hat{\underline{\theta}})$ is to compute interval estimates by finding appropriate contours on the likelihood (or sum of squares) surface. Another approach is "jackknifing." Both are costly computer approaches if more than a few parameters are involved. There are possibly some clues to improved practical approaches in the Princeton robustness work, but that point is nebulous.

A similar problem arises when one decides upon one form for $\underline{\phi}(\underset{\sim}{x})$ from among a permissible class of forms. Methods of selecting predictors now widely used illustrate the matter.

Ordinarily $\underset{\sim}{\phi}(\underset{\sim}{x})$ is of the form $\underset{\sim}{x}\ \underline{\theta}$, $\underline{\theta}$ being, say, q x 1. One version is to decide which is best from among the set, $\underline{\theta}_1$, $\underline{\theta}_2$, . . . , $\underline{\theta}_m$, of all vectors represented by $\underline{\theta}$ but having only $p < q$ nonzero elements. One estimator examines all possibilities and picks the one that minimizes some function of the residual mean square. Even though the $\hat{\underline{\theta}}$ obtained is in some sense the best estimate, the confidence placed in it as being the truly best $\underline{\theta}$, or somewhere near the truly best, is clearly small, except under very favorable conditions that are seldom met in practice in my experience. Unfortunately, many people using this approach seem not to recognize that fact and place tragically unwarranted confidence in the results. To date, the only way I have seen to assess the dispersion of the estimator in practice is some version of jackknifing.

Selection of predictors would be much improved, I believe, if the programmed procedures were not used so blindly; in most cases I've seen, data as taken are simply run through the computer. It seems certain that the alternative model forms to be decided between could be reduced in number and complexity if previous knowledge about the system under study were intelligently utilized.

I now come to the choice of model forms. Again a single example will illustrate my concern.

In the substantive fields of interest to me the trend in recent years has been toward the study of response surfaces, not simply comparison of treatments. Not surprisingly, I suppose, in view of the general availability of computer programs to fit quadratic models, compute their contours, etc., such models are being widely used. Animal and plant responses in general, however, are not well described by quadratic models, except over limited domains of input factors (x-variables). Actual responses can have horizontally asymptoting characteristics on the right, or on both left and right, or they can have long gently sloping or slowly curving regions in between sharply curving regions, etc.

In a couple of papers I've refereed, sufficient information was included so the authors could be shown, by plots of the data and the fitted curves, that the quadratic model was a bad choice. It was easy to suggest linear models, containing no more parameters than the quadratic, that would curve like the data did.

Further, the models were those that would, considering the nature of the experimental materials, behave well outside the x-domains employed in the studies.

If graphical procedures were generally integral and routine parts of data analysis, as they used to be, using poor model forms and some other worrisome practices would not be so common. I suppose that high-speed computing and some packaged programs have contributed to the decline of graphical analysis, but that should not be. High-speed computers, via the plotters and the CRT terminals (often with hard-copy devices) now generally available, can immensely expedite the use of graphical procedures. Many sets of data I have seen would have been better analyzed by using graphical methods combined with relatively simple numerical approaches often called "quick and dirty" or "approximate."

A matter very disconcerting to me about choice of models is that there often exists one (or sometimes more) good theoretical model, with meaningful parameters, based on insight in the substantive field. Such models should serve as a base for designing studies and analyzing data. During the past decade, however, I have seen several instances where such models are available, yet the investigator chose central composite or similar designs, and based analysis of the resulting data on the quadratic model. This is very puzzling to me. It seems as though many investigators do not realize that the tie between statistical analysis and the real world is the model. I think that statisticians have a grave responsibility to help correct this situation. Even in cases where a theoretical model is not immediately available, the statistician working intimately with the research man can arrive at something reflecting the behavior of the system under study much better, and having more meaningful parameters, than do the textbook models, so widely used. This matter is, of course, closely related to the selection of predictor problem.

Before entering the closing phase of my discussion, I feel it worthwhile to say something about skewed (asymmetric) data. The distributions in mind are not those encompassed by well behaved distributions such as the binomial, log normal, Poisson, gamma, exponential, etc. They are more a mixture of two distributions having different means. The proportion of mix as well as the means can be related to $\underline{x}$.

Such situations are found in animal and plant growth data, where runts occur with some frequency, especially under certain conditions. Although not a universal practice, it is common for researchers to eliminate the runts before analyzing their data. In general, this is a dangerous practice. The fraction of runts is, of course, a relatively high variance item, so showing a definite relation to $\underline{x}$ often requires accumulation of much data. Until sufficient data are obtained, one may wish to analyze both with and without the runts included and draw tentative conclusions. Generally, I suspect that such skewed distributions are not adequately describable by two parameters although much can be done by relating the variance as well as the mean to $\underline{x}$.

As noted early in this discussion, I have not been nearly as concerned about the robustness of estimators as I have about getting good estimates of error for whatever estimates are obtained. Despite the very impressive Princeton work, my concern has not changed. One can certainly compromise on robustness of estimators if ancillary information can be used in design and analysis to reduce the error of estimates sufficiently, and if computation of less robust estimates and estimates of their errors is enough easier.

Nevertheless, I believe that there is a strong need for robust estimators. Many important decisions in most walks of life must be made, sometimes quickly, on the basis of whatever data are already available or can be quickly obtained. Such data often are very heterogeneous and/or otherwise nonideal in character, and ancillary information is scarce or too difficult to utilize in design or analysis. Hence, methods of analysis are needed that do not mislead yet do extract as much information as is feasible within the bounds of time and resources the anxieties of administrators and/or society permit.

I have no alternative, therefore, except to highly commend Dr. Tukey and the Princeton work and to encourage continued intense effort. I do feel, however, that the context should be broadened to include studies on the particularly troublesome matter of obtaining good indices of dispersion for $\underline{\phi}$ in the nonlinear and in the selection of predictor cases. Also, I should like to encourage work dealing with skewed distributions. In general I feel that we have focused too much on estimators of location and we must face the general problem of relating two or more parameters of distributions to factors under study ($\underline{x}$).

I have said little about Dr. Tukey's paper per se, mainly because I have little criticism. He clearly indicates that his analysis is not a final one — that he is mainly trying to illustrate some reasonable approaches to analyzing data that have highly misbehaving distributions. Two thoughts did occur to me, however. One is that some of the things learned from the asymptotic aspects of the Princeton study might have been invoked in choosing his $\underline{\phi}(\underset{\sim}{x})$. The other is that, apparently more by accident than by design, the data being analyzed had considerable true replication (i.e., duplicate results for the same situations). As best I can tell, he did not invoke that fact to learn something about the basic noisiness of his data, which I think would be very helpful in assessing how far he can go in borrowing strength.

CHAPTER 6

ANALYZING A SERIES OF SOIL FERTILITY EXPERIMENTS FOR PREDICTION

F.B. CADY
Biometrics Unit, Cornell University

R.L. ANDERSON
Department of Statistics, University of Kentucky

and

D.M. ALLEN
Department of Statistics, University of Kentucky

Soil scientists base their prediction of crop yields on values of soil, climate and management variables. A good prediction depends on the determination of important variables and the parameter estimation of a prediction equation that is useful in calculating fertilizer requirements.

In many areas corn yield is a function of nitrogen fertility. During 1962-65 a series of 82 experiments was conducted in a nonirrigated western part of the El Bajio area in central Mexico. Each series was designed to study four levels of applied nitrogen replicated in a randomized complete block design.

Before the experiments, a large number of variables that could not be controlled at a single or differing levels, but could be "measured" at each site were considered. Some (though perhaps important) were eliminated due to limitations on available resources; others were found to be unimportant in previous studies or in the general literature. Other site variables, including several based on laboratory tests and not requiring field observations during the growing season, were measured but then eliminated if a sufficient range and uniform distribution of the measured values were not obtained, or if extremely high associations with the retained site variables were observed.

The mean (of four replications) corn yields and measured site variables are on file at the Health Sciences Computing Facility (HSCF), University of California, Los Angeles. A Data Set Description, including a complete listing in file format, is presented at the end of this chapter. Only the results from 72 experiments appear in the Data Set Description. The other ten were eliminated on the basis of poor population stands, unexpected site conditions during the experimentation period or extreme within-field variability, usually resulting from microclimatic environments.

The scales and indices used for the site variables were developed from past experience. For example, the drought index was calculated by summing the products of the number of days of wilting during different parts of the growing season by the estimated reductions in yield per day, based on several experiences reported in the literature. Each value of a scale involves a category of field conditions defined so an approximate linear relationship exists between yield and the scale.

Each experiment was analyzed according to a randomized complete block design model. The experimental errors from each experiment were tested for homogeneity of error. A quadratic polynomial equation between yield and applied nitrogen was then calculated and the estimated response curves compared. Part of the observed variation among the curves is due to different levels of soil or endogenous nitrogen at the various sites; i.e., the true response surface between yield and total nitrogen available through a given time period might be the same for all sites but the responses between yield and applied nitrogen are estimated using different portions of the total available nitrogen abscissa. In addition, other site variables affect the observed response between yield and applied nitrogen, resulting in additional variability in individual experiment response surfaces.

Historically, a combined analysis of variance would be calculated after examination of the experiment analyses.

Source of variation	d. f.
Sites (S)	$(s-1)$
Blocks/sites	$s(b-1)$
Applied nitrogen levels (L)	$(\ell-1)$
S x L	$(s-1)(\ell-1)$
Combined experimental error	$s(b-1)(\ell-1)$

Usually, as for the present data set, the site by applied nitrogen level mean square is significant and the problem in these analyses is to identify characteristics of sites so the interaction of variables can be interpreted. More recently, agricultural scientists have been able to quantitate a number of potentially important site variables, making their measurement more practical. Consequently, in addition to the response variable — the yield of corn at controlled levels of applied (or fertilizer) nitrogen) — the investigators in Mexico had available the following site variables that were measured but not controlled:

- total soil nitrogen, percentage by weight X 100;
- excess moisture, 0 - 6 scale;
- drought, weighted index based on days of plant wilting;
- depth of rooting zone, centimeters;
- soil slope, percent X 10;

- soil texture, 1 - 5 scale;
- previous crop, 10 - 25 scale;
- hail, 0 - 6 scale;
- blight (H. Turcicum), 0 - 9 scale; and
- weeds, 0 - 6 scale.

This increased ability to measure site variables coincided with the advent of high speed digital computers and software, e.g., the BMD statistical package, that could adequately handle large multiple regression problems. A model, linear in the parameters, was formulated based on three sources of nitrogen: applied nitrogen, soil nitrogen and previous crop nitrogen. The relationship was believed to be approximated by a quadratic polynomial including the linear by linear interactions of the three nitrogen variables. Based on soil science knowledge, it was decided that certain site variables (depth of rooting zone, soil slope and soil texture) would not interact with nitrogen but that a quadratic effect was expected given the range of soil texture. The interaction between the other site variables and the various sources of nitrogen was hypothesized. As a result, the 33 independent variables shown in table 1 were developed and called the full model.

The least squares estimation procedure was used to estimate the parameters in the full multiple linear regression model. Applied nitrogen was coded, dividing by 40. Variables AN, B^2, CA, CB, DB, HA, JA, and KA were coded, multiplying by 0.1 and A^2, BA and DA by 0.01. The symbols are defined and the estimated coefficients given in table 1, reproduced from Cady and Allen (1972).

The signs of the intercept and the linear effects of soil nitrogen, excess moisture, drought, depth of rooting zone, and hail were bothersome to one trying to interpret the estimated coefficients. The magnitudes of the linear effects of certain variables also

Table 1. The independent variables and the estimated partial regression coefficients for the full and reduced models. (Reproduced with modifications from Cady and Allen, 1972.)

Independent Variable	Symbol	Estimated Regression Coefficients		
		Full	Stepwise	SPA
Constant	K	-0.3170	-0.5446	1.5780
Applied nitrogen (linear)	N	1.8410	1.8050	1.4540
Applied nitrogen (quadratic)	N^2	-0.1552	-0.1547	-0.1528
Total soil nitrogen (linear)	A	-0.0290		0.0098
Total soil nitrogen (quadratic)	A^2	0.0150	0.0032	
A x N	AN	-0.0396	-0.0406	
Previous crop (linear)	B	0.2220		
Previous crop (quadratic)	B^2	-0.0813	-0.0176	
B x N	BN	-0.0014		
B x A	BA	0.0771	0.0711	
Excess moisture	C	0.1066	-0.2656	-0.2436
C x N	CN	-0.0374		
C x A	CA	-0.0217		
C x B	CB	-0.0794		
Drought	D	0.0309		
D x N	DN	-0.0096	-0.0091	-0.0091
D x A	DA	-0.0023		
D x B	DB	-0.0259		
Depth of rooting zone	E	-0.0054		
Soil slope	F	-0.0124	-0.0111	-0.0086
Soil texture (linear)	G	1.2800	1.2740	
Soil texture (quadratic)	G^2	-0.1591	-0.1630	
Hail	H	0.5556	0.2651	-0.2737
H x N	HN	-0.0003		
H x A	HA	-0.0802	-0.0694	
H x B	HB	-0.0159		
Blight (H. Turcicum)	J	-1.0890	-0.2733	-0.2677
J x N	JN	0.0183		
J x A	JA	0.0611		
J x B	JB	0.0139		
Weeds	L	-1.7750	-0.9231	
L x N	LN	-0.0004		
L x A	LA	0.1111	0.0757	
L x B	LB	0.0458	0.0183	

were not in agreement with agronomic expectation, e.g., the linear effect of blight and weeds seems large. Looking only at the linear effect can, of course, be misleading. By considering all four variables involving hail and using average values for the three sources of nitrogen, a reasonable overall estimate of the effect of hail can be deduced. However, the net effect of the four blight variables is higher than agronomic expectation.

Despite the attempt to include only important site variables and to depend on the estimator of experimental error to be used in hypothesis testing as discussed in Laird and Cady (1969), more than one third of the variables were not statistically significant from zero using a Type I error rate of .05. This relatively large number of nonrejections plus nonappealing signs and magnitudes of several estimates led to an attempt to reduce the full model. The stepwise regression program (Draper and Smith, 1966) yielded an equation with the 17 variables shown in table 1. Even though all the variables were then significant at a Type I error rate of .05, other bothersome events appeared, such as A^2 and AN in the model, but A not in the model.

The question of comparing the reduced model with the full model arose. The R^2 for the two were nearly the same, indicating that the fit to the data was equally well-handled by the reduced model as by the full model. Remembering that the primary objective was to develop a prediction equation, it seemed fruitful to compare the two models on the basis of how well the two estimated models would predict observations not included in the least squares estimation procedure.

To some, it may appear that the full model would do better than the reduced model on the basis that the extra variables must help, or could not harm, since the effect of a near zero estimate on the predicted values would be minimal. Thus, the data set was divided into halves. The full model and the reduced model (previously selected by stepwise regression) were then estimated on each half. The resulting prediction equations were used to predict the other half and the squared deviations between observed and predicted were added over both halves and called the prediction sum of squares.

As reported in Anderson, et al (1972), the half and half procedure was repeated four times with various modifications. The average mean square, calculated by dividing the prediction sum of squares by the number of observations, for the full model was 2.01 and for the reduced model 0.74. These calculations included some data not used in the present study, but the 2.01 and 0.74 are comparable to 0.38 and 0.39, respectively, the usual residual mean squares.

Other divisions of the data, including estimating on n - 1 observations and predicting for the nth, were calculated with the same general results. The poor performance of the full model is surprising and indicates that criteria used to arrive at a good prediction equation should be rethought.

Data analysts do not give sufficient attention to the fact that the variance of a predicted response cannot decrease, and usually increases, with the addition of a variable to the prediction equation. However, not including important variables gives a biased predicted value. Therefore, in an estimated prediction equation some balance between variance and bias is desired. It was clear from the previous results that the full model was fine for predicting those observations used in the estimation, but too many variables were included for it to be a good prediction equation for observations not used in the estimation procedure. To a lesser extent the same conclusion could be made for the reduced model selected by stepwise regression. The problem here is primarily in the usual stopping rule that stops the selection procedure when a pseudo-F statistic is less than an arbitrarily chosen percentage point of the F distribution.

A natural extension to the activity of comparing models is the use of a criterion incorporating the ability of the prediction equation into the variable selection procedure. Another is the development of a criterion which would give different relative weights to the variance and bias of a predicted value. The C_p criterion was used recently and is discussed in Daniel and Wood (1970).

We adopted the Prediction Sum of Squares (PRESS) criterion as developed in Allen (1971) and Cady and Allen (1972). To obtain PRESS, each observation is "predicted" using all the other observations. The resulting "errors of prediction" are squared and summed to form PRESS. PRESS is appealing because it simulates prediction. It does not use an observation to aid in the prediction of itself. The Sequential PRESS Algorithm (SPA) presented in Allen (1971) is used to calculate PRESS for any given subset of variables and to identify the additional variable that will result in the largest reduction of PRESS.

Using the 33 potential variables and the 72 experiments presented earlier, the prediction sum of squares decreases rapidly with the first few variables to enter, followed by several variables with small increases before a minimum is reached and then concluding with an increase in the prediction sum of squares. Strictly adhering to the prediction sum of squares criterion, variables are added to the prediction equation until the minimum is reached; however, the shape of the curve resulting from plotting the prediction sum of squares against the order of the entering variables is such that a practical decision can be made to stop bringing variables into the prediction equation earlier. The resulting prediction equation, from using the prediction sum of squares (PRESS), is given in the last column of table 1.

The SPA procedure yields a prediction equation containing all main effects except for the drought by applied nitrogen interaction. In addition, the signs and magnitudes of the estimates are agronomically reasonable leading to a straightforward interpretation of each selected variable. One cannot expect to interpret the data by viewing the estimated coefficients in the full model; and it is almost as difficult with the variables selected by stepwise regression. Interactions are undoubtedly important in a complete understanding of the basic underlying relationships involving yield determining variables. The last column of table 1 selects variables and gives estimates leading to a reasonable partial interpretation concerning kind and relative size of variables important in yield determination.

The question of which model in table 1 will give the best predictions can only be answered by additional experiments in future years. However, one approach is to use the first three years of data, determine the important variables, calculate estimates, and predict the corn yields for the fourth year using the already known numerical values for applied nitrogen and appropriate site variables. Therefore, three prediction equations, based on the full model and two reduced models selected by stepwise and SPA, were calculated using part of the data, 228 observations from the first three years, and predicting the remaining 60 observations from the fourth year. The residual mean squares based on the 228 observations were 0.35, 0.38 and 0.42 for the full, stepwise and SPA procedures, respectively. The "residual mean squares" based on the predictions of the 60 observations not used in the estimation were 1.12, 0.71 and 0.51. Again the poor performance of the full model is noted, and the PRESS criterion has given the smallest increase when predicting for observations not included in the estimation.

DATA SET DESCRIPTION

DATA - EL BAJIO, MEXICO, SOIL FERTILITY EXPERIMENT
(F. CADY, R. ANDERSON AND D. ALLEN)

CASES AND VARIABLES

72 CASES (EXPERIMENT SITES)
16 VARIABLES - 14 MEASUREMENT, 2 IDENTIFICATION

BRIEF DESCRIPTION

VAR	NAME
***	****
1-2	IDENTIFICATION
3-6	CORN YIELD
7-16	SITE FIELD CONDITIONS

PURPOSE OF STUDY

THE DATA CONSIST OF CORN YIELDS OBTAINED WITH FOUR LEVELS OF APPLIED NITROGEN IN 72 DISTINCT FERTILIZER TRIALS CONDUCTED IN CENTRAL MEXICO. TEN SITE VARIABLES DESCRIBE THE CHARACTERISTICS OF EACH FERTILIZER TRIAL AREA. THE PURPOSE OF THE EXPERIMENT WAS TO PREDICT YIELD AS A FUNCTION OF APPLIED NITROGEN WITH ADJUSTMENT FOR MEASUREABLE BUT NONCONTROLLABLE SITE VARIABLES.

REFERENCE

LAIRD, R. J., ALVARO RUIZ B., HORACIO RODRIGUEZ G., AND FOSTER B. CADY. COMBINING DATA FROM FERTILIZER EXPERIMENTS INTO A FUNCTION USEFUL FOR ESTIMATING SPECIFIC FERTILIZER RECOMMENDATIONS. RESEARCH BULLETIN 12, INTERNATIONAL MAIZE AND WHEAT IMPROVEMENT CENTER, LONDRES 40, MEXICO 6, D. F., 1969.

DETAILED DESCRIPTION

VAR	COL	FORMAT	DESCRIPTION
***	***	******	***********
1	1-2	XX.	SEQUENCE NUMBER
2	4-6	XXX.	EXPERIMENT NUMBER
3	10-12	XXX.	YIELD (TONS/100 HECTARES) FOR 0 KG/HECTARE OF APPLIED NITROGEN
4	14-16	XXX.	YIELD (TONS/100 HECTARES) FOR 40 KG/HECTARE OF APPLIED NITROGEN
5	18-20	XXX.	YIELD (TONS/100 HECTARES) FOR 80 KG/HECTARE OF APPLIED NITROGEN
6	22-24	XXX.	YIELD (TONS/100 HECTARES) FOR 120 KG/HECTARE OF APPLIED NITROGEN
7	28-30	XXX.	TOTAL SOIL NITROGEN (PERCENT BY WEIGHT X 100)
8	32	X.	EXCESS MOISTURE (0-6 SCALE)

```
 9  34-36    XXX.    DROUGHT (WEIGHTED INDEX BASED ON DAYS OF PLANT
                     WILTING)
10  38-39     XX.    DEPTH OF ROOTING ZONE (CENTIMETERS)
11  41-42     XX.    SOIL SLOPE (PERCENT SLOPE X 10)
12     44      X.    SOIL TEXTURE (1-5 SCALE BASED ON CLAY
                     CONTENT)
13  46-47     XX.    PREVIOUS CROP (10-25 SCALE)
14     49      X.    HAIL (0-6 SCALE)
15     51      X.    BLIGHT (H. TURCICUM, 0-9 SCALE)
16     53      X.    WEEDS (0-6 SCALE)

FORMAT IS (F2.0,1X,F3.0,3X,4(F3.0,1X),2X,F3.0,1X,F1.0,1X,F3.0,1X,
2(F2.0,1X),F1.0,1X,F2.0,3(1X,F1.0))      N = 72
```

LOCATION OF DATA

CARD IMAGE - FS.C073.CCADY1

LISTING OF DATA (THE FIRST TWO LINES ARE COLUMN GUIDES)

```
0         1         2         3         4         5
12345678901234567890123456789012345678901234567890123

 SITE       CORN YIELD            FIELD CONDITIONS
     2          4         6        8     10  12  14  16
 1          3         5        7     9     11  13  15

 1 202   186 339 452 515    87 0  35 70  1 5 17 0 2 0
 2 204   182 371 480 548   120 0   5 90 80 5 17 1 0 2
 3 206   156 339 349 368    87 3  26 50 30 5 17 0 1 3
 4 207   231 363 386 464    73 1  28 80 30 4 10 1 1 0
 5 208   314 441 541 575   120 0   0 99  1 5 20 1 3 0
 6 209    59 229 362 426    76 2   0 99 20 5 25 0 0 0
 7 210    80 208 342 375    76 0   0 90 20 5 10 5 0 1
 8 211   132 283 397 394    78 0  26 25 80 3 17 0 0 1
 9 212    55 282 419 552    55 2   0 99  2 5 20 0 1 1
10 213   174 315 401 417    54 2   0 60  1 5 25 2 1 0
11 216   198 318 421 525    78 2   0 99  1 5 13 0 1 0
12 217   314 404 371 438    95 0  23 90  2 5 13 0 0 0
13 218   143 300 469 518    94 0  10 75  3 5 25 0 0 0
14 219   112 263 362 413    95 2  44 55  5 5 13 0 1 0
15 221   348 447 588 622    99 1   0 99  1 5 25 0 0 2
16 222    98 275 392 444    69 0  17 60 20 5 10 2 0 1
17 227   269 379 408 432    60 0  34 65 40 4 13 0 0 0
18 228   175 305 399 422    81 1  27 50  1 5 13 0 0 1
19 229   104 205 265 268    78 5   0 45  1 5 13 0 0 3
20 232   257 294 292 278    87 0  59 50  2 4 13 0 0 0
21 302   199 245 215 218    98 0  60 38 26 3 13 0 1 0
22 303   178 268 264 332   103 0  40 35 11 2 17 0 1 0
23 304   185 392 446 475    75 0  20 50 28 4 17 0 1 0
24 306   220 387 423 481    90 1   0 80  5 4 21 0 1 0
25 307   224 350 451 480    85 0   0 60 21 3 13 0 1 0
26 308   125 266 369 443    83 3   0 55  4 5 13 0 1 2
27 309   110 225 363 452    63 0   0 99  5 5 20 0 2 0
28 310    76 195 331 386   137 1   0 55 74 5 10 1 1 1
29 311    41 139 303 386    85 2   0 75 26 5 13 1 1 1
30 312    47 103 244 302   100 2  15 55 17 5 13 1 4 0
31 313    60 145 283 374   120 0   0 75 17 5 10 2 5 1
```

SITE		CORN YIELD				FIELD CONDITIONS									
1	2	3	4	5	6	7	8	9	10	11	12	13	14	15	16
32	314	117	236	296	389	90	1	3	90	2	5	13	0	1	0
33	315	36	146	243	319	100	4	0	80	35	5	25	3	1	4
34	316	52	166	276	306	73	3	0	99	16	5	25	0	5	3
35	318	51	106	159	255	100	0	0	99	3	4	13	1	8	1
36	319	29	138	271	352	92	0	10	99	35	5	25	0	5	0
37	320	23	127	298	400	90	0	3	90	20	5	10	1	3	0
38	321	78	218	363	453	100	0	3	99	20	5	20	1	2	0
39	323	83	185	283	389	113	1	0	65	17	5	10	1	1	0
40	402	98	179	267	282	46	0	112	40	13	2	13	1	0	0
41	405	191	318	372	415	72	0	45	60	5	2	10	1	0	0
42	406	67	181	217	248	56	1	55	60	18	2	17	0	0	0
43	408	100	160	178	187	64	0	54	50	8	3	13	4	0	1
44	409	264	360	393	394	75	0	30	65	13	2	13	1	0	1
45	410	247	390	432	425	95	0	31	70	12	3	17	0	0	1
46	411	257	289	264	288	113	1	71	32	23	3	13	0	1	0
47	412	99	253	356	361	46	0	32	35	10	3	10	0	1	1
48	413	139	252	266	329	60	0	25	55	11	2	17	0	1	1
49	414	31	195	344	467	47	0	10	90	4	5	10	1	2	0
50	415	112	311	434	511	59	0	3	99	7	5	13	0	2	0
51	416	156	267	392	488	72	0	3	99	2	5	10	0	1	0
52	417	88	230	386	482	61	0	5	90	7	5	10	0	1	0
53	418	167	391	475	507	81	0	0	60	13	5	10	0	2	0
54	419	137	286	456	588	129	0	0	55	22	5	13	0	1	1
55	420	155	317	365	442	75	0	13	80	2	5	25	0	2	0
56	422	179	282	402	427	84	0	15	75	12	5	13	0	1	0
57	423	455	474	585	594	123	0	10	42	16	5	13	0	1	1
58	501	216	341	377	400	87	2	0	99	1	2	13	0	0	0
59	503	446	464	501	486	124	1	0	72	2	4	13	1	0	0
60	504	345	414	430	447	101	0	0	85	4	3	25	0	0	0
61	505	99	280	311	332	99	1	0	55	10	2	15	1	0	1
62	506	238	355	482	517	69	0	0	60	22	3	10	0	1	0
63	507	322	371	476	501	105	2	0	73	10	3	17	0	1	1
64	508	173	290	373	437	94	3	0	99	2	5	13	0	3	0
65	509	137	211	326	378	77	3	0	99	1	5	13	0	3	0
66	510	93	167	282	421	58	3	0	80	3	4	13	0	3	0
67	511	92	179	328	410	54	3	0	70	14	4	13	0	2	0
68	512	89	174	309	418	74	3	0	80	5	5	13	1	2	0
69	513	151	221	312	383	111	3	0	55	10	4	25	1	2	0
70	514	190	226	385	463	125	2	0	99	5	4	10	0	1	0
71	515	129	226	336	418	84	3	0	60	15	5	25	0	1	0
72	516	100	186	269	378	67	3	0	90	5	5	10	0	1	1

REFERENCES

Allen, D.M. (1971). The prediction sum of squares as a criterion for selecting predictor variables. Technical Report No. 23, Department of Statistics, University of Kentucky.

Anderson, R.L., D.M. Allen and F.B. Cady (1972). Selection of predictor variables in linear multiple regression. In *Statistical Papers in Honor of George W. Snedecor*, T.A. Bancroft, Ed., Iowa State University Press.

Cady, F.B. and D.M. Allen (1972). Combining experiments to predict future yield data. *Agron. J.* 64, 211-214.

Daniel, C. and F.S. Wood (1970). *Fitting Equations to Data.* New York, Wiley.

Draper, N.R. and H. Smith (1966). *Applied Regression Analysis.* New York, Wiley.

Laird, R.J. and F.B. Cady (1969). Combined analysis of yield data from fertilizer experiments. *Agron J.* 61, 829-834.

DISCUSSION OF THE CADY-ANDERSON-ALLEN CHAPTER

R. Elashoff

The problem

This paper continues the analysis of corn yield data F.B. Cady began a few years ago. It seems that the problem of interest is to identify the important site variables to explain the site by nitrogen interaction. Presumably, this is to be solved by developing a prediction equation for corn yield from applied nitrogen levels and characteristics of the site which makes agronomic sense.

The main issue in this paper is how to choose a model or prediction equation relating corn yield to those predictor variables.

Brief remarks on the authors' problem definition

The problem as stated by the authors is their scientific problem. A practical agronomist has this problem: My site has these characteristics, A, B, etc. How much nitrogen should I apply to maximize corn yield? This practical problem, I seem to recall, was formulated by John Tukey at the Conference. This practical problem should be solvable from the prediction equation developed from the analyses. For each site, then, the authors should find the nitrogen level to maximize corn yield. The final version of their paper does not include the nitrogen values. The paper would have more usefulness to agronomists if the results were included. For example, what rule of thumb can we give to agronomists about how much nitrogen to use in sites with certain patterns of characteristics?

The authors' solution

The authors develop prediction equations based on an agronomic model, a stepwise procedure, and their own PRESS routine available in the literature (idea due to R. L. Anderson). They find that, based on a reasonable criterion, either the stepwise procedure in BMD02R or their own PRESS routine seems to give the best results. BMD02R gives some anomalies which might be overcome by using orthogonal polynomials. I suspect that PRESS was not developed with this example in mind. I am not surprised that the full agronomic model did poorly considering the relatively small number of observations to the number of parameters to be estimated.

Concluding remarks

The main thrust of this paper is to compare three techniques of estimating a prediction equation. I regard the paper in two ways: first, it brings two statistical techniques to the attention of agronomists to help them choose multiple regression equations; second, it is an illustration of a new technique of choosing variables in multiple regression — say an example in a paper that introduces such a new technique.

The paper, then, discusses a technique applied to a problem. It is very much like the Dickey and Walrath paper in this respect. Both papers are interesting, informative and useful. Neither paper is a data analysis, except at the simplest level.

J. Tukey

(preconference comments)

This paper reveals a perennial problem. I feel we have made a significant part of it for ourselves by not facing up to what kinds of data we are free to bring in from outside. We feel relatively happy about which variables are to be given zero coefficients because they are entirely left out of the picture. We are not equally willing, though I see no reason why we should not be, to accept values other than zero based on evidence and insight external to the data, either as forced by, or only to be overridden by, significant evidence drawn from the data. Intervals of allowed values should also, it seems to me, be given equally strong consideration. "If this be Bayes, make the most of it."

(after reading final manuscript)

After seeing the results and participating in the initial discussion, I was strongly impressed by the apparent potential of PRESS. (I look forward to its use in connection with resistant/robust techniques of regression. Someone has to learn how to make the combination!)

As the later discussion indicates, I was glad to stress (at the presentation) the extent to which such economically-oriented regressions (in a broad sense of "economic" this includes regressions involving comparative clinical treatments, for example) have, of necessity, to be focussed on changes — on differential effects. It should make us all feel better that the data of the example were strong enough for analysis of the differences between adjacent treatments to come out clearly.

The whole approach — both the paper and the discussion — illustrates the gains to be found by avoiding naive overconsistency. At least in this example, the effective calculation proceeds as follows:

- minimizing RESS (residual SS) to fix the value of coefficients actually fitted,
- minimizing PRESS to choose which coefficient to consider adding next, and
- minimizing 3PRESS minus 2PRESS — or perhaps 2PRESS minus RESS — to decide how far to go.

Followers of the will-o'-the-wisp of apparent naive consistency would, I fear, argue that we ought to use the same criterion for all three. How happy it is that we are sophisticated enough to dismiss such a misleading argument! We need to try — carefully, thoughtfully, and with much empirical guidance — splitting criteria in other situations, too.

Even in the wonderful world of mixed criteria, problems such as those placed by the author among "other bothersome events," specifically A^2 and AN in the model but A out, are still likely to occur. At least two kinds of attack on this particular class of problem are reasonable:

- Making sure that A^2 is orthogonal to A, and thus not likely to come in in its stead, and

- restricting the order in which a stepwise procedure can proceed, perhaps by requiring A to come in with A^2, if not in already.

CHAPTER 7

EVOKED OFF RESPONSE TO AN AUDIO TONE

D. C. MARTIN
Departments of Biostatistics and Psychiatry, University of Washington

and

H. L. LUCAS
Biomathematics Program, North Carolina State University

The analysis presented here is based on an experiment performed by D. C. Martin, P. L. Hein, Jr., and J. C. Martin* on the response of gross electrical activity of the brain to an audio stimulus. This experiment was selected for two reasons:

- Many of the problems of analysis, typical of a large class of statistical problems are found in this experiment. In particular, the data reported include short nonstationary time series responses with a significant covariance structure, repeated measures on the same individuals, and several sources of variation.

- The evoked responses and analysis of these data are of interest in their own right.

---

* The experiment is unpublished to date. It was a joint effort between D. C. Martin, Biomathematics Program, North Carolina State University and P. L. Hein, Jr. and J. C. Martin, Department of Psychiatry, Duke University.

A SHORT DESCRIPTION OF THE EXPERIMENT

The experiment examined the evoked electrical potentials of the brain to an audio stimulus. The stimulus was a 500 Hz square wave signal and the points of interest are the responses to the onset and to the offset of the tone.

Surface electrodes were placed on the subject's scalp (C_z and mastoid) and the electrical potentials were amplified by a standard Grass Model 7 Polygraph with 7T5 preamplifiers. These responses, along with the tone on/off signal were recorded on an FM tape recorder and a conventional ink strip chart recorder. The tones were presented by headphones to the subjects in a semidarkened soundproof anechoic chamber.

Two treatments were used. In the first (A), 1/3 second intervals of silence were presented with the tone on at all other times. In the second (B), a 1/3 second interval of tone on was presented and the time between presentations varied from approximately two to nine seconds. The period of interest is the electrical activity immediately after the tone on or tone off. Figures 1 and 2 show a short period of the electrical activity from the strip chart recorder for both treatments.

The usual method of analysis is to average a number of responses to the stimulus. An example of such an average curve with a two standard deviation confidence interval is shown in figure 3. These data are from a pilot experiment with a 1000 Hz sine wave stimulus.

PURPOSE OF ANALYSIS

The analysis of data from this experiment was undertaken because the data are typical of several statistical problems that appear frequently in data analysis. In particular:

- The basic unit of response is a segment of a nonstationary time series. This is common to both time series problems and to many problems where the

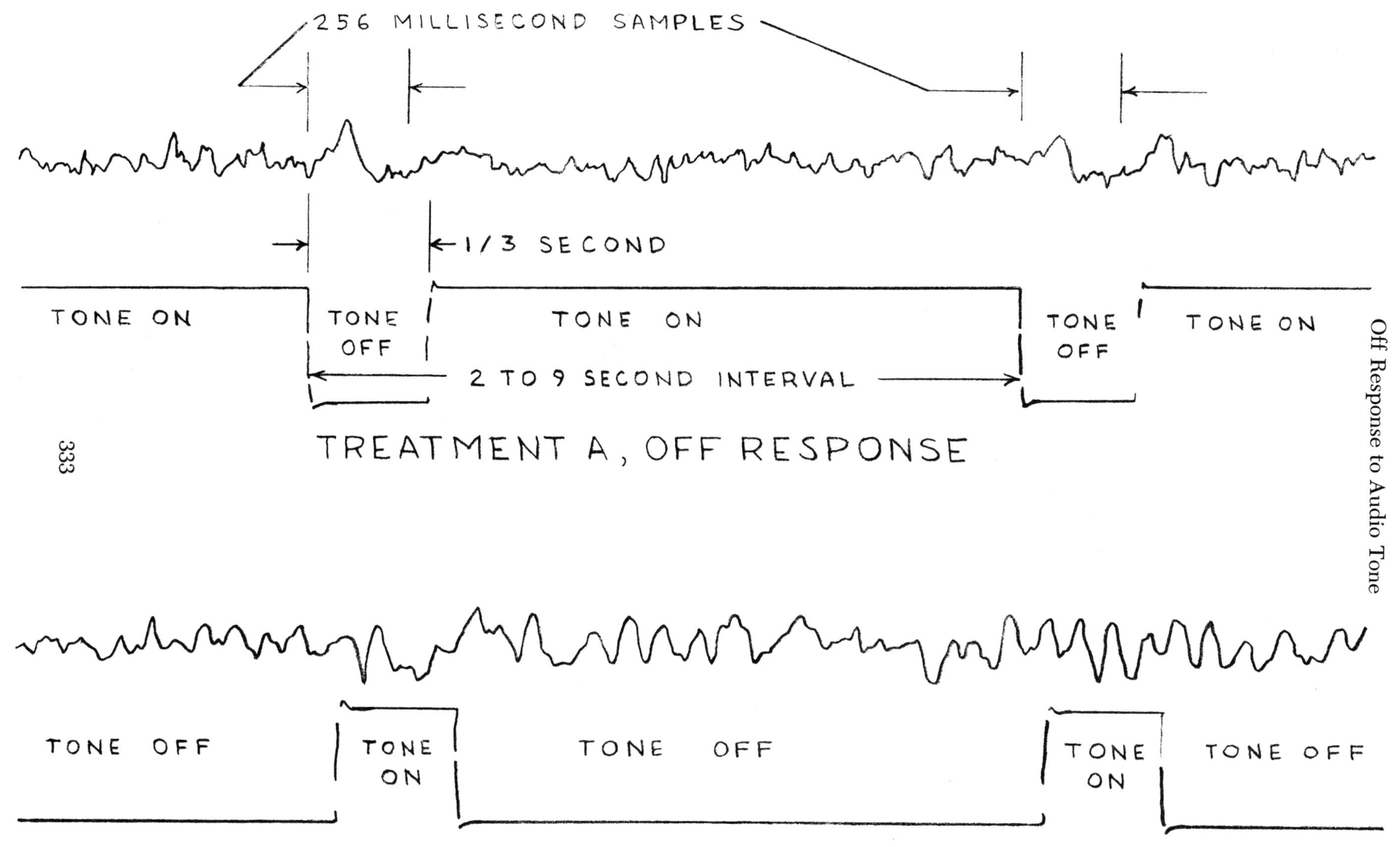

Figures 1 and 2. Electrical activity of treatment from strip chart recorder showing (A) off response and (B) on response.

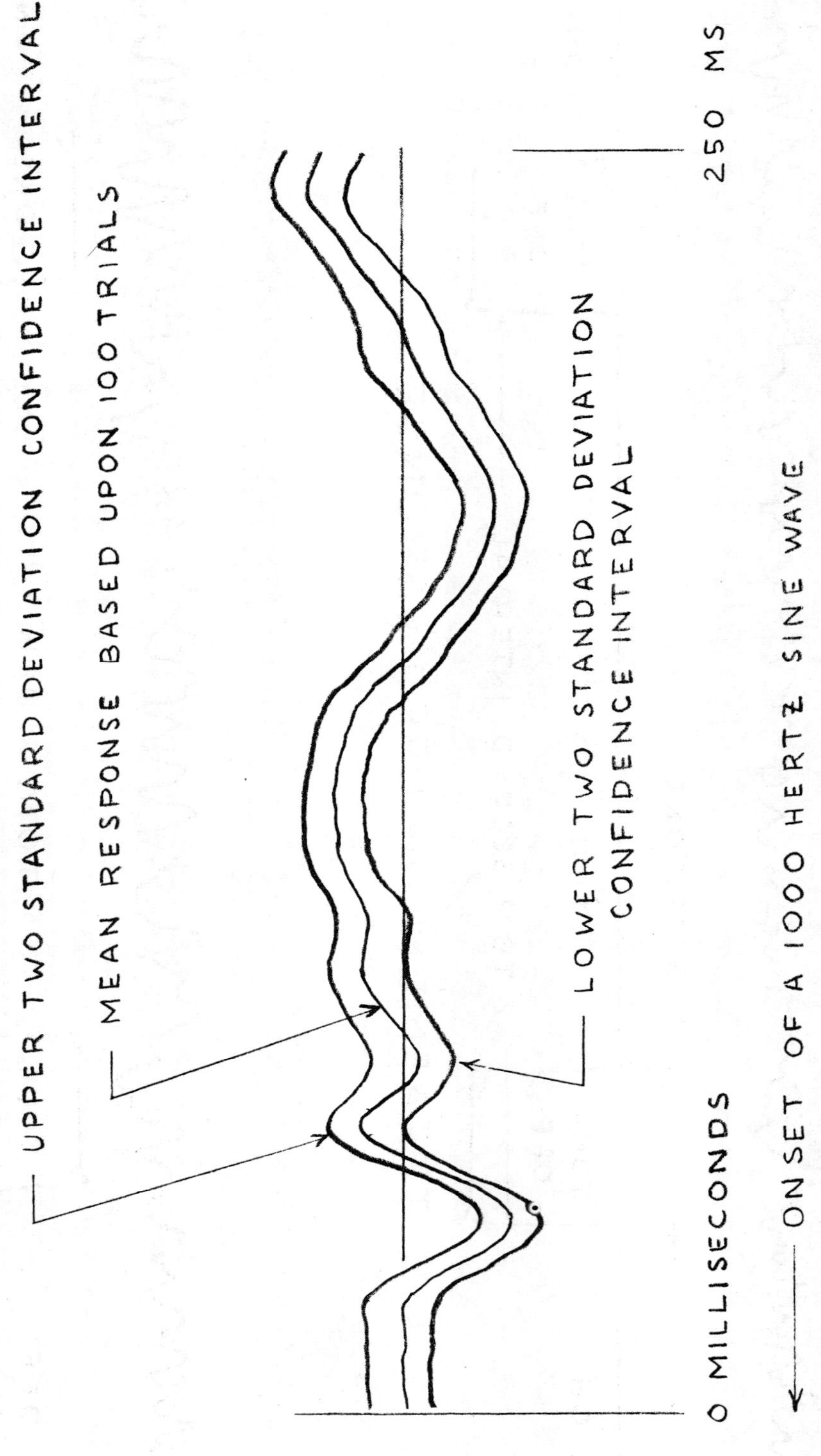

Figure 3. Average response curve with a two standard deviation confidence interval (1000 Hz sine wave stimulus).

response is a curve.

- Standard multivariate procedures used here are difficult because of possible carry-over effects, the large number of variates in the observation vector, and possible changes in the error structure due to variance components.

- The error structure is relatively interesting. There are at least two sources of error. The major error component in the signal averaging analysis is due to EEG activity that is not synchronized with the stimulus presentation (background EEG) — the low frequency signal leads to very high correlations in the error structure, and some electrical noise comes from various other sources. We also expect that variances and perhaps responses may vary between subjects.

- Some assumptions about the relationships of the trials within a run and the runs within a subject need to be investigated. The assumption proposed here is to treat the runs and trials within a run as independent, or at least to have zero covariances; and the only effects are trends due to fatigue both within and between runs and possible carry-over from one treatment to the next.

The problems mentioned in the last two items are common in repeated split plot and repeated measures designs. These data supply a reasonably large base for the development of both methods of analysis and computational algorithms.

Specialized methods of analysis and hypothesis tests are also needed for evoked responses. Donchin (1969) has a discussion of the methods of analysis. Schwartz (1970) criticizes a paper for using an unspecified number of t tests in comparing two mean curves. The authors' reply to Schwartz (Bartlett and John, 1970) has several interesting points about the types of variability that are typical in evoked response experiments. Although a number of interesting analyses have been performed on different experiments, it appears that there is still a need for more general methods of analysis in evoked responses.

A specific goal of this analysis is the investigation of the off response. Evoked responses were first noted in 1941 and various signal averaging systems have been available since the early sixties. As a result, there is a very large literature on evoked responses. Surprisingly, there seems to have been very little work on the off response. Sandel and Kiang (1961) worked with the off response in cats. They concluded, for anesthetized cats, that the off response is in fact an on response to the new frequency components introduced in the stimulus when the tone is turned off. Kiang and Sandel (1961) then modified their conclusions for unanesthetized cats and concluded that a second mechanism exists.

The audio stimulus for this experiment was designed to introduce no variation in the spectrum due to turning off the stimulus. Needless to say, the Fourier representation differs when an off period is included in the interval. However, the spectral components introduced by turning the stimulus off are always present because a square wave, 5 volt and 0 volt peaks, that was shut off on the 0 volt cycle was used. (It will be of interest to see how the results of this experiment compare with those of Sandel and Kiang. One hypothesis in this experiment is that there is no off response; this would agree with Sandel and Kiang's work on anesthetized cats.)

Another specific goal is to test several hypotheses about the off response relative to the usual evoked (on) response. The most interesting of these hypotheses is based on Freeman's (1964) model of the evoked response. His model is based on feedback loops in a linear system with a step function input. It is conjectured that this model predicts an off response that is the negation of the on response. This assumes that the stimulus has been present for a long enough period for the transient responses to the onset to damp out, and that the cessation of the stimulus is a step function back to the original level.

The analyses described here are not yet complete.

BASIC STATISTICAL DESIGN

The basic unit of response is a short time series of approximately 1/4 second that consists of 256 observations taken at one millisecond intervals. There are two treatments:

A. Tone normally on with 1/3 second intervals of silence. The response measured is to the offset of the tone.

B. Tone normally off with 1/3 second intervals of tone on. The response measured is to the onset of the tone.

A trial consists of the onset or offset of the tone and the digitized response of 256 points. Each run was six minutes in length and 75 trials were taken from the approximately 85 trials available in each run. This is a total of 1575 (= 3·7·75) trials and 403,200 observations. The time between trials varied from approximately two to nine seconds. A sequence of ten intertrial intervals was used throughout each run. This served two purposes in the design. It made it difficult for the subject to anticipate the tone and it reduced the possibility of systematic effects being carried over from one trial to the next.

The trials were arranged in a changeover design. Each subject had three six minute runs with a two minute rest between each run. The treatments were:

Subject	Run 1	Run 2	Run 3
1	B	A	A
2	A	B	B
3	A	B	B
4	B	A	A
5	B	A	A
6	A	B	B
7	A	B	B

The seventh subject was run as a backup for subject two (subject two had an EKG artifact and a history of high blood pressure). However, it was decided to leave subject two in the analysis and use the data from subject seven in the selection of various types of analysis of the response within subjects.

Data from the study are available from the Health Sciences Computing Facility, University of California, Los Angeles on a standard labeled 9-track tape labeled DSLIB1. A sample of the data (two trials from run 1, subject 1) is included in the Data Set Description at the end of this chapter.

ASSUMPTIONS, NOTATION AND MODEL

The proposed preliminary analysis is contingent on several of the preliminary steps. We begin with the preliminary model:

$$\widetilde{Y}_{rst}(\tau) = R_{rs}(\tau) + T_{rs}(\tau) + C_{rs}(\tau) + \widetilde{A}_{rst}(\tau) + \widetilde{\varepsilon}_{rst}(\tau)$$

where

$\tau = 1, \ldots, 256$ is the time in milliseconds

$r = 1, 2, 3$ is the run number for a subject

$s = 1, \ldots, 7$ is the subject number

$t = 1, \ldots, 75$ is the trial number.

The functions of time in the above model are:

$\widetilde{Y}_{rst}(\tau)$ the observed response;

$R_{rs}(\tau)$ the evoked response for subject s for 75 trials on run r;

$T_{rs}(\tau)$ a time trend over trials and runs within a subject;

$C_{rs}(\tau)$ a carry over effect from the previous treatment;

$\tilde{A}_{rst}(\tau)$ the background EEG activity that is not synchronized with the stimulus, assumed random or mixed; and

$\tilde{\varepsilon}_{rst}(\tau)$ wide band (white) noise due to electrical noise, tape recording and analog to digital conversion, assumed to be completely random.

The preliminary analysis assumes that $T_{rs}(\tau) \equiv 0$ and $C_{rs}(\tau) \equiv 0$, that is, that trend and treatment carry over are negligible.

No functional forms are specified at this time since the selection of these forms is a part of the analysis. Note, since we are always dealing with 256 values over a 256 millisecond interval we can use vectors instead of functions when it is convenient.

Let

$$\underline{\tilde{Y}}'_{rst} = [\tilde{Y}_{rst}(1), \ldots, \tilde{Y}_{rst}(256)] \text{ and define } \underline{R}_{rs}, \underline{T}_{rs}, \underline{C}_{rs},$$

$\underline{\tilde{A}}_{rst}$ and $\underline{\tilde{\varepsilon}}_{rst}$ in an analogous manner.

The following assumptions will be used:

A1 $E(\underline{\tilde{A}}_{rst}) = E(\underline{\tilde{\varepsilon}}_{rst}) = 0$

A2 $\text{Cov}(\underline{\tilde{\varepsilon}}_{rst}) = \sigma^2_{rs}\underline{I}$

A3 $\text{Cov}(\underline{\tilde{A}}_{rst}) = \underline{\Sigma}_s$

A4 $\underline{\tilde{A}}_{rst}$ and $\underline{\tilde{\varepsilon}}_{r's't'}$ are independent for all r, s, t, r', s' and t'

A5 $\underset{-}{\tilde{A}}_{rst}$ and $\underset{-}{\tilde{A}}_{r's't'}$ are $\left\{\begin{matrix}\text{uncorrelated}\\ \text{independent}\end{matrix}\right\}$ if $r \neq r'$ or $s \neq s'$ or $t \neq t'$

A6 $\underset{-}{\tilde{\varepsilon}}_{rst}$ and $\underset{-}{\tilde{\varepsilon}}_{r's't'}$ are $\left\{\begin{matrix}\text{uncorrelated}\\ \text{independent}\end{matrix}\right\}$ if $r \neq r'$ or $s \neq s'$ or $t \neq t'$

A7 $\underset{-}{\tilde{A}}_{rst}$ and $\underset{-}{\tilde{\varepsilon}}_{rst}$ are multivariate normal random deviates.

In general, the weaker uncorrelated form of A5 and A6 are used in parameter estimation and A7 can be omitted. The stronger independence assumption and A7 are used in confidence intervals and hypothesis tests.

A1 is the usual assumption in evoked response analysis. This is essentially the assumption that the expected value of the response, $\underline{R}_{rs}$, does not change over a six minute interval. Some alternatives are a random effects response or a time dependent response. Neither of these alternative assumptions appears attractive although they have been discussed in the evoked response literature.

A2 is reasonable if $\underline{\tilde{\varepsilon}}$ is due to high frequency or wide band electrical noise. Unfortunately, any low frequency components will be confounded with $\underline{A}$, the background EEG. It is desirable to check for 60 Hz and harmonics before accepting this assumption. The possible change in variance is proposed for two reasons. The electrode resistance varies widely between subjects, and an unidentified high frequency noise appeared between runs 1 and 2 on subject 4 and may have been recorded on the following runs.

In most experiments the high frequency or white noise effects would be minimized by the use of low pass electrical filters. We have not done this because the filters change the shape of the response and increase the correlations between observations, and the low frequency components are further confused with the low frequency background EEG. Since we wished to estimate the variability due to this source, we set the cutoff frequencies to relatively high values.

Assumption A3 is that each subject has the same covariance for the background EEG for a 22 minute period. In practice this means either no change, or a slow change in the nature of this activity. This assumption is perhaps stronger than it should be but is still much more realistic than the usual weak (covariance) stationary process assumptions. It is also subject to at least a partial test since covariance estimates can be computed for different runs, or even parts of runs, for the same subject.

Assumption A4 is reasonable if the sources of variation $\underline{\tilde{A}}_{rst}$ and $\underline{\tilde{\varepsilon}}_{rst}$ are correctly identified.

Assumption A5 is an approximation that assumes the autocovariance function of the background EEG is sufficiently small after the two to nine second intervals between trials to be neglected. This assumption is strengthened by the partial randomization of the intertrial intervals because it is reasonable to assume that any dependency would be time dependent.

Assumption A6 is reasonable if $\underline{\tilde{\varepsilon}}_{rst}$ is due to wide band or white electrical noise.

The normality assumption A7 is made with the hope that it will not be needed because all of the tests are sufficiently robust.

OUTLINE OF THE ANALYSIS

Steps 1 to 3 are preliminary analyses. Steps 1 and 2 could be based on the intertrial interval data so that no data are used again in the next stages of the analysis. However, this did not appear to justify the additional programming effort since the sampled data used in later steps is less than 10% of the data used in these two steps.

Step 1: Power spectrum estimates will be computed from several intervals on the analog tape to check for 60 Hz contamination.

Step 2: The lag zero autocovariance, or RMS power of the signal, will be computed for one or more intervals for each subject. If these appear to differ appreciably between subjects we may then consider scaling the amplitudes so that the RMS powers are the same.

Step 3: Preliminary estimates of the $\underset{\sim}{\tilde{\varepsilon}}_{rst}$ variance will be computed by a variate difference method for each of the 1575 (= 7·3·75) trials.

Step 4: The average evoked response will be computed for each of the 21 runs by

$$\hat{R}_{rs}(\tau) = \frac{1}{75} \sum_{t=1}^{75} \tilde{Y}_{rst}(\tau)$$

and plotted. A variance estimate and a Durbin-Watson statistic may also be computed.

Step 5: The dimensionality of the $\underset{\sim}{\tilde{Y}}_{rst}$ vectors will be reduced by a linear transformation. This transformation will be based upon the (possibly weighted) least squares parameter estimates found by fitting a linear model of the form

$$L(\tau) = \sum_{i=1}^{n} B_i \phi_i(\tau)$$

to each of the observations.

The basis functions $\phi_1(\tau), \ldots, \phi_n(\tau)$ will be selected by trying several sets of functions on the data collected from subject 7. We have more ideas for basis functions than we are likely to be able to try. The first two will be the usual sine-cosine Fourier and a spline-like segmented polynomial that is being developed. Other candidates are: average values for intervals, square waves (Walsh functions), polynomials (probably Chebyshev), impulse function (Haar functions), exponentially damped sine waves (Cautz basis functions), etc. The selection procedure will start with the average response curves from subject 7 computed in step 4. The poor fits will be eliminated and the remaining models will be tried on a sampling of runs on subject 7. Note that the residual sum of squares will allow a second estimate of the variance of $\underline{\tilde{\varepsilon}}_{rst}$ that can be compared with the estimates computed in step 3.

When a satisfactory model has been found, the parameter estimates will be computed for each trial on the remaining 18 runs of 75 trials. This procedure also acts as a low pass digital filter to reduce the effect of $\underline{\tilde{\varepsilon}}$ and possibly of 60 Hz line noise.

Step 6: The mean and sample covariance matrix $\underline{\bar{B}}_{rs}$ and $\mathrm{cov}(\underline{\hat{B}}_{rst})$ will be computed over the 75 trials in each run for the parameter estimates. The basis functions will be evaluated with the mean parameter estimates and the plots compared with those computed in step 4 as a check on the choice of functions. The parameter estimate vectors,

$$\underline{\hat{B}}_{rst} = \underline{\Gamma}\, \underline{\tilde{Y}}_{rst}$$

where $\underline{\Gamma}$ is the linear transformation given by the least squares fit in step 5, will be used as the raw data for the remaining steps.

Step 7: Given the assumption of no carry over effects, the hypothesis that $\underline{R}_{15} = -\underline{R}_{25}$ can be tested for each subject with T^2 statistic. However, this omits the third run for each subject. A second analysis based upon a multivariate regression procedure will then be used with a more elaborate model.

Although further analysis is planned, the methods are not yet to the point where they can be simply described. The previous analyses are essentially done with subjects. Plans are being made to analyze the data with between subject variability taken into account as well as carry over and trend effects. Further study of the residuals, especially across trials within runs, is also planned.

RESULTS TO DATE

Dr. Tukey suggested that a pre-stimulus sample of 512 milliseconds should be taken before the tone on or off transition. The sampling program was rewritten and the new data digitized from the FM analog tape records. The analysis was based upon these data.

Several of the proposed steps were modified. The power spectrum estimates in steps 1 and 2 were computed from the pre-stimulus sample rather than resampling the analog tape. These showed some 60 Hz contamination, as expected. An unexpected 60 Hz sine wave artifact appeared in the averaged responses. This was due to the device used to control the time intervals between trials. It was a digital electronics device that obtained its time pulses by counting down from the 60 Hz line frequency. This inadvertently synchronized our samples with the power line. As a result, a 60 Hz contamination appears in the averaged responses.

The third step was to calculate variate difference statistics (Morse and Grubbs, 1947). The pre-stimulus sample was treated as two blocks of 256 milliseconds each. The first ten difference estimates of dispersion were computed on the two pre-stimulus samples thus obtained and on the post-stimulus sample for all of the data. This led to 4650 sets of dispersion statistics. These statistics were examined and the data were plotted for all unusually large values as well as control plots of adjacent trials. This editing process was to check for gross errors, equipment failures, etc. Although some of the trials had large dispersion estimates, no gross errors or failures were identified and all the data were retained. The 60 Hz sine wave component was not expected to affect the higher order difference estimates of dispersion. (See Table II of Morse and Grubbs and their discussion of the sensitivity of this type of statistic to sine wave trends.)

The average evoked responses were computed as described in step 4 for both the pre- and post-stimulus samples. The standard deviation, two standard deviation, and two standard deviation of the mean estimates of the confidence interval were computed also. These data were then used to generate 21 plots, one for each run of 75 trials. Samples of these plots are shown in figures 4, 5 and 6. The 75 consecutive trials in each of the 21 runs were then subdivided into three blocks of 25 trials. The means and standard deviations were computed for each time point in these blocks. (These data will be used for a rough check for time trends in a run. If such trends exist, they are expected to appear as a decrease in the amplitude of the evoked response. These test statistics have not yet been computed.)

The investigation of various linear transforms to reduce the dimensionality of the observation vectors as described in step 5 is only partially complete. The Fourier approximation has been investigated in some detail. It was selected as the initial choice for several reasons:

- it is in wide use in various forms and has been quite successful in other similar applications,
- very efficient computational algorithms are widely available (this was not completely trivial because even in this relatively small experiment we have to process about 1.2 million observations), and
- the orthogonality of the harmonic sine-cosine basis makes it easy to compute an ANOVA table that displays the reduction in the sum of squares of the residuals for each harmonic.

Each block of 256 points was transformed separately; transforms for both the raw data and windowed data were computed. The

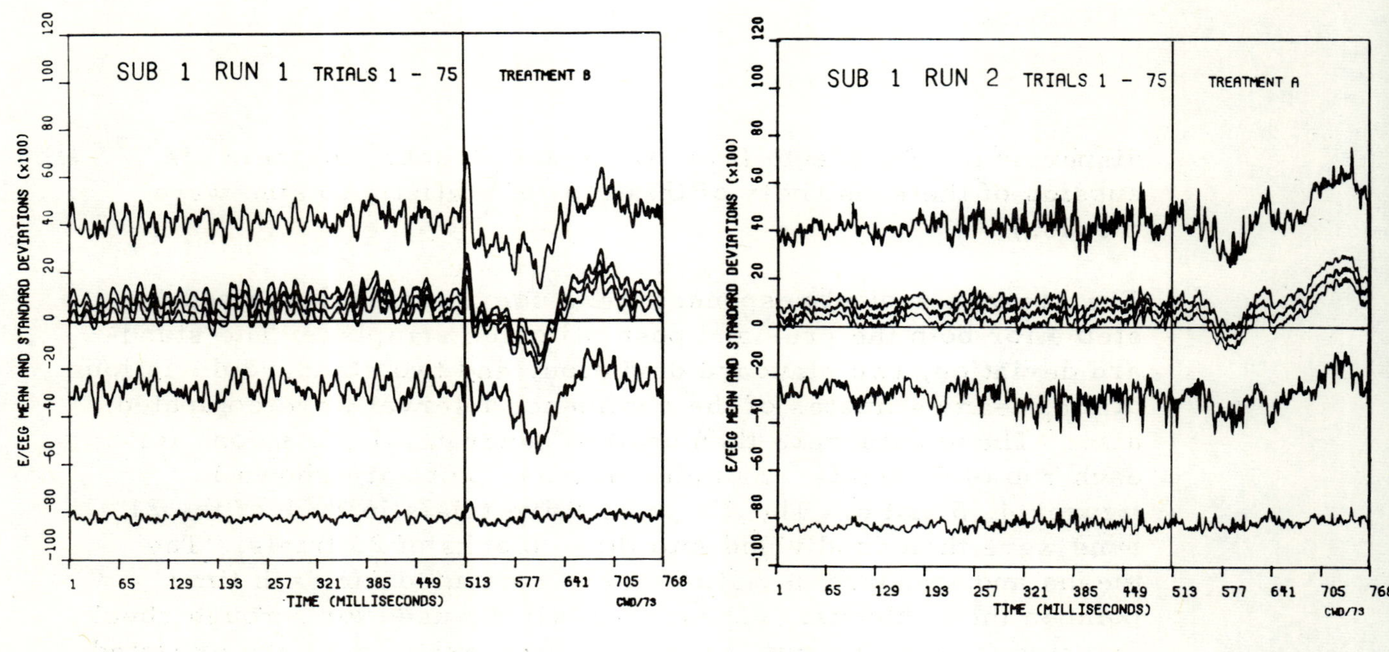

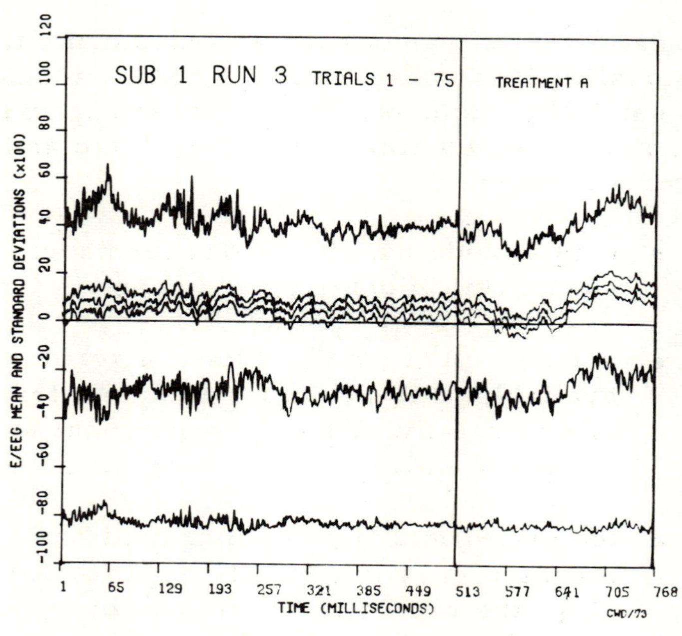

Figures 4, 5 and 6. Average evoked response plots showing standard deviation, two standard deviation and two standard deviation of the mean estimates of the confidence interval.

windowed data had the mean linear trend removed so the mean and slope of the 256 points were zero. A data window was used that applied a cosine bell to the first and last 10% of the data. (The computation of this data window is given in Section VII of Bingham, Godfrey and Tukey, 1967, in their discussion of "interim computational recipes.")

The approximations were computed and plotted for a number of harmonics and trials. On the basis of these plots and the sum of squares of residuals, it was decided that 21 coefficients were more than adequate to approximate the responses. In most curves, this approximation accounted for over 90% of the sum of squares. A sample plot showing the reconstructions based upon Fourier coefficients is shown in figure 7. The truncated Fourier approximation is very insensitive to the 60 Hz component. Therefore no special additional procedures are required to reduce the effect of this signal.

A few of the responses have been approximated by a least squares fit of high degree Chebyshev polynomials. It was pointed out at the Conference that the Chebyshev polynomial fit was related to the Fourier fit. In particular, the Fourier cosine series can be related to the Chebyshev polynomials through the trigonometric form of the Chebyshev recursion relations. Our current work in this area is as yet inadequate to suggest any comparisons.

The other basis functions mentioned in step 5 have not yet been explored. It seems likely that either Fourier or Chebyshev approximations can reduce a vector of 256 observations to a vector of length 21 or smaller while retaining a very good approximation of the average response.

The multivariate analysis of variance procedures described in steps 6 and 7 have not yet been computed. We have computed the pre-stimulus samples and found the Fourier coefficients but we have not yet tried adjusting the post-stimulus samples for these measurements. We also intend to compute the usual univariate least squares contrasts to estimate treatment effects, carry over, etc., and to plot these.

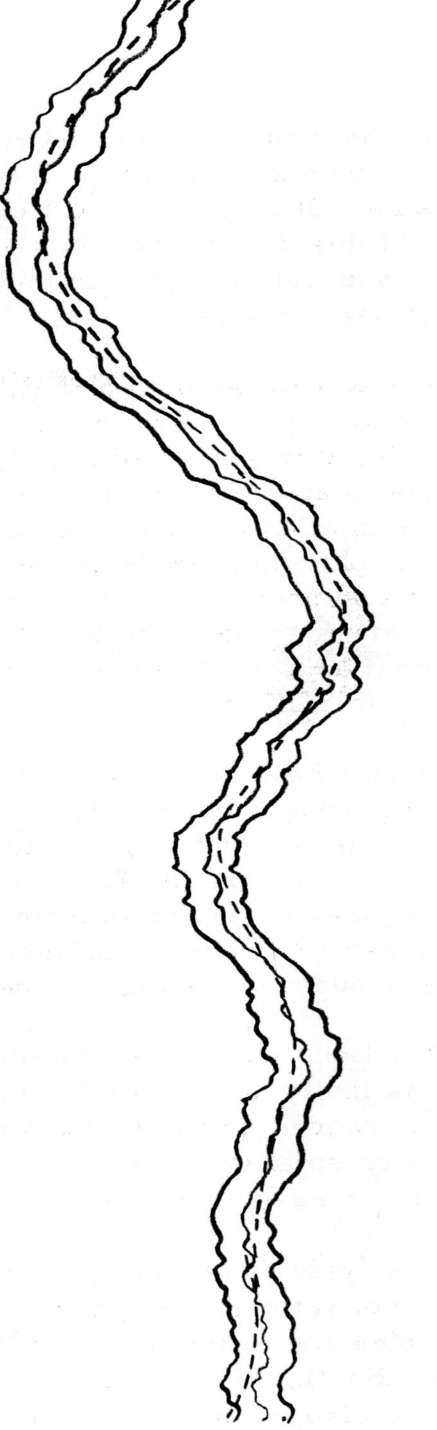

Figure 7. Average evoked response and two standard deviation confidence interval (solid lines) with an eleven term truncated Fourier approximation (dashed lines).

ACKNOWLEDGEMENTS

This study was supported in part by NIH Grant Computer use in the Health Sciences, 5 PO7 PR00011-09.

Mr. A. Angelone designed and built the square wave tone stimulus generator, the switching logic so that the wave was cut off at the end of a full cycle and the intertrial interval time control unit. He then loaned us his HiFi stereo headphones. Greater love of knowledge is indeed rare!

Mr. J. C. Christopher wrote the program to digitize the data and the software to generate the IBM 360 compatible data tape. He also digitized the data.

DATA SET DESCRIPTION

DATA - EVOKED OFF RESPONSE TO AN AUDIO TONE (D. C. MARTIN AND H. L. LUCAS)

CASES AND VARIABLES

THERE ARE SEVEN SUBJECTS WITH THREE RUNS EACH. EACH RUN CONSISTS OF 75 TRIALS WITH 256 SAMPLES PER TRIAL.

THE DATA ARE AVAILABLE ON A STANDARD LABELED 9-TRACK TAPE.

BRIEF DESCRIPTION

VAR	NAME
***	****
1	SUBJECT NUMBER AND RUN NUMBER
2	NUMBER OF POINTS IN TRIAL
3	STAT PACKAGE HEADER WORD
4	STAT PACKAGE HEADER WORD
5	STAT PACKAGE HEADER WORD
6	PAD
7	PAD
8	TRIAL NUMBER
9	NORMAL (TONE-ON) OR INVERTED (TONE-OFF)
10	NUMBER OF INTERRUPTS
11	SAMPLE RATE
12-267	DATA POINTS

PURPOSE OF STUDY

THE STUDY EXAMINED EVOKED ELECTRICAL POTENTIALS OF THE BRAIN TO AN AUDIO STIMULUS. ELECTRICAL POTENTIALS WERE AMPLIFIED BY A STANDARD GRASS MODEL 7 POLYGRAPH WITH 7T5 PREAMPLIFIERS. THE DATA WERE COLLECTED ON AN FM AMPEX ANALOGUE FR 1300 PORTABLE TAPE RECORDER AND A CONVENTIONAL INK STRIP CHART RECORDER. THREE CHANNELS OF DATA WERE COLLECTED. ONE CHANNEL WAS A MARKER CHANNEL TO INDICATE WHEN TO SAMPLE THE EEG. ONE CHANNEL WAS A TONE GENERATED BY THE STIMULUS GENERATOR. THE LAST WAS AN EEG CHANNEL. THE DATA RUNS ARE APPROXIMATELY SIX MINUTES LONG. THERE ARE SEVEN SUBJECTS WITH THREE RUNS EACH. EACH RUN CONSISTS OF 75 TRIALS WITH 256 SAMPLES PER TRIAL.

REFERENCE

MARTIN, D. C. AND H. L. LUCAS. PUBLISHED IN THIS VOLUME FOR THE FIRST TIME.

DETAILED DESCRIPTION

VAR	COL	FORMAT	DESCRIPTION
***	***	******	***********

```
  1     2-10   XXXXXXXXX.  SUBJECT NUMBER AND RUN NUMBER
                             FIRST DIGIT = SUBJECT NUMBER (1-7)
                             SECOND DIGIT = RUN NUMBER (1-3)
  2    11-19   XXXXXXXXX.  NUMBER OF POINTS IN TRIAL (256)
  3    20-28   XXXXXXXXX.  STAT PACKAGE HEADER WORD (327702)
  4    29-37   XXXXXXXXX.  STAT PACKAGE HEADER WORD (50)
  5    38-46   XXXXXXXXX.  STAT PACKAGE HEADER WORD (4194320)
  6    47-55   XXXXXXXXX.  PAD (0)
  7    56-64   XXXXXXXXX.  PAD (0)
  8    65-73   XXXXXXXXX.  TRIAL NUMBER (1-75)
  9    74-82   XXXXXXXXX.  NORMAL (TONE-ON) OR INVERTED (TONE-OFF)
                             1 = NORMAL (TONE-ON)
                             2 = INVERTED (TONE-OFF)
 10    83-91   XXXXXXXXX.  NUMBER OF INTERRUPTS (1 MS)/6
 11    92-100  XXXXXXXXX.  SAMPLE RATE 1000 HZ

 12     2-7       XXXXXX.  FIRST DATA POINT
...    ...        ...      ...
 30   110-115     XXXXXX.  NINETEENTH DATA POINT

                           THE NEXT 13 RECORDS ARE IDENTICAL IN FORMAT
                           TO THE PRECEDING ONE EXCEPT THAT THE LAST
                           RECORD CONTAINS ONLY NINE DATA POINTS.

FORMAT IS (1X,11F9.0/13(1X,19F6.0/),1X,9F6.0)
```

LOCATION OF DATA

THE DATA ARE AVAILABLE ON A STANDARD LABELED 9-TRACK TAPE LABELED DSLIB1.

DATA ****	DSNAME ******	LABEL *****
SUBJECT 1, RUN 1	SUB1RUN1	14
SUBJECT 1, RUN 2	SUB1RUN2	15
SUBJECT 1, RUN 3	SUB1RUN3	16
SUBJECT 2, RUN 1	SUB2RUN1	17
SUBJECT 2, RUN 2	SUB2RUN2	18
SUBJECT 2, RUN 3	SUB2RUN3	19
SUBJECT 3, RUN 1	SUB3RUN1	20
SUBJECT 3, RUN 2	SUB3RUN2	21
SUBJECT 3, RUN 3	SUB3RUN3	22
SUBJECT 4, RUN 1	SUB4RUN1	23
SUBJECT 4, RUN 2	SUB4RUN2	24
SUBJECT 4, RUN 3	SUB4RUN3	25
SUBJECT 5, RUN 1	SUB5RUN1	26
SUBJECT 5, RUN 2	SUB5RUN2	27
SUBJECT 5, RUN 3	SUB5RUN3	28
SUBJECT 6, RUN 1	SUB6RUN1	29
SUBJECT 6, RUN 2	SUB6RUN2	30
SUBJECT 6, RUN 3	SUB6RUN3	31
SUBJECT 7, RUN 1	SUB7RUN1	32
SUBJECT 7, RUN 2	SUB7RUN2	33
SUBJECT 7, RUN 3	SUB7RUN3	34

LISTING OF DATA FOR SUBJECT 1, TRIAL 1

```
(THE FIRST FIVE LINES ARE COLUMN GUIDES)

STUDIES WITH A TREMENDOUS AMOUNT OF DATA, SUCH AS THIS ONE, MUST
UTILIZE TAPE OR DISK.  THE SAMPLE OF DATA LISTED BELOW CONSISTS OF
TWO TRIALS FROM RUN 1 FOR SUBJECT 1.  THE ENTIRE FILE FOR THIS
STUDY IS ALMOST 800 TIMES AS LARGE AS THIS SAMPLE.

0         1         2         3         4         5         6
1234567890123456789012345678901234567890123456789012345678901234567890123456

                                                1         1
          7         8         9         0         1
789012345678901234567890123456789012345678901234567890123456789012345

       11      256   327702        50   4194320        0        0
                          1         2      5089     1000
  -1425    -463     831    1971    1303    1129    2679    2659    1489    -163      51
                            319    -257    -571    -449    -335   -1285    -667     -77
   -837   -1633    -545    -287   -1331    -219    -129   -2717   -3143   -2177   -1817
                          -2373   -1639   -1089   -1769   -2689   -3339   -3775   -2993
  -3329   -3681   -3813   -4007   -3495   -2967   -2353   -2129   -1861   -1921   -2389
                          -2097   -2273   -2243   -2467   -2113   -1247   -1537    -915
   -561   -1415   -2089   -2247   -2431   -2641   -2195   -2183   -2539   -1567   -2049
                          -3137   -4017   -2855   -2651   -2319   -1249   -2357   -4353
  -6159   -6947   -5633   -5421   -5857   -6321   -4743   -4783   -4961   -5557   -5639
                          -5645   -5361   -4401   -5001   -5311   -5559   -4893   -5105
  -4407   -4657   -4439   -3783   -3889   -3885   -3985   -3377   -3265   -3073   -2621
                          -2401   -2593   -2801   -2305   -2033   -2177   -2001   -1873
  -2369   -2733   -2959   -2817   -2913   -2539   -2205   -2115   -2087   -1921   -2129
                          -1711    -599    -199    -971   -1153   -1373   -1039   -1587
  -2689   -1867    -469    -185   -1217   -1619    -823    -385    -413   -1483   -1121
                           -701   -1137    -809    -579    -719    -341    -257    -283
    511     343    1487    2927    4591    4847    5311    4031    3953    4095    4351
                           4479    4847    5183    4479    4415    4095    4463    5439
   4991    4209    3315    2921    3441    3001    2623    2225    2337    2111    2401
                           1407    1613    2175    2047    1623    1745    1779    1469
   1493    1881    2047    2047    2105    1407    1249    1585    3071    2559    2389
                           3967    5231    3839    3041    2047    1319    1131    2429
   3199    3543    2673    3519    3761    3199    2047    1883    2407    1913    1599
                           1727    1919    2227    3199    4095    3763    3519    4095
   4095    4451    3427    3389    3823    4479    5567    5887    5103    3481    1919
                           1969    3627    4863    5487    4735    4991    4991    4479
   2869    1515    1913    2559    1663     313     383     255     511

       11      256   327702        50   4194320        0        0
                          2         2      5551     1000
   1521    2047    2639    4607    7599    6271    3199    -979    -577    1023    1279
                           1023    -481    -913    -965    -151     369     767     975
    745    1203    1991    1651    1727    2297    1981    1347    2175    2291    2375
                           2473    2687    2641    3455    3967    3525    2675    1787
   2175    2175    1887    1241    1407    1641    1383    1151     639     191    -157
                           -691    -845   -1231   -1541   -1715   -1725   -2183   -2583
  -2735   -3131   -3613   -3521   -3985   -3765   -3985   -4865   -4953   -4681   -4815
                          -5123   -4509   -4331   -4043   -3873   -3905   -4289   -4033
  -3377   -2999   -1927   -1667   -2889   -3045   -2775   -2707   -2457   -2239   -2063
                          -2545   -3161   -3521   -4145   -4503   -4395   -4487   -4273
  -3483   -2753   -2997   -3137   -3313   -3527   -3249   -3201   -3249   -3025   -3593
                          -3343   -2017   -1409   -1689   -2619   -2757   -1921   -2101
  -1889   -1551    -455     747    2639    3275    2337    2273    3135    3017    1983
                           2673    2793    2943    3429    2815    2303    2685    2765
```

Off Response to Audio Tone

3839	3839	4575	5359	5183	3125	2815	3041	3583	4479	4543
			4095	3957	3647	3665	3709	4351	4847	3929
3281	3581	2647	2417	1851	2559	3041	3391	2877	3071	2547
			2673	2397	2485	1625	751	1279	1247	1633
2785	2943	1879	1239	1535	1373	1535	1585	1747	1791	1023
			895	1279	1791	2453	1407	1225	1619	1279
1897	2343	1735	1149	637	−129	63	−65	−205	213	187
			−17	437	369	−1	615	625	633	−519
−385	−485	−641	−305	481	1407	2289	3433	3327	2267	1279
			895	733	1535	1971	3007	2813	1621	2673
3261	2929	1645	431	639	383	337	79	−1	1717	2257
			1649	995	255	−705	−371	73	767	1087
497	115	319	207	823	735	613	959	1215		

REFERENCES

Bartlett, F. and E.R. John (1970). Reply to Schwartz's comments. Science 169, 304-305.

Bingham, C., M.D. Godfrey and J.W. Tukey (1967). Modern techniques of power spectrum estimation. I.E.E.E. Transactions on Audio and Electroacoustics, AU15, No. 2, 56-66.

Donchin, E. (1969). Data analysis techniques in average evoked potential research. In Average Evoked Potentials, E. Donchin and D.B. Lindsley, Eds., Washington, D.C. Govt. Printing Offices, NASA SP-191, 199-217.

Freeman, W.J. (1964). A linear distributed feedback model for prepyriform cortex. Exp. Neurol. 10, 525-547.

Kiang, N.Y.S. and T.T. Sandel (1961). Off-responses from the auditory cortex of unanesthetized cats. Arch. Ital. Biol. 99, 121-134.

Morse, A.P. and F.E. Grubbs (1947). The estimation of dispersion from differences. Ann. Math. Statist. 18, 194-214.

Sandel, T.T. and N.Y.S. Kiang (1961). Off-responses from the auditory cortex of anesthetized cats. Arch. Ital. Biol. 99, 105-120.

Schwartz, M. (1970). Means and variances of average-response wave forms. Science 169, 303-304.

DISCUSSION OF THE LUCAS AND MARTIN CHAPTER

R. Mickey

Professors Martin and Lucas have presented us with a large data set from a carefully planned experiment. It even appears that they applied to themselves the advice they would give to others, particularly with respect to advance planning of the analysis. Unfortunately — in one sense at least — time did not permit the plans to come to fruition. On the other hand, as illustrative data for analysis this incompleteness leaves us with free scope to speculate on what might be interesting to try, without being inhibited by knowing what worked out well. I agree with Martin and Lucas that the problem is representative of a substantial class; it perhaps has the advantage of being more "neat" than many similar problems in that it may not present as many side issues to divert our attention.

One question that arises for me is whether or not there is in fact a response. I feel a bit uncomfortable in asking the question, since those in the evoked response business doubtless have methods of determining this that are acceptable to the community. Or is this possibility one of the questions for us to treat in the data analysis? The question of response may not be sensible to ask of a given trial or run, although it would seem natural to say of a given run that the subject showed a response, particularly if the response was large in some sense. One of the reasons for raising the question at all comes from wondering if one doesn't need pre-stimulus data to give an answer. Perhaps the question is a side issue.

What is available to look at in the way of results is the set of tracings of post-stimulus average potential for each of the three runs for each of the seven test subjects. By way of subjective analysis, I removed the treatment and run number identification and presented the tracings to a couple of subjects, telling them that for each test two of the tracings followed one stimulus and the third tracing followed a second stimulus; the subjects were asked to pick the one that was different. The task appeared to present no difficulties and both subjects correctly sorted the tracings for each of the seven tests. The replicate tracings seem to be quite similar and are sufficiently different from the third that there was no point in continuing with additional subjects. The replicate tracings are always the second and third runs, so it is possible that the distinction is an order rather than a treatment effect. The subjects were then asked to sort the tracings into two groups to correspond to the two treatments. One subject sorted the tracings in complete correspondence with treatment and the other had a few errors. The second task has substantially more uncertainty than the first because of the variability among test subjects, but nevertheless does not seem to present a guessing situation. My conclusion is that at least one of the stimuli elicits a response and, although there is quite a bit of variation among responders, the response patterns among test subjects are similar enough to generally distinguish the two stimuli.

I also tried ordering the test subjects by response patterns. This was done separately for the on response and the off response. My object was to group together patterns that seemed quite similar and to see if there was a gradation among the patterns. I did this separately for the on response and the off response. There was no rank correlation between the two orderings of the subjects. While there were some tracings that seemed relatively quite similar, (for example, the on response for subjects 1 and 4) in general I doubt that my orderings would be very reproducible were I to repeat the process after a suitable time lapse.

The results of one such ordering are shown in figure 1. I don't think there is any particular significance to the ordering, or even that attempting to order is particularly instructive. I do think the figure brings out similarities among patterns and may be useful for thinking about numerical analysis. For example, there appears to be an important amplitude variation among subjects. Also there is some suggestion that subjects vary in

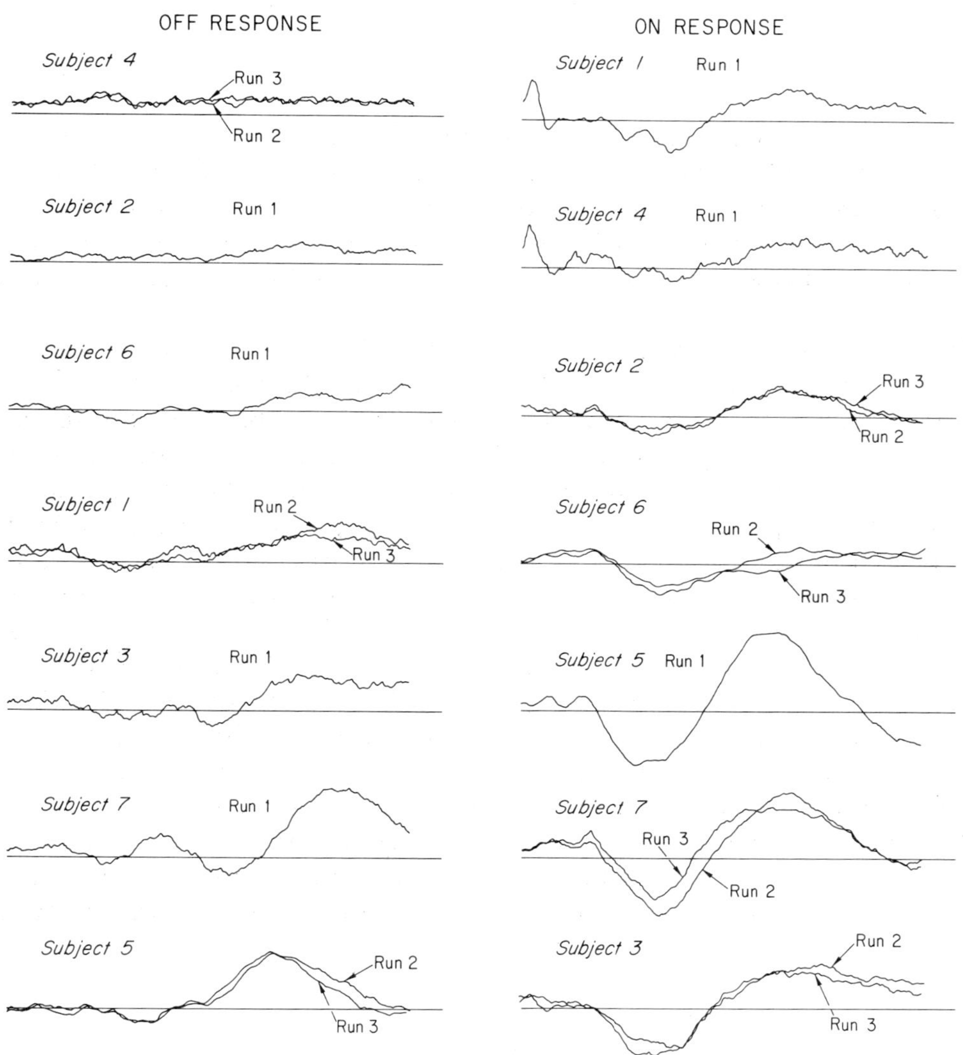

Figure 1. Evoked off and on responses to an audio tone arranged according to gradation of response. Tracings are from those supplied by Lucas and Martin. There is no difficulty in sorting out the responses of a given subject to the repeated stimulus, and little to moderate difficulty in distinguishing between on and off response in an unlabeled presentation of the response curves.

their time scale. Representations that explicitly allow for these variations might be more useful than representations that do not. I take the results as shown in figure 1 to indicate that there probably are some nicely informative ways of analyzing the data, and I look forward to the completed analysis by Martin and Lucas.

One of the difficulties with this type of problem seems to be that despite all of the numbers that enter the data base, there are not many degrees of freedom when it gets down to developing the final conclusions. This aspect of the analysis seems more severe if one were to take into account other attributes (such as hearing sensitivity, pitch discrimination ability, and perhaps indicators of nervous sensitivity) that might plausibly relate to the response. Since covariate type information is not available we do not need to consider how it might affect the analysis. Noting the possibility does raise the question of the extent to which the data analysis needs to be based on ideas of physiological interest. I will hazard the opinion that data analysts can develop findings of physiological interest with a minimum of guidance from physiology and that it is quite appropriate to proceed on our own. In order to contribute effectively to the scientific substance, however, I think that a second, third, etc. pass at the data will ordinarily be needed, and that these will require appropriate collaboration.

J. Tukey

(preconference comments)

The main issues here seem to be the extent to which background interference (mainly from alpha rhythms) can be allowed for, and what ought to be done to look more clearly into the data, especially in view of the possibility of entrainment of alpha rhythms.

Let us suppose that we have been able to digitize a long piece of record for many cases. For definiteness I suppose details that may not be essential, specifically that we have data for adjacent time intervals as follows:

Interval A = 256 points at 8 milliseconds

Interval B = 256 points at 1 millisecond

(stimulus change)

Interval D = 256 points at 1 millisecond

The first step is to analyze intervals A and B together, seeking for a good predictor at each point of interval B based on the values in interval A. As a practical matter it may well suffice to find regressions for points of interval B spaced every 8 to 16 milliseconds and interpolate between. Finding a smoothed mean spectrum for interval A will guide us to fairly good linear predictors, and stepwise adjustment (using both raw and locally smoothed values as possible carriers) can polish these predictors up easily.

Let now interval C consist of 256 points at 8 millisecond spacing running up to just before the stimulus change. It will contain

the last 256 - 32 = 224 points of interval A

every 8th point, 32 in all, of interval B.

We can now apply the same predictors to interval C that we found did well when applied to interval A. They may or may not do well in suppressing background over interval D. (If we had not changed the stimulus, they would have done well.)

Since we are in doubt as to how these predictors will work, it is natural to treat them as covariates, and ask with what coefficients they will do their best. These coefficients might well change as we move across interval D. If the pattern of their change is clear enough, this pattern ought to tell us more about what is really going on.

Plausibility arguments can be made for trying both "predictor value" and "smoothed rate of change of predictor value with time predicted for" as covariates. (An empiricist might also be willing to dump in the square of the latter.)

In any event, covariance, across trials but within individuals, ought to be tried as a way to get better data for the final analysis.

Another, more complex hypothesis, would be that the stimulus change entrains the alpha rhythms. A way to try to use this would be the following:

- make predictions through interval D as before;
- find their Fourier transforms;
- rotate each (cosine coeff, sine coeff) 2-vector in two ways, so the resulting covariate (a) has a + maximum at the stimulus change, (b) has a + maximum slope there;
- transform each of (a) and (b) back; and
- use three covariates, the original predictor and the retransforms of (a) and (b).

If this does better, it is the residuals from this that ought to go to further analysis.

What further analysis? It seems to me that a DFT (discrete Fourier transform) should come first. We know that different frequency bands behave quite differently. We ought to have a look at what we can do about this first.

While it would be possible to DFT the 256 values (which may have been adjusted by covariance) directly, which corresponds to using a rectangular data window, and while it may be worthwhile to do this for comparison, past experience suggests strongly that this will not turn out to be the better thing to do. DFT using a data window of the form

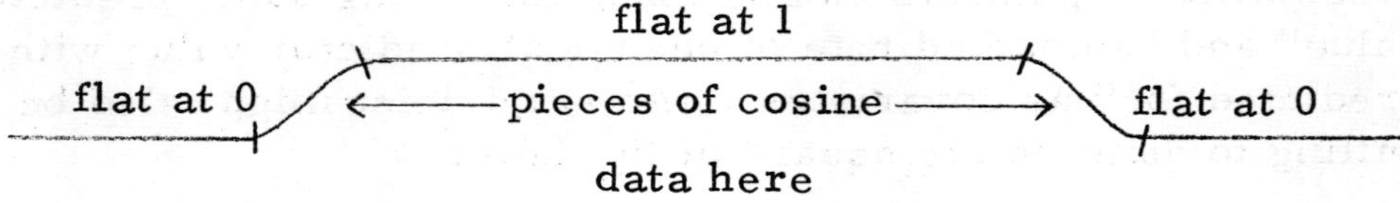

with the ends about 20 milliseconds long is much more likely to give clean results.

What do we do with our sine and cosine coefficients? Having obtained them for many runs for a single individual, we can think of plotting the sine-and-cosine-coefficient vectors for each frequency. We will get some kind of point cloud whose characteristics will change from frequency to frequency.

One simple approach that could be taken would be to ask what shrinking of the mean vector of each point cloud leads to minimum mean square error. After finding the shrunken values, they can be retransformed to a time function that might be a better typical pattern for an individual than the actual time means (themselves a retransformation of the unshrunken mean vectors).

Another approach would be to relate the DFT components for interval D with those for interval B. Here relationship could mean:

- ordinary (vector) regression, which would take care of unusual dependences of phase shift on frequency;
- regression on (or also on) the amplitude of the earlier vector; and
- sorting out, say into 3 or 4 equinumerous parts, in accord with the values of the earlier amplitude, followed by looking at spreads to see if weighting by some function of the earlier amplitude should be helpful.

Whether or not some time-side function expansion is to be used, it seems likely that an initial excursion to the frequency side will be well worthwhile.

Another possibility for the prediction game, which would be more compact, and might be more effective, would be to predict short period (interval B or D) Fourier coefficients from long period (interval A or C) Fourier coefficients. Here only about 8 or 10 (or maybe 16 or 20) 2-vectors would naturally appear as carriers for a given 2-vector, and the sample size could often be doubled by 90° rotations. Differences in regression coefficients for B on A and D on C could be assigned to "entrainment" and might throw additional light on the problem.

(later comments, after reading final version of paper)

1. We have not yet seen either middle or later stages of analysis for this problem.

2. I would like, were I analyzing the data, to see plots of "variance about mean" against "time after stimulus" and against "slope of mean." (And the latter for all but the first part of the response.) Hard looks at these might lead us to some insights as to plausible models (e. g., a time stretch-shrink component in the response ought to show up as increased variance near high slopes).

3. All of us — unwise statisticians that we are — would like to look at the full principal component analysis of dimension 256. (Except that one would need about 400 runs, one could think of doing this. I have done eigenvalues in 104 dimensions.) Why not plan to do something less than this, but more flexible than fitting chosen sets of functions?

One approach would be to take a stepwise approach. Consider times 16h + 1, 16h + 2, . . . , 16h + 16 for h = 0, 1, 2, . . . , 15. For each h separately, look at the deviations of the 75 runs from their mean (or the three sets of 25, each from its own means, etc.) and do a full principal components analysis. Hopefully 0 to 3 components will be outstanding at each h. If 0, we will include the largest available, giving us 1 to 3.

The second step is to combine the 16 to 40 resulting components in another principal component analysis. The truly outstanding components here, if any, can be taken as reasonable leads.

Given one such, we can separate the trials into 3 to 5 zones based on the value of this component. The means of the zones — more precisely the differences of, or appropriate contrasts among, these means — can give us then an even better picture of which components seem to show large variance (within subject).

An alternative to 16 blocks of 16 laid end to end is 15 blocks of 32 laid "16 to weather" (in shingle terminology), that is $16h + 1, 16h + 2, \ldots, 16h + 32$ for $h = 0, 1, \ldots, 15$.

In either case, the stepwise character of the principal components would keep us from looking at two weird components. (The limited results of the stepwise eigening may well prove safer than a full 256-dimensional analysis for fewer than, say, 2000 trials. Who knows?)

4. The tracing of 60 Hz synchronism to the stimulus control offers a simple design opportunity for later experiments. A reversing switch in the power lead for the control only should allow locking-in the 60 Hz in reversed phase on half the trials, so that its effect will cancel out in the mean.

5. Dr. John Hartigan's comments from "the pattern recognition point of view" tacitly assume that the features of importance will show most clearly in a time history. I see no reason to assume this — and, I guess, no reason to assume the contrary. It is an ever-present danger of data analysis to assume that, if the data would like to show us something, we can see it in the form first presented to us. (The great usefulness of polynomial fits, and row-PLUS-column analyses of 2-way tables of responses are prominent denials of any such assumption.)

CHAPTER 8

THE USE OF DENSITY ESTIMATES BASED ON ORTHOGONAL EXPANSIONS

RICHARD A. KRONMAL
Department of Biostatistics, University of Washington, Seattle

and

MICHAEL TARTER
Division of Biostatistics, University of California, Berkeley

In this chapter we discuss the application of orthogonal polynomial estimates of the univariate and bivariate densities as well as the cumulative for a set of data contrasting coronary heart disease patients with age matched "normals." We are concerned primarily with the use of these techniques for description of the data, particularly their use as possible replacements for the histogram, two variable plot and step function.

A description of the study is given below, and a data set description taken from a published paper describing the results of the study, is given at the end of the chapter.

The results of the use of these methods on the data are described, with particular attention given to the estimation of the cumulative and of the bivariate distributions of several of the variables. The general theory for orthogonal polynomial density estimation for the orthonormal system based on the complex trigonometric functions is also given.

DESCRIPTION OF THE STUDY

This analysis is based on a study conducted by Kasser and Bruce (1969) comparing a group of coronary heart disease patients with a group of age matched normals. The description given below is taken from the published paper with sections omitted for brevity.

> Coronary heart disease exhibits marked variability in its clinical manifestations. While some patients are disabled by angina pectoris, others surviving myocardial infarction pursue their usual activities with little or no symptoms. Nevertheless, hemodynamic studies indicate that most coronary patients have evidence of impaired myocardial contractility. When stressed by exercise or isoproterenol infusion, they show subnormal increments in cardiac output, stroke volume, and systolic ejection rate. More advanced disease may manifest overt left ventricular failure with elevated resting left ventricular end-diastolic and pulmonary capillary pressures.
>
> In a parallel manner, advancing age also adversely affects cardiovascular function. Whereas older individuals show only slight abnormalities at submaximal work loads, their maximal exercise performance is significantly limited. Oxygen consumption, heart rate, stroke volume, and cardiac output are all reduced at high work loads. Consequently, most middle-aged coronary patients suffer from the added limitations of both disease and increasing age.
>
> Many clinicians limit exercise testing of coronary patients to simply detecting myocardial ischemia from postexertional S-T depression. Yet with slightly more effort, an objective assessment of each patient's functional capacity can be obtained. In this regard, maximal exercise on a multistage treadmill has proved to be a useful test. In less than 12 min., asymptomatic as well as a class IV cardiac patients (New York Heart Association classification) can be exercised using the same test procedure, and their performance compared quantitatively to that of normal subjects.

This study was undertaken to define the maximal exercise capacity of normal men and coronary patients and assess the relative contributions of aging and disease to the overall functional limitations of an individual patient. . . .

One hundred seventeen (117) men, whose ages ranged from 34 to 73 years with a mean age of 52 years were selected because of a clinical diagnosis of definite coronary heart disease. The diagnosis was based on (1) recurrent exertional chest pain typical of angina pectoris, (2) a Q wave of at least 0.04-sec duration on a resting electrocardiogram, or (3) a clinical episode compatible with acute myocardial infarction substantiated by elevated serum enzymes or myocardial injury recorded on the electrocardiogram, or both. All men were in sinus rhythm, and none were receiving digitalis. No patient was exercised earlier than 2 months following a myocardial infarction.

For comparison, 117 age-matched normal men, whose ages ranged from 33 to 72, with a mean age of 51 years, were chosen from a group of healthy faculty and YMCA volunteers. In addition, a third group of 62 normal young men, whose ages ranged from 15 to 35 years, with a mean of 25 years, were evaluated in the outpatient clinic. Both these groups of men were judged to be normal by history, physical examination, chest x-rays, and 12-lead electrocardiogram. All normal subjects were normotensive according to World Health Organization criteria, with a resting blood pressure of less than 160/95 mm Hg. Data were also available on 79 of the older normal men who had been tested previously.

The multistage exercise test involves an uninterrupted series of work loads on a motor driven treadmill; the initial submaximal load (stage 1) requires walking slowly on a 10% grade. Since there is no increase in oxygen uptake after 3 min. of submaximal exertion, the speed and grade are increased every 3 min. Each person continues exercise until a self-determined end point of exhausting fatigue, marked

> dyspnea, aching or weakness in the legs, dizziness, chest pain, or various combinations of these....
>
> Each subject was examined before and after the test by a physician who remained in attendance during the entire procedure.
>
> ....Heart rates were recorded at rest and during each minute of exercise and recovery, while blood pressures were obtained at rest and on initial recovery as well as 3 and 6 min. after exertion.
>
> Exercise performance was evaluated by the following parameters of cardiac function: (1) total duration of exertion, (2) maximal heart rate, and (3) maximal systolic blood pressure....Resting heart rate and blood pressure were obtained while each subject was sitting just before the start of the exercise.

The aim of the study was primarily descriptive and exploratory. Thus the authors were interested in techniques that would describe their data well and allow them to develop hypotheses for future studies. The use of orthogonal polynomial nonparametric density estimation provided them with a useful tool for these purposes.

SOME EXAMPLES OF THE USE OF ORTHOGONAL POLYNOMIAL DENSITY ESTIMATES

Although there are many uses of density estimates, particularly in the area of multivariate analysis, the application described here is restricted to the description of data. The focus is on estimation of the cumulative, the univariate density, and bivariate density (Tarter and Kronmal, 1970; Kronmal and Tarter, 1968; and Cencov, 1962). We illustrate this application by examples from the data described earlier as well as with some Monte Carlo simulation data.

The use of a new technique for the description of data must be justified on at least three grounds: its mathematical statistical properties, its acceptability to applied statisticians, and most importantly its usefulness to the researcher.

Although there are a number of papers on the mathematical aspects of orthogonal polynomial estimation, there is little to suggest the use of these methods for descriptive purposes. However, this is to be expected for a topic as new and as mathematically difficult as this one because of its complexity and the lack of a clear definition as to what constitutes a good description of the data.

The large Monte Carlo studies described by Anderson (1969) seem to support the view that for sample sizes as small as 50, excellent estimates of the underlying probability density are obtained through the use of the orthogonal polynomial estimates. Figures 1 and 2 show the estimates of a density and cumulative respectively from a sample of size 200 from the normal distribution with mean 0 and variance 1. Figure 3 shows a graph of the approximate probability contours for a bivariate normal with means equal to zero and variances equal to 1 and correlation equal to 0.69 based on a sample size of 100. The program generating the contours is still in the developmental stage, so these plots appear on a line printer and thus are somewhat distorted. Although this program was unavailable at the time the Kasser-Bruce study was analyzed, we use it to illustrate the potential value of graphical displays of bivariate contours. One can see from figures 1, 2 and 3 that the estimates generated by these methods seem to be quite adequate for describing the populations from which the data were generated.

To our knowledge, little use has been made of any of the methods for nonparametric density estimation (with the possible exception of its use in classification). In a statistical system of computer programs (CCSS) for the maintenance and description of complicated data files, a program is included that produces estimates of the univariate density and cumulative (Kronmal et al, 1970). Other than this package of programs and its use by

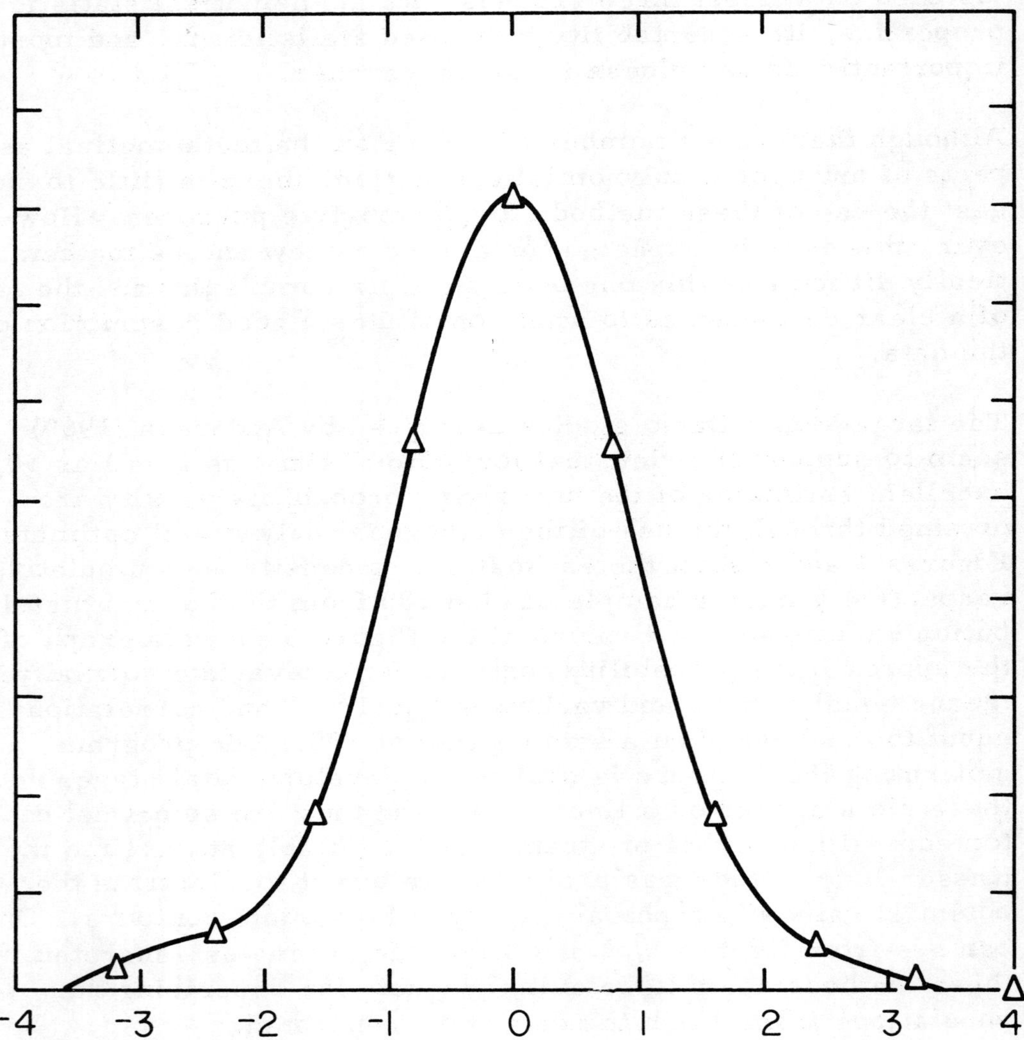

Figure 1. Estimate of density for sample of size 200 from N(0, 1).

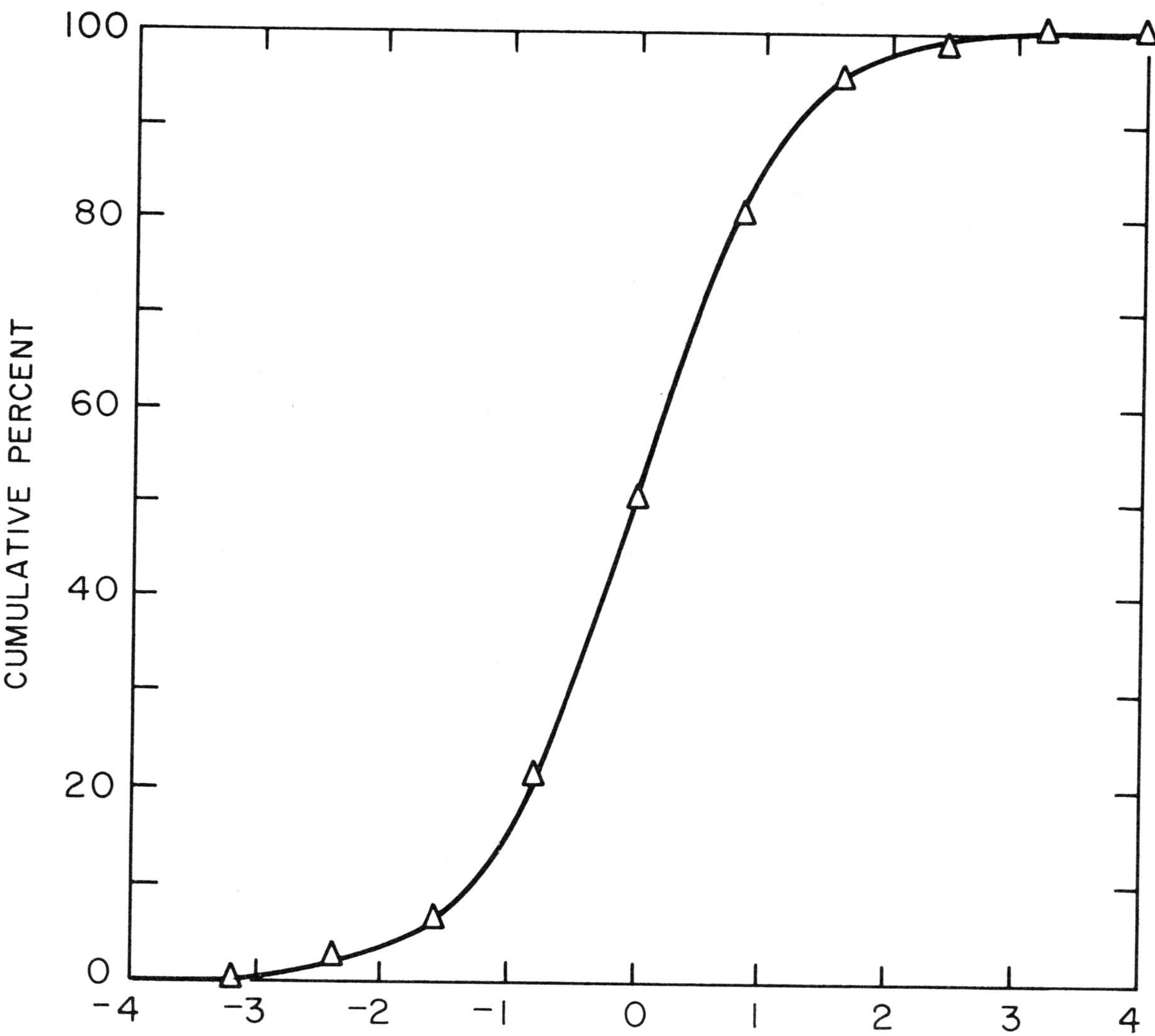

Figure 2. Estimate of cumulative distribution for sample size 200 from N(0, 1).

FIGURE 3. ESTIMATE OF CONTOURS OF PROBABILITY DENSITY FOR N(0,1,0,1,.8). THE HEIGHTS OF THE CONTOURS ARE SCALED TO THE APPROXIMATE MAXIMUM HEIGHT OF THE DENSITY. THE SYMBOLS 2, 4, 6, 8, M REPRESENT 20+5%, 40+5%, 60+5%, 80+5%, 100-5% RESPECTIVELY OF THE MAXIMUM HEIGHT OF THE DENSITY. FOR EXAMPLE, HEIGHTS FROM 15% TO 25% OF THE MAXIMUM ARE SHOWN AS A 2 ON THE PLOT. NUMBER OF OBSERVATIONS = 100.

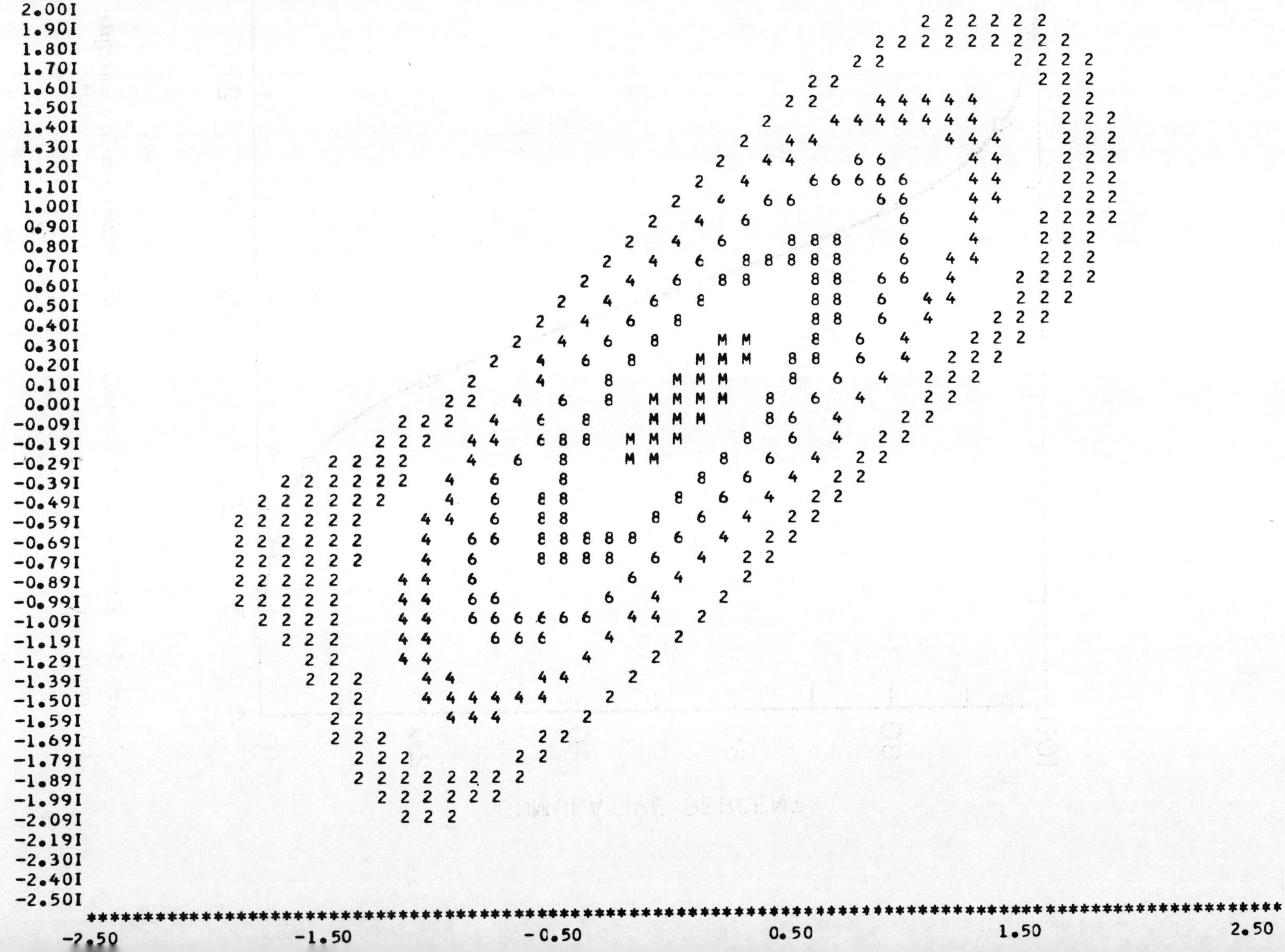

the authors in recent months we are unaware of others using these methods.

This technique is not yet well known among statisticians but it has found its way into research in cardiovascular disease. Kasser and Bruce (1969) used orthogonal polynomial cumulative distributions to describe cardiovascular properties of disease and matched normal groups. In this application, the sample step function would have been an acceptable substitute for the orthogonal polynomial estimates. Yet most medical investigators faced with a sample step function would probably have had them smoothed by an artist, probably resulting in a less adequate estimator than would be given by the methods described here. Of course, as in any descriptive techniques, one must use a certain amount of restraint in interpreting the results of such displays.

Tarter et al (1967) discussed the comparison of the histogram to the nonparametric density estimate:

> It is not too surprising that a graphical descriptive procedure which is well suited for sample description is not well suited for displaying an estimate of the population distribution. The very features which sharpen the histogram as a method of viewing isolated and extreme data points make it less powerful as a tool for the detection of such features of the population distribution as skewness, kurtosis, truncation or the existence of superposed component distributions.
>
> Too often the excess of visual cues present in a histogram can lead an experimenter to the wrong conclusion about the presence or absence of some of these features and in many cases the picture yields such a bewildering array of gaps, plateaus and spikes that the experimenter disregards the histogram entirely. Of course the rough edges of the histogram or frequency polygon may be used by the experimenter to remind himself that only a limited sample of information is available. However, there may also be a

subconscious tendency to interpolate across these rough edges which in turn may be dangerous if the interpolated picture is made to look the way the experimenter wants it to look.

We used the same data to illustrate density estimation by our methods. Figures 4 and 5 show the density estimates for the resting and maximum heart rate for the patient group. Note particularly that there seems to be a bimodal distribution for the maximum heart rate. Figure 6 shows the bivariate contours for these same two variables. Again note the bimodal appearance of this distribution. Figure 7 shows the bivariate contours for the same two variables but in this instance for the normal individuals. The scale for this plot is somewhat different than that of the previous contour plot, and an examination of these plots shows there is a considerable difference in both the shape and location of the two distributions.

Figure 8 gives the bivariate density estimate for the diastolic and systolic blood pressure of the coronary patients. To see what type of a contour picture would be presented by data from a normal distribution, a sample was generated with the same means, variances and correlation as the patient data (figure 9). Although these two distributions differ somewhat, it is remarkable how similar they appear.

One might be suspicious of a method of estimation of probability densities that gave "pretty" pictures on all samples of real data, since it is most likely that some proportion of real data follows very irregular distributions. Figure 10 shows the estimated contours for the normals for diastolic versus systolic blood pressure. As seen from this figure, these contours lack the regular features exhibited by the previous figures. A glance at figure 11, which is a plot of the systolic versus diastolic blood pressure in these individuals, shows the irregularities in the pattern of the data.

FIGURE 4. DENSITY ESTIMATE FOR RESTING HEART RATE FOR THE PATIENT GROUP.

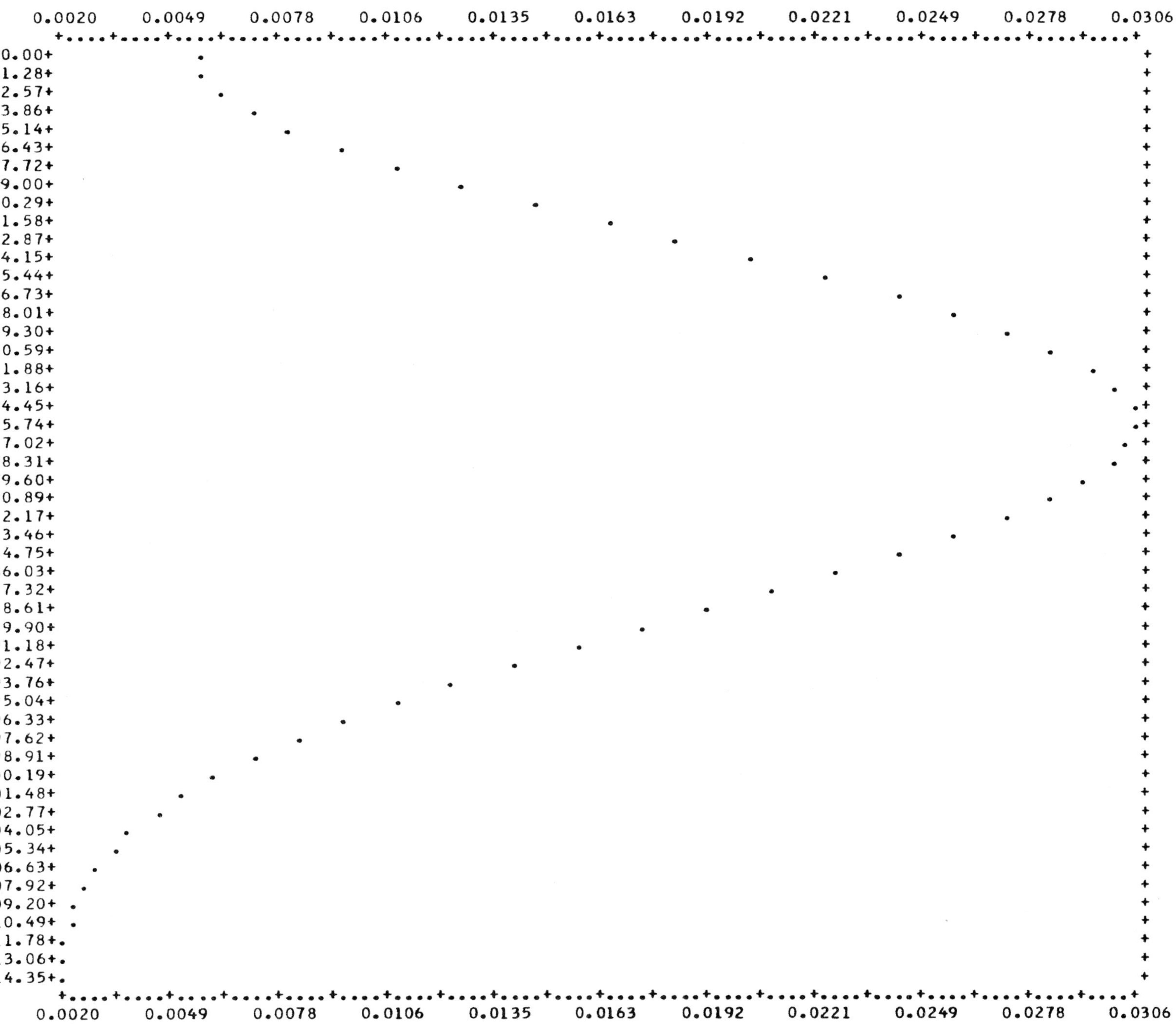

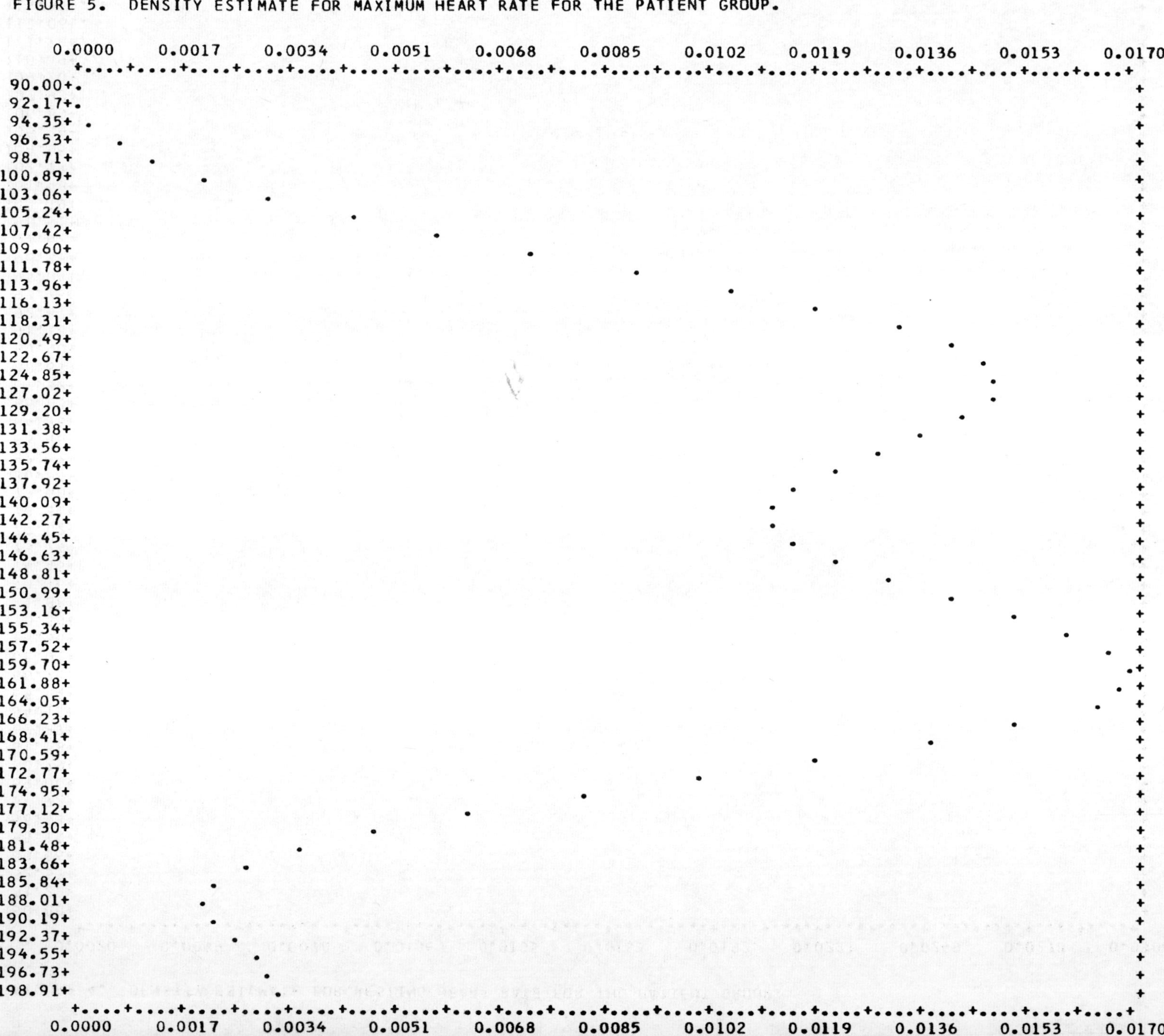

FIGURE 5. DENSITY ESTIMATE FOR MAXIMUM HEART RATE FOR THE PATIENT GROUP.

FIGURE 6. ESTIMATE OF BIVARIATE DENSITY OF MAXIMUM AND RESTING HEART RATE IN THE PATIENT GROUP. SEE FIGURE 3 FOR EXPLANATION OF CONTOURS. NUMBER OF OBSERVATIONS = 117.

```
MAXHR

200.00I
197.80I
195.60I
193.40I
191.20I
189.00I
186.80I
184.60I
182.40I
180.20I
178.00I
175.80I
173.60I
171.40I
169.20I
167.00I
164.80I
162.60I
160.40I
158.20I
156.00I
153.80I
151.60I
149.40I
147.20I
145.00I
142.80I
140.60I
138.40I
136.20I
134.00I
131.80I
129.60I
127.40I
125.20I
123.00I
120.80I
118.60I
116.40I
114.20I
112.00I
109.80I
107.60I
105.40I
103.20I
101.00I
98.80I
96.60I
94.40I
92.20I
90.00I
     50.00        63.00        76.00        88.99        101.99        114.99

                                RESTHR
```

FIGURE 7. CONTOURS OF THE ESTIMATED PROBABILITY DENSITY FOR MAXIMUM AND RESTING HEART RATE IN THE NORMAL GROUP. SEE FIGURE 3 FOR EXPLANATION OF CONTOURS. NUMBER OF OBSERVATIONS = 117.

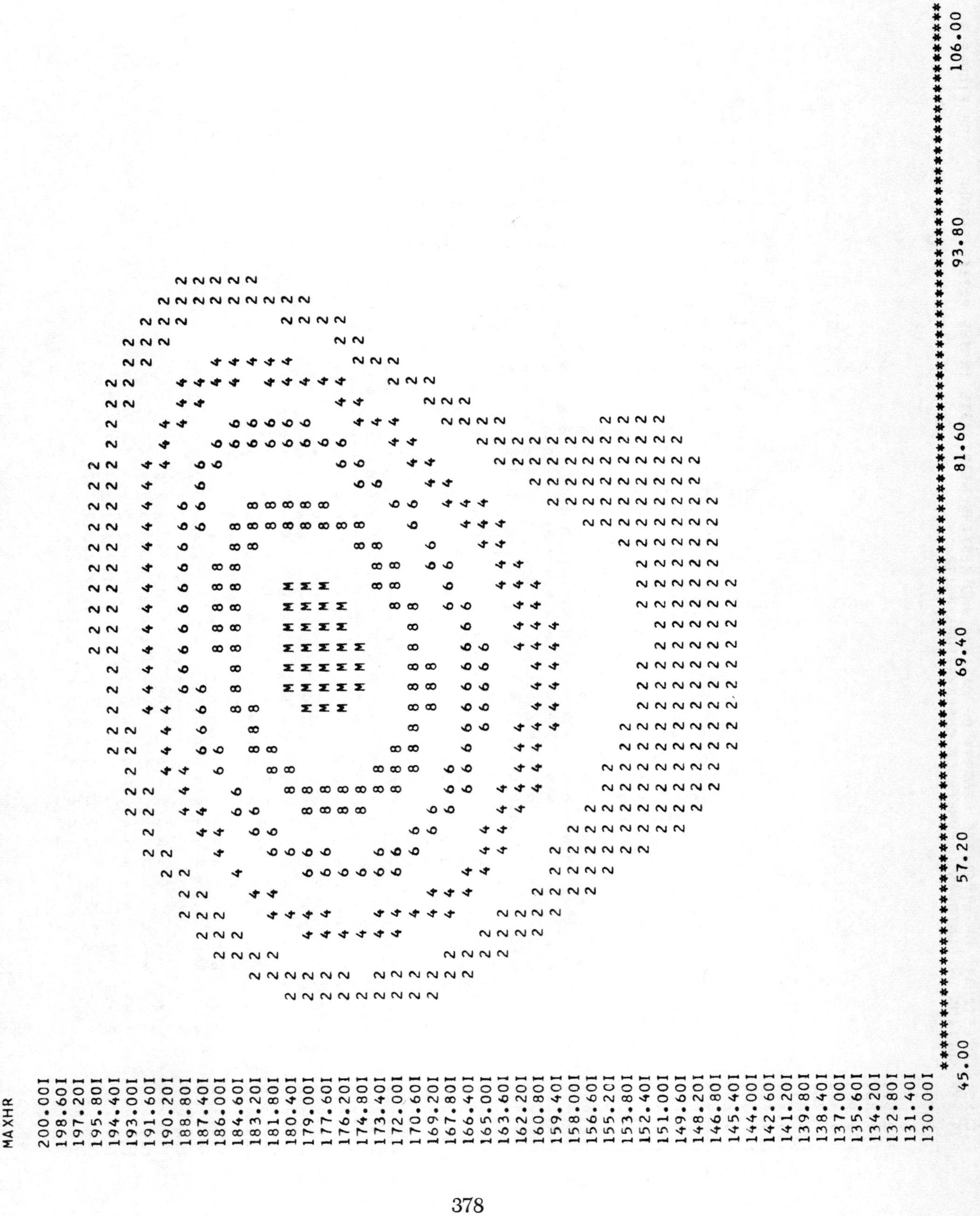

MAXHR

200.00I
198.60I
197.20I
195.80I
194.40I
193.00I
191.60I
190.20I
188.80I
187.40I
186.00I
184.60I
183.20I
181.80I
180.40I
179.00I
177.60I
176.20I
174.80I
173.40I
172.00I
170.60I
169.20I
167.80I
166.40I
165.00I
163.60I
162.20I
160.80I
159.40I
158.00I
156.60I
155.20I
153.80I
152.40I
151.00I
149.60I
148.20I
146.80I
145.40I
144.00I
142.60I
141.20I
139.80I
138.40I
137.00I
135.60I
134.20I
132.80I
131.40I
130.00I

45.00 57.20 69.40 81.60 93.80 106.00

RESTHR

FIGURE 8. CONTOURS OF THE ESTIMATED PROBABILITY DENSITY FOR SYSTOLIC AND DIASTOLIC BLOOD PRESSURE FOR THE PATIENT GROUP. SEE FIGURE 3 FOR EXPLANATION OF CONTOURS. NUMBER OF OBSERVATIONS = 117.

```
PDISBP

125.00I
123.50I
122.00I
120.50I
119.00I
117.50I
116.00I
114.50I
113.00I
111.50I
110.00I
108.50I
107.00I
105.50I                                                        2 2 2 2 2 2
104.00I                                                  2 2 2 2 2 2 2 2 2 2
102.50I                                              2 2 2               2 2 2 2
101.00I                                          2 2 2         4 4           2 2 2
 99.50I                                        2 2     4 4 4 4 4 4 4 4 4       2 2 2
 98.00I                                    2 2     4 4               4 4 4       2 2 2
 96.50I                                2 2     4 4     6 6 6 6 6       4 4 4     2 2 2 2
 95.00I                              2     4     6 6 6         6 6 6     4 4     2 2 2 2
 93.50I                          2 2   4 4   6 6                 6 6     4 4     2 2 2 2 2
 92.00I                        2     4   6 6     8 8 8 8 8 8     6 6     4 4     2 2 2 2
 90.50I                    2 2   4     6     8 8         8 8 8   6 6     4       2 2 2
 89.00I                  2     4   6 6   8 8               8     6 6   4 4     2 2 2
 87.50I              2 2   4 4   6     8       M M M     8 8   6 6   4 4     2 2 2
 86.00I            2 2   4 4   6   8 8     M M M M M     8     6   4 4     2 2
 84.50I          2 2   4     6   8 8     M M M M M     8     6   4 4     2 2
 83.00I        2 2   4     6   8 8     M M M M M     8     6   4 4   2 2
 81.50I    2 2 2   4 4   6     8     M M M M       8   6 6   4     2 2
 80.00I    2 2   4 4   6     8 8               8 8   6     4   2 2
 78.50I  2 2     4     6     8 8           8 8     6   4 4   2 2
 77.00I  2     4 4   6 6     8 8 8 8 8 8 8     6     4     2
 75.50I  2     4     6 6     8 8 8 8 8     6 6   4 4   2 2
 74.00I  2     4     6 6               6 6     4     2
 72.50I  2     4 4     6 6 6     6 6 6     4 4   2 2
 71.00I  2     4 4       6 6 6 6 6     4 4     2
 69.50I  2     4 4 4               4 4     2 2
 68.00I  2       4 4 4 4   4 4 4 4       2 2
 66.50I  2 2       4 4 4 4 4 4       2 2
 65.00I  2 2                     2 2 2
 63.50I  2 2 2               2 2 2
 62.00I  2 2 2 2         2 2 2 2
 60.50I  2 2 2 2 2 2 2 2 2 2 2
 59.00I  2 2 2 2 2 2 2 2 2
 57.50I    2 2 2 2 2 2
 56.00I    2 2 2 2
 54.50I
 53.00I
 51.50I
 50.00I
      ********************************************************************************************************
      95.00          115.00          135.00          155.00          175.00          195.00

                                    PSYSBP
```

FIGURE 9. ESTIMATE OF CONTOURS FOR SAMPLE FROM BIVARIATE NORMAL WITH MEANS, VARIANCES AND CORRELATION EQUAL TO THE VALUES OF THESE FOR THE DATA SHOWN IN FIGURE 8. SEE FIGURE 3 FOR EXPLANATION OF CONTOURS. NUMBER OF OBSERVATIONS = 117.

```
125.00I
123.50I
122.00I
120.50I
119.00I
117.50I
116.00I
114.50I
113.00I
111.50I
110.00I
108.50I                                                                             2 2 2 2
107.00I                                                                       2 2 2 2 2 2 2
105.50I                                                           2 2 2 2 2 2 2 2 2 2 2 2 2
104.00I                                                     2 2 2 2 2 2 2 2 2 2 2 2 2 2 2 2
102.50I                                               2 2 2 2                       2 2 2 2
101.00I                                             2 2                               2 2 2
 99.50I                                         2 2     4 4 4 4 4 4 4 4 4 4           2 2 2
 98.00I                                       2     4 4 4             4 4 4 4 4       2 2 2
 96.50I                                   2 2   4 4       6 6 6           4 4 4       2 2 2
 95.00I                                 2     4     6 6 6 6 6 6 6 6       4 4 4       2 2
 93.50I                             2 2   4 4   6 6             6 6 6       4 4     2 2 2
 92.00I                         2 2     4     6     8 8 8 8       6 6       4 4     2 2 2
 90.50I                     2 2 2     4   6 6   8 8 8 8 8 8 8     6 6     4 4       2 2
 89.00I             2 2 2 2 2 2   4 4   6     8 8         8 8 8   6 6     4 4     2 2 2
 87.50I         2 2 2 2 2 2     4 4   6     8               8     6 6   4 4       2 2
 86.00I       2 2 2 2 2 2     4 4   6     8     M M M       8     6     4 4     2 2 2
 84.50I     2 2 2 2 2 2     4 4   6     8     M M M M     8 8   6     4 4     2 2 2
 83.00I     2 2 2 2 2     4 4   6 6   8     M M M M M     8   6 6   4 4     2 2 2
 81.50I     2 2 2 2       4     6   8 8     M M M M     8   6 6   4 4     2 2 2
 80.00I     2 2 2 2     4 4   6 6   8     M M M M     8     6   4 4     2 2 2
 78.50I     2 2 2       4 4   6     8               8     6   4 4     2 2 2
 77.00I     2 2 2       4     6     8           8 8   6 6   4     2 2 2
 75.50I     2 2 2     4 4     6     8 8 8   8 8 8   6 6   4     2 2 2
 74.00I     2 2 2     4 4     6     8 8 8 8 8     6     4     2 2
 72.50I     2 2 2       4     6 6             6 6   4 4   2 2
 71.00I       2 2 2     4 4     6 6 6     6 6     4     2 2
 69.50I       2 2 2     4 4       6 6 6 6     4 4   2 2
 68.00I       2 2 2       4 4             4 4     2 2
 66.50I         2 2 2       4 4 4 4 4 4 4 4     2
 65.00I           2 2 2       4 4 4 4 4     2 2
 63.50I           2 2 2 2               2 2 2
 62.00I             2 2 2 2 2       2 2 2
 60.50I                 2 2 2 2 2 2 2 2
 59.00I                     2 2 2
 57.50I
 56.00I
 54.50I
 53.00I
 51.50I
 50.00I
       ********************************************************************************************************
     95.00              115.00              135.00              155.00              175.00              195.00
```

FIGURE 10. ESTIMATED CONTOURS OF THE PROBABILITY DENSITY FOR SYSTOLIC AND DIASTOLIC BLOOD PRESSURE IN THE NORMAL GROUP. SEE FIGURE 3 FOR EXPLANATION OF CONTOURS. NUMBER OF OBSERVATIONS = 117.

```
PDISBP

110.00I
109.00I
108.00I
107.00I
106.00I
105.00I
104.00I
103.00I
102.00I
101.00I                                    2 2 2 2 2 2 2 2 2 2 2
100.00I                                2 2 2 2 2 2             2 2 2 2 2 2
 99.00I                                      2 2 2 2 2 2 2 2 2 2 2 2 2 2 2 2 2
 98.00I
 97.00I
 96.00I
 95.00I
 94.00I
 93.00I
 92.00I                    2 2 2 2 2 2 2 2 2 2 2 2 2 2 2
 91.00I        2 2 2       4 4 4 4             4 4 4       2 2
 90.00I          2 2     4 4     6 6 6 6 6 6 6 6 6     4 4     2 2
 89.00I            2 2 2     4 4       6 6 6 6         4 4     2 2 2
 88.00I                  2 2         4 4 4 4 4 4         2 2 2
 87.00I                      2 2 2 2 2 2 2 2 2 2
 86.00I                            2 2 2
 85.00I                          2 2 2 2 2 2
 84.00I                      2 2 2 2 2 2 2 2 2 2 2
 83.00I            2 2       4 4 4 4 4 4 4 4       2 2 2
 82.00I  2 2     4 4     6 6             6 6     4 4     2 2
 81.00I      4   6 6   8 8               8 8     6   4 4   2 2                                             2
 80.00I  2   4 4   6   8 8     M M M M M     8 8   6     4     2 2                                         2
 79.00I    2 2   4     6     8 8 8 8 8 8 8     6 6     4 4     2 2 2
 78.00I        2 2     4 4 4             4 4 4         2 2 2 2 2 2 2
 77.00I          2 2 2 2 2     2 2 2 2 2
 76.00I              2 2 2 2
 75.00I
 74.00I
 73.00I
 72.00I              2 2 2 2 2 2
 71.00I      2 2 2 2 2 2 2 2 2 2 2 2 2
 70.00I      2 2 2 2 2 2 2 2 2 2 2 2 2 2
 69.00I          2 2 2 2 2 2 2 2 2
 68.00I
 67.00I
 66.00I
 65.00I
 64.00I
 63.00I
 62.00I
 61.00I
 60.00I
      ****************************************************************************************************
     95.00              114.00              133.00              151.99              170.99              189.99

                                                   PSYSBP
```

```
FIGURE 11.  SCATTERGRAM OF SYSTOLIC VS. DIASTOLIC BLOOD PRESSURE FOR THE NORMAL GROUP.  THE SYMBOLS 2, 3, ..., A
REPRESENT THE NUMBER OF POINTS PLOTTED AT APPROXIMATELY THE SAME COORDINATES (A=10).  NUMBER OF POINTS = 113.
PDISBP

125.00+
123.50.
122.00.
120.50.
119.00.
117.50+
116.00.
114.50.
113.00.
111.50.
110.00+                                                *                  *
108.50.
107.00.
105.50.
104.00.
102.50+
101.00.
 99.50.                             2  *   *   * *          2
 98.00.
 96.50.                                            *
 95.00+
 93.50.                             *                *
 92.00.                                          *
 90.50.      *   2  *   * 2 2       5  *     3         *        *   *     *                  *
 89.00.
 87.50+               *   *         *        *
 86.00.                         *            *
 84.50.    *     2      2 2         *                  *
 83.00.
 81.50.            *     2     *      *   *
 80.00+2          42 *   * 7   3 *   A        3         *        *
 78.50.            *       *
 77.00.
 75.50.           *      * *                        *   *
 74.00.         *
 72.50+                    *
 71.00.
 69.50.           3          *                          2
 68.00.           **
 66.50.            *                      *
 65.00+
 63.50.
 62.00.
 60.50.
 59.00.
 57.50+
 56.00.
 54.50.
 53.00.
 51.50.
 50.00+
       +....+....+....+....+....+....+....+....+....+....+....+....+....+....+....+....+....+....+....+....+
    100.00    110.50    121.00    131.50    142.00    152.50    163.00    173.50    184.00    194.50    205.00
                                                PSYSBP
```

THEORETICAL RESULTS

In this section we present the theoretical results necessary to understand the use of orthogonal series expansions for the estimation of probability densities. We restrict the choice of an orthogonal series to the trigonometric polynomials. Most of the results given here can be found in Tarter and Kronmal (1970). For other results on the same subject we refer the reader to Kronmal and Tarter (1968), Cencov (1962) and Anderson (1969).

The following notation, given by Zygmund (1959) will be used in this section: The symbols j and k represent p-tuples of integers. The sets M and N and their complements $\overline{M}$ and $\overline{N}$ are composed of such p-tuples. The symbol χ is used to represent a p-tuple of real variables and B_k and $\hat{B}_k$ represent complex valued scalars. The scalar $\hat{B}_k$ is a statistic formed from n I. I. D. random variates. The complex value functions $\psi_k(\chi)$ are defined by

$$\psi_k(\chi) = e^{2\pi i k' \chi} \tag{1}$$

Let X_s $(s = 1, \ldots, n)$ be n p-vectors of real valued I. I. D. random variates with multivariate density f(x) with respect to Lebesgue measure on R where it is assumed that $f \varepsilon L_2$ and R represent the p-dimensional unit hypercube in p-dimensional Euclidean space, E^p. B_k is defined as the k-th coefficient in the Fourier expansion of f,

$$B_k = \int_R f(x)\, e^{-2\pi i k' x} dx \tag{2}$$

$\hat{B}_k$ will be defined as the following estimate of B_k

$$\hat{B}_k = \left(\frac{1}{n}\right) \sum_{s=1}^{n} e^{-2\pi i k' X_s} . \tag{3}$$

It is important to note the obvious fact that

$$E\,\hat{B}_k = B_k\,. \tag{4}$$

From the conditions given above we can express f(x) in terms of its Fourier series expansion as

$$f_N(x) = \sum_{k\varepsilon N} B_k\,\psi_k(x)\,. \tag{5}$$

We further define $\hat{f}_M$, our estimate of f, as

$$\hat{f}_M(x) = \sum_{k\varepsilon M} \hat{B}_k\,\psi_k(x) \tag{6}$$

where $M \subseteq N$.

Theorem 1 shows that not only is the first moment of the statistic $\hat{B}_k$ expressible in terms of the Fourier coefficients of f, as would be true if any orthogonal ψ_k were substituted into expressions (2) and (3), but the second ordinary moments as well as the product moments of $\hat{B}_k$ are likewise easily expressible by equation (7).

<u>Theorem 1.</u> <u>For all</u> j <u>and</u> k

$$E[(\hat{B}_k - B_k)(\overline{\hat{B}}_j - \overline{B}_j)] = \frac{1}{n}[B_{k-j} - B_k B_{-j}] \tag{7}$$

which in turn equals

$$\frac{1}{n}[B_{k-j} - B_k \overline{B}_j] \tag{8}$$

<u>Proof</u>:

$$E[(\hat{B}_k - B_k)(\overline{\hat{B}}_j - \overline{B}_j)] = E\,\hat{B}_k\overline{\hat{B}}_j - B_k\overline{B}_j. \tag{9}$$

$$E\hat{B}_k\bar{\hat{B}}_j = E\left(\frac{1}{n}\right)^2\left\{\sum_{s=1}^{n} e^{-2\pi i[(k-j)/X_s]} + \sum_{s\neq t} e^{-2\pi i(k/X_s - j/X_t)}\right\} \quad (10)$$

$$= \left(\frac{1}{n}\right)\left\{Ee^{-2\pi i(k-j)/x} + (n-1)(Ee^{-2\pi ik/x})(Ee^{2\pi ij/x})\right\} \quad (11)$$

Expression (7) follows from the substitution of expression (2) and (11) into (9) and expression (8) results from the identity $\bar{B}_j = B_{-j}$ which is apparent from expression (2). We define the M.I.S.E. (mean integrated square error) as

$$J(f_N, \hat{f}_M) = E\int_R [f_N(\chi) - \hat{f}_M(\chi)]\,[f_N(\chi) - \hat{f}_M(\chi)]\,d\chi. \quad (12)$$

Substituting for $\hat{f}_M$ and f_N their orthogonal expansions (5), (6) and letting $j = k$ in expression (8) yields the following theorem.

Theorem 2.

$$J(f_N, \hat{f}_M) = \left(\frac{1}{n}\right)\sum_{k\epsilon M}[1 - B_k\bar{B}_k] + \sum_{k\epsilon(N\bar{M})} B_k\bar{B}_k\,. \quad (13)$$

If the set M^+ is defined as $M\{k_0\}, k_0 \notin M$, we get:

Corollary 1.

Define the error increment (ΔJ_{k_0}) due to adding a term k_0 as

$J(f_N, \hat{f}_M+) - J(f_N, \hat{f}_M)$, then

$$\Delta J_{k_0} = 2\left\{\frac{1}{n} - \frac{n+1}{n} B_{k_0}\bar{B}_{k_0}\right\}. \quad (14)$$

Note that the sign of ΔJ_{k_0}, equals the sign of $\{\frac{1}{n+1} - B_{k_0}\bar{B}_{k_0}\}$ or in other words, the inclusion of the k_0-th term to form estimate f_{M^+}, as opposed to the estimate $\hat{f}_M$, is indicated whenever

$$B_{k_0} \bar{B}_{k_0} > \frac{1}{n+1} . \tag{15}$$

As suggested by Cencov, "it is then possible to select n, 'here k_0,' according to results of observations, restricting oneself... only to such terms whose coefficients '$\hat{B}_{k_0}$' are essentially greater than their experimental mean square error." However the substitution of the estimate $\hat{B}_{k_0}$ for B_{k_0} in (15) will result in a biased rule that would select too many terms. It will be shown using Corollary 2 that the appropriate inclusion rule would indicate the use of the term k_0 whenever

$$\hat{B}_{k_0} \bar{\hat{B}}_{k_0} > \frac{2}{n+1} \tag{16}$$

rather than

$$\hat{B}_{k_0} \bar{\hat{B}}_{k_0} > \frac{1}{n+1} .$$

<u>Corollary 2.</u>

<u>Let</u>

$$\hat{\Delta J}_{k_0} = 2 \{ \frac{1}{n} - (\frac{n+1}{n}) [\frac{n}{n-1} \hat{B}_{k_0} \bar{\hat{B}}_{k_0} - \frac{1}{n-1}]\} . \tag{17}$$

<u>Then</u>

$$E(\hat{\Delta J}_{k_0}) = \Delta J_{k_0} . \tag{18}$$

Corollary 2 follows from expression (11). Now from (17) we see that the sign of $\hat{\Delta J}_{k_0}$, the unbiased estimator of ΔJ_{k_0}, equals the sign of

$$\{\frac{2}{n+1} - \hat{B}_{k_0}\overline{\hat{B}}_{k_0}\}$$

i.e., we use expression (16) for our inclusion rule.

It is worth noting that $\hat{f}_M$ will be a real valued function if the term selection rule given in formula (16) is used since if $\hat{B}_k$ is selected then $\hat{B}_{-k}$ will also be selected and this has the effect of cancelling out the imaginary portion of $\hat{f}_M$.

CONCLUSION

We would like to emphasize that we profess no special insight into the usefulness or value of these methods. We do feel, however, that such methods may find a place in the tools available to the statistician and investigator for the description of data.

ACKNOWLEDGEMENT

Thanks to Mr. S. Raman, a candidate for a Ph. D. in Biostatistics at Berkeley, for the use of his program for the contours estimation. We would also like to express our appreciation to Drs. Bruce and Kasser for permission to reprint portions of their paper and for the use of their data.

This research was supported in part by PHS Research Career Award Grant 1-KO3-GM38645-01.

DATA SET DESCRIPTION

DATA - KASSER AND BRUCE CORONARY HEART DISEASE STUDY (R. KRONMAL)

CASES AND VARIABLES

234 CASES (SUBJECTS)
34 VARIABLES - 16 MEASUREMENT, 16 CATEGORICAL, 2 IDENTIFICATION

BRIEF DESCRIPTION

VAR	NAME
***	****
1	SUBJECT NUMBER
2	CARD NUMBER
3	SOURCE
4	SEX
5	AGE
6	FUNCTIONAL CLASS
7	ACTIVE
8	DIGITALIS TREATMENT
9	12-LEAD ECG RESULTS
10	RHYTHM
11	SYSTOLIC BLOOD PRESSURE
12	RESTING HEART RATE
13	MAXIMUM HEART RATE
14	DURATION ON TREADMILL
15	SYSTOLIC BLOOD PRESSURE AT PHYSICAL EXAM
16	DIASTOLIC BLOOD PRESSURE AT PHYSICAL EXAM
17	MAXIMUM SYSTOLIC BLOOD PRESSURE ON TREADMILL
18	MAXIMUM DIASTOLIC BLOOD PRESSURE ON TREADMILL
19	HISTORY OF MYOCARDIAL INFARCTION
20	HISTORY OF ANGINA PECTORIS
21	ECG
22	XRAY
23	HISTORY OF HIGH BLOOD PRESSURE
24	SURVIVAL STATUS
25	MONTHS
26	NUMBER OF PAST MYOCARDIAL INFARCTIONS
27	ANGINA PECTORIS
28	CONGESTIVE HEART
29	HEART RATE AT STAGE ONE ON TREADMILL
30	CHANGE IN HEART RATE
31	SYSTOLIC BLOOD PRESSURE TIMES HEART RATE
32	VOLUME OF OXYGEN PREDICTED FOR HEALTHY MAN WITH SIMILAR CHARACTERISTICS
33	ACTUAL ESTIMATED VOLUME OF OXYGEN FOR THIS MAN
34	FUNCTIONAL AEROBIC IMPAIRMENT

PURPOSE OF STUDY

THE PURPOSE OF THE STUDY IS TO EXPLORE THE RELATIVE CONTRIBUTIONS OF AGE AND CORONARY HEART DISEASE TO THE IMPAIRMENT OF FUNCTIONAL CAPACITY. THE SAMPLE CONSISTS OF 117 MALE CORONARY PATIENTS AND 117 AGE-MATCHED HEALTHY MIDDLE-AGED MEN.

REFERENCE

KASSER, IRWIN S. AND ROBERT A. BRUCE. COMPARATIVE EFFECTS OF AGING AND CORONARY HEART DISEASE AND SUBMAXIMAL AND MAXIMAL EXERCISE. CIRCULATION 39, 1969, 759-774.

```
DETAILED DESCRIPTION

  VAR   COL   FORMAT   DESCRIPTION
  ***   ***   ******   ***********

   1    1-3    XXX.    SUBJECT NUMBER
                         001 - 364 = CONTROL GROUP
                         401 - 517 = CORONARY GROUP
   2      4      X.    CARD NUMBER
                         1 = CARD 1
   3      5      X.    SOURCE
                         1 = CLINIC
                         5 = FACULTY
                         7 = YMCA
   4      6      X.    SEX
                         1 = MALE
   5    7-8     XX.    AGE (YEARS)
   6      9      X.    FUNCTIONAL CLASS
                         0 = NONE
                         1 = MINIMAL
                         2 = MODERATE
                         3 = MODERATE TO SEVERE
                         4 = SEVERE
   7     10      X.    ACTIVE
                         0 = UNKNOWN
                         1 = VERY
                         2 = NORMAL
                         3 = LIMITED
   8     11      X.    DIGITALIS TREATMENT
                         0 = NO
                         1 = YES
   9     12      X.    12-LEAD ECG RESULTS
                         0 = UNKNOWN
                         1 = NORMAL
                         3 = ABNORMAL
  10     13      X.    RHYTHM
                         0 = UNKNOWN
                         1 = NORMAL
                         2 = ATRIAL FIBRILLATION
                         3 = OTHER
  11  14-16    XXX.    SYSTOLIC BLOOD PRESSURE
                         0 = UNKNOWN
  12  17-19    XXX.    RESTING HEART RATE
  13  20-22    XXX.    MAXIMUM HEART RATE
  14  23-25    XXX.    DURATION ON TREADMILL (SECONDS)
                         999 = UNKNOWN
  15  26-28    XXX.    SYSTOLIC BLOOD PRESSURE AT PHYSICAL EXAM
  16  29-31    XXX.    DIASTOLIC BLOOD PRESSURE AT PHYSICAL EXAM
                         999 = UNKNOWN
  17  32-34    XXX.    MAXIMUM SYSTOLIC BLOOD PRESSURE ON TREADMILL
                         999 = UNKNOWN
  18  35-37    XXX.    MAXIMUM DIASTOLIC BLOOD PRESSURE ON TREADMILL
                         999 = UNKNOWN
  19     38      X.    HISTORY OF MYOCARDIAL INFARCTION
                         0 = NONE
                         1 = POSSIBLE
                         2 = DEFINITE
                         9 = CONTROL GROUP
  20     39      X.    HISTORY OF ANGINA PECTORIS
                         0 = NONE
                         1 = POSSIBLE
                         2 = DEFINITE
                         9 = CONTROL GROUP
  21     40      X.    ECG
```

```
                    0 = NORMAL
                    1 = POSSIBLE MYOCARDIAL INFARCTION
                    2 = PROBABLE MYOCARDIAL INFARCTION
                    3 = OTHER
                    9 = UNKNOWN OR CONTROL GROUP
22      41     X.   XRAY
                    0 = NORMAL
                    1 = ENLARGED
                    9 = UNKNOWN OR CONTROL GROUP
23      42     X.   HISTORY OF HIGH BLOOD PRESSURE
                    0 = NORMAL
                    1 = HYPERTENSIVE
                    9 = CONTROL GROUP
24      43     X.   SURVIVAL STATUS
                    1 = ALIVE
                    2 = DEAD
                    8 = CONTROL GROUP
                    9 = UNKNOWN
25   44-45    XX.   MONTHS
                    99 = UNKNOWN OR CONTROL GROUP
26      46     X.   NUMBER OF PAST MYOCARDIAL INFARCTIONS
                    1 = 1
                    2 = 2
                    3 = NONE
                    8 = CONTROL GROUP
                    9 = UNKNOWN
27      47     X.   ANGINA PECTORIS
                    1 = SLIGHT
                    2 = SIGNIFICANT
                    3 = NONE
                    8 = CONTROL GROUP
                    9 = UNKNOWN
28      48     X.   CONGESTIVE HEART
                    1 = YES
                    3 = NONE
                    8 = CONTROL GROUP
                    9 = UNKNOWN
29   49-51   XXX.   HEART RATE AT STAGE ONE ON TREADMILL
                    999 = UNKNOWN
30   52-54   XXX.   CHANGE IN HEART RATE (MAXIMUM HEART RATE
                    RESTING HEART RATE)
31   55-57   XXX.   SYSTOLIC BLOOD PRESSURE TIMES HEART RATE
                    999 = UNKNOWN
32   58-60   XXX.   VOLUME OF OXYGEN PREDICTED FOR HEALTHY MAN WITH
                    SIMILAR CHARACTERISTICS
33   61-63   XXX.   ACTUAL ESTIMATED VOLUME OF OXYGEN FOR THIS MAN
                    999 = UNKNOWN
34   64-66   XXX.   FUNCTIONAL AEROBIC IMPAIRMENT
                    999 = UNKNOWN

FORMAT IS (F3.0,3F1.0,F2.0,5F1.0,8F3.0,6F1.0,F2.0,3F1.0,6F3.0)

N = 234
```

LOCATION OF DATA

```
CARD IMAGE - FS.C073.CKASS1
```

LISTING OF DATA (THE FIRST TWO LINES ARE COLUMN GUIDES)

```
0         1          2         3         4      5           6
12345 6789 0123456789 012345678 901234567 89012 345678 901 234 567 890 123 456

  117 1480 1031 80 76 180577120  90200 80 99999 899888 114 104 359 388 359   7
  217 1490 1011 80 68 176785120  80200 80 99999 899888  95 108 351 385 461 -19
```

317	1410	1011	48	70	174775132	82180	60	99999	899888	91	104	313	407	456	-12
417	1520	1031	46	57	170624144	92190	60	99999	899888	98	113	322	377	382	-1
617	1520	1011	40	66	170567110	68150	60	99999	899888	96	104	254	377	354	6
1417	1480	1011	10	60	174720130	90120	60	99999	899888	96	114	208	388	429	-10
1617	1450	2011	38	84	190570112	82150	60	99999	899888	126	106	284	365	355	2
2117	1510	2031	84	68	182517 96	62180	30	99999	899888	106	114	327	349	329	5
2517	1590	2011	70	64	148558120	80190	50	99999	899888	88	84	281	328	349	-6
2917	1440	1011	180	72	194607110	70290	60	99999	899888	106	122	562	399	373	6
3017	1590	1011	94	75	180583116	88210	80	99999	899888	106	105	377	359	362	0
3117	1440	1011	0	57	184765190	90190	60	99999	899888	106	127	349	399	451	-13
3317	1620	1011	78	76	152523112	66190	80	99999	899888	126	76	288	351	332	5
3417	1420	1011	36	63	190688146	96110	60	99999	899888	102	127	208	404	413	-2
3617	1520	1011	76	74	152496104	84180	80	99999	899888	100	78	273	377	319	15
3817	1460	2011	80	65	174445120	80200	70	99999	899888	106	109	347	362	294	18
4417	1510	1011	68	54	150635122	70190	70	99999	899888	86	96	284	380	387	-1
4517	1460	1011	40	62	170630130	80170	70	99999	899888	95	108	288	393	385	2
4817	1660	1011	60	70	156476150	76210	60	99999	899888	116	86	327	340	309	8
4917	1390	1011	140	64	194585140	80280	60	99999	899888	110	130	543	412	363	12
5117	1520	1011	50	45	158705140	90190	70	99999	899888	70	113	300	377	421	-11
5317	1450	1011	70	66	170675150	70220	80	99999	899888	96	104	373	396	407	-2
5517	1540	1011	0	68	184695150	90220	80	99999	899888	108	116	404	372	417	-11
5617	1540	1011	160	67	190740120	80280	999	99999	899888	108	123	531	372	439	-17
5717	1490	1011	70	60	152523150	80220	80	99999	899888	90	92	334	385	332	13
5917	1560	1011	110	68	176675130	80240	60	99999	899888	94	108	422	367	407	-10
6017	1450	1011	64	55	160555 96	80160	70	99999	899888	104	105	255	396	348	12
6217	1600	1011	56	84	184555130	80186	70	99999	899888	118	100	342	356	348	2
6617	1480	1011	26	66	174676114	80140	80	99999	899888	108	108	243	388	407	-4
6717	1500	1031	90	58	180750130	100220	100	99999	899888	85	122	395	383	443	-15
6817	1600	1011	30	77	174602110	80140	50	99999	899888	106	97	243	356	371	-4
7017	1410	1011	80	63	180708110	90190	70	99999	899888	90	117	341	407	423	-3
7117	1600	1011	46	65	180590144	100190	80	99999	899888	100	115	341	356	365	-2
7217	1520	1011	110	60	166624130	90240	50	99999	899888	90	106	398	377	382	-1
7317	1480	1011	54	80	166681156	100210	90	99999	899888	116	86	348	388	410	-5
7417	1550	1031	56	78	170647120	78176	60	99999	899888	100	92	299	369	393	-6
7917	1590	1031	38	68	184578112	78150	60	99999	899888	112	116	275	359	359	0
8217	1390	2011	50	65	180575110	76160	50	99999	899888	106	115	287	381	358	6
8417	1720	1031	30	80	130218170	90200	90	99999	899888	120	50	259	324	183	43
9017	1550	2031	76	88	182573136	82212	80	99999	899888	114	94	385	338	357	-5
9217	1390	1031	20	78	180610130	80150	70	99999	899888	118	102	269	412	375	9
9417	1540	1011	88	64	184660138	100236	90	99999	899888	100	120	434	372	399	-7
10317	1520	1011	30	56	156640130	80160	50	99999	899888	86	100	249	377	390	-3
10517	1560	1011	24	64	168580130	90154	80	99999	899888	92	104	258	367	360	1
10817	1460	1011	84	80	176560126	80210	40	99999	899888	108	96	369	393	350	10
10917	1390	1011	40	66	180742140	90180	999	99999	899888	95	114	323	412	440	-6
11117	1380	1011	20	68	170623120	88140	70	99999	899888	100	102	237	415	381	8
11717	1610	1011	62	88	150262118	82180	80	99999	899888	138	62	269	353	204	42
11917	1460	2011	50	80	190580130	80180	60	99999	899888	112	110	341	362	360	0
12017	1490	2011	22	54	182585118	82140	70	99999	899888	92	128	254	354	363	-2
12217	1530	2011	62	106	190596148	94210	96	99999	899888	130	84	398	344	368	-7
12517	1380	2011	58	56	184575122	90180	80	99999	899888	108	128	331	384	358	6
13117	1600	2011	48	75	160425112	80160	60	99999	899888	114	85	255	325	284	12
13217	1490	2033	82	64	168550108	74190	60	99999	899888	106	104	319	354	345	2
13417	1500	2011	76	80	190575164	90240	80	99999	899888	116	110	455	352	358	-1
13617	1510	2011	20	72	190610130	100150	90	99999	899888	102	118	284	349	375	-7
13717	1570	2011	60	88	200455140	86200	80	99999	899888	132	112	399	333	299	10
14117	1390	2031	60	86	200610120	90180	92	99999	899888	136	114	359	381	375	1
14917	1520	2011	20	62	184618110	80130	70	99999	899888	102	122	239	346	379	-9
15217	1480	1011	62	76	176540118	90180	60	99999	899888	128	100	316	388	341	12
15317	1530	2011	60	76	184615140	80200	70	99999	899888	999	108	367	344	377	-9
15517	1530	2011	102	80	184624118	76220	999	99999	899888	120	104	404	344	382	-11
15717	1480	2011	34	75	184583106	90140	60	99999	899888	100	109	257	357	362	-1
16317	1420	1011	50	56	176790130	84180	80	99999	899888	96	120	316	404	463	-14
16817	1520	2011	40	80	170443100	80140	80	99999	899888	122	90	237	346	293	15
17417	1610	1011	10	68	148487130	90120	70	99999	899888	999	80	177	353	315	10
17617	1530	2011	0	84	194575130	80130	70	99999	899888	124	110	252	344	358	-4
17917	1630	1011	52	58	156465134	100186	60	99999	899888	106	98	290	348	304	12
18017	1470	1011	26	72	180551124	80150	40	99999	899888	104	108	269	391	346	11

18417	1480	2011	30	78	180545120	72150	50	99999	899888	100	102	269	357	343	3
18817	1520	1011	20	56	184605140	88160	60	99999	899888	103	128	294	377	372	1
19017	1400	2011	80	64	164450118	84198	60	99999	899888	100	100	324	378	296	21
19117	1480	1011	52	62	176650118	80170	60	99999	899888	104	114	299	388	394	-1
19317	1540	1011	76	62	184720134	90210	80	99999	899888	102	122	386	372	429	-15
19517	1390	2011	32	62	184630110	70142	50	99999	899888	94	122	261	381	385	0
19817	1450	1011	54	64	190710136	66190	70	99999	899888	110	126	360	396	424	-7
19917	1570	2011	18	80	130245124	80142	70	99999	899888	999	50	184	333	196	41
20017	1480	1011	40	60	164605160	80200	60	99999	899888	94	104	327	388	372	4
20117	1420	1011	36	74	186595100	80136	60	99999	899888	114	112	252	404	368	9
20317	1390	2011	28	75	146450112	80140	50	99999	899888	999	71	204	381	296	22
20617	1490	1011	80	72	174478150	70230	50	99999	899888	102	102	400	385	310	19
20717	1720	1031	82	65	160360124	82206	90	99999	899888	156	95	329	324	252	22
20917	1390	1011	110	74	194763110	90220	60	99999	899888	98	120	426	412	450	-9
21017	1440	2011	30	72	180550110	80140	40	99999	899888	100	108	251	368	345	6
21717	1391	1011	94	72	176598156	100250	60	99999	899888	104	104	439	412	369	10
21817	1430	1011	60	84	178628130	90190	60	99999	899888	114	94	338	401	384	4
22117	1540	1011	90	80	174593170	110260	60	99999	899888	96	94	452	372	367	1
22417	1550	1011	60	58	180660120	80180	80	99999	899888	94	122	323	369	399	-8
22517	1420	2011	20	68	184590110	84130	70	99999	899888	110	116	239	373	365	2
30115	1610	2031	10	68	160465114	90105	80	99999	899888	108	92	167	322	304	5
30715	1510	2011	66	86	164414130	88196	60	99999	899888	114	78	321	349	279	20
31315	1630	1011	46	55	180610130	80176	70	99999	899888	96	125	316	348	375	-7
31715	1610	2011	30	72	158401130	80160	90	99999	899888	106	86	252	322	272	15
31915	1520	2001	38	66	172564122	90160	82	99999	899888	102	106	275	346	352	-1
32015	1550	2011	90	68	164568120	84210	70	99999	899888	74	96	344	338	354	-4
32315	1710	2011	24	61	160360120	84144	60	99999	899888	122	99	230	295	252	14
32415	1380	2011	30	88	200637150	84140	70	99999	899888	150	112	279	384	388	-1
32615	1630	2031	20	74	170405130	94150	80	99999	899888	116	96	254	317	274	13
32715	1640	2001	48	68	140357120	80168	72	99999	899888	105	72	235	314	251	20
32815	1500	2011	80	68	200600110	84190	100	99999	899888	114	132	379	352	370	-5
32915	1580	2031	46	74	166530140	90186	80	99999	899888	124	92	308	330	336	-1
33315	1580	1011	62	66	172448110	70172	70	99999	899888	112	106	295	361	296	18
33415	1380	0001	88	65	174477112	68200	70	99999	899888	999	109	347	384	310	19
33715	1390	2032	66	86	190646140	999206	80	99999	899888	126	104	391	381	393	-2
33815	1570	2011	40	58	168596120	80160	80	99999	899888	104	110	268	333	368	-10
33915	1500	1011	54	76	180500150	110210	90	99999	899888	130	104	377	383	321	16
34815	1610	2011	38	84	188508142	100180	80	99999	899888	134	104	338	322	325	0
35215	1680	2033	40	56	175577140	80180	80	99999	899888	100	119	314	303	359	-18
35315	1410	1011	44	74	174605 96	70140	60	99999	899888	102	100	243	407	372	8
35415	1550	1011	50	76	168497130	80180	90	99999	899888	104	92	302	369	320	13
35515	1500	2011	12	76	180512118	84130	40	99999	899888	999	104	233	352	327	7
35615	1540	2011	44	82	180550126	86170	70	99999	899888	114	98	305	341	345	-1
35715	1480	1011	66	63	180590124	80190	60	99999	899888	110	117	341	388	365	5
36015	1550	2031	40	70	152438120	76160	70	99999	899888	106	82	243	338	291	14
36115	1620	2011	74	72	154510146	76250	78	99999	899888	104	82	384	320	326	-1
36315	1710	2001	30	60	160450160	90190	70	99999	899888	114	100	303	295	296	0
36415	1480	2031	50	85	180637110	80160	90	99999	899888	112	95	287	357	388	-8
40111	1422	2031	56	88	166545104	70160	80	12200	141123	122	78	265	373	343	8
40211	1662	2001	56	90	156345150	90190	90	12200	999999	122	66	296	309	245	20
40311	1562	0000	9	90	153 90104	80130	80	22000	212199	153	63	198	336	120	64
40411	1552	2001	20	82	130410100	80128	80	22900	999999	110	48	166	338	277	18
40511	1412	2001	30	84	160385170	120210	90	12211	145999	135	76	335	376	265	29
40611	1620	0000	50	82	160240150	90200	80	20201	148333	152	78	319	320	194	39
40711	1462	2001	40	75	130415150	100190	98	12001	154313	130	55	246	362	279	22
40811	1442	1001	0	83	135710150	90190	85	02001	2 9123	112	52	256	399	424	-6
40911	1501	2031	40	55	160565140	90180	100	02201	223131	78	105	287	352	353	0
41011	1733	3031	10	92	140135120	75130	75	02210	221199	135	48	181	290	142	50
41111	1482	2031	10	65	145330130	85140	85	22200	124929	110	80	202	357	238	33
41211	1532	2001	10	56	120275130	80140	80	22210	229313	102	64	167	344	211	38
41311	1513	1031	18	70	118315104	80140	85	12301	220123	100	48	165	380	230	39
41411	1590	0000	60	82	154390165	100225	125	02311	146331	125	72	346	328	267	18
41511	1543	3031	45	76	100268195	105240	120	12301	2 8129	97	24	239	341	207	39
41611	1412	2031	58	61	164565130	90188	105	21201	122313	108	103	308	376	353	6
41711	1562	2031	0	76	130100130	100999	999	20211	2 4199	130	54	999	336	125	62
41811	1380	2001	50	70	162710150	90200	110	02001	148323	97	92	323	384	424	-10
41911	1403	3031	20	85	160250110	70130	90	22200	1 8313	138	75	207	378	198	47

```
42011 1421 2031 74 82 170400110  80184 90 12200 111323 144  88 312 373 272  27
42111 1511 2011 60 75 175510140  80200 90 02000 111313 116 100 349 349 326   6
42211 1521 0001  0 73 173999100  90999999 21210 171313 113 100 999 346 999 999
42311 1370 1011 60 56 190675130  80160 80 02000 1 5313  97 134 303 418 407   2
42411 1481 2131 10102 172305130  88140 70 20210 1 5333 150  70 240 357 225  36
42511 1350 0000 30 75 130430130  70160 90 12210 141323 108  55 207 392 287  26
42611 1351 1031110 98 180628140  95250100 20200 999999 134  82 449 423 384   9
42711 1483 3031 10 84 160360105  70130 84 02001 1 8323 144  76 207 357 252  29
42811 1522 2011 48 74 166470122  85170 95 02001 119323 100  92 282 346 306  11
42911 1462 3011  0 70 150130150  88190 90 02001 1 6999 140  80 284 362 140  61
43011 1513 0001 12107 141135168  98178100 02300 172313 141  34 250 349 142  59
43111 1500 0000 10 58 125405150  80170 80 02990 156313 125  67 212 352 274  21
43211 1723 3011 35106 126 30170  85205 90 01201 112333 116  20 258 293  91  68
43311 1563 3031 35 65 148420125  78160 90 22200 139323  98  83 236 336 282  16
43411 1563 3031 25 56 114140105  60130 80 22300 114323 114  58 148 336 145  56
43511 1632 2011 45 77 162390130  90175 90 12200 120323 116  85 283 317 267  15
43611 1531 2011 45104 146235105  65150 60 22010 999999 134  42 218 344 191  44
43711 1530 1000 58 76 132390100  70158 86 20200 157333 100  56 208 375 267  28
43811 1573 2031 75 65 120260125  80200 80 02210 118121 114  55 239 333 203  38
43911 1571 2031 20 56 132265190  78210 85 02001 120333 104  76 277 333 206  38
44011 1622 2031 34 85 154270180  90214 96 11200 142123 999  69 329 320 208  34
44111 1732 2031152 60 104204152  88186104 02390 999999 100  44 193 290 176  39
44211 1442 0000 14 68 128280116  72130 70 02000 999999 118  60 166 368 213  41
44311 1633 2001 20 75 134180140  85160 80 22200 999999 128  59 214 317 164  48
44411 1591 1001 20 68 110270140  80160 80 02000 153313  98  42 175 359 208  41
44511 1511 0000 44 62 150410116  78160 70 20210 161333  88  88 239 349 277  20
44611 1523 0001 20 72 122 85140  90160110 22200 166323 122  50 195 346 118  65
44711 1640 0000 40 98 156475154  80194 80 10200 151333 124  58 302 314 309   1
44811 1532 2031 40 70 160421150  90190 80 20200 999999 115  90 303 344 282  17
44911 1581 2011 24 96 162590116  74140 84 02000 999999 116  66 226 330 365 -10
45011 1530 2031 40 75 156420160 110200 94 22201 123313 118  81 311 344 282  17
45111 1582 2011 38 52 105135140  80178 85 02100 122313 103  53 186 330 142  56
45211 1451 2031 70 82 150412150  90220110 22001 999999 107  68 329 365 278  23
45311 1423 2011 45 78 120110105  60150 90 02000 130999 120  42 179 373 130  65
45411 1602 2031 15 72 138213115  70160110 02211 210199 136  66 220 325 180  44
45511 1341 0001 50 80 200420130  80180100 10201 999999 138 120 359 395 282  28
45611 1642 2000 30 58  94210110  75120 80 22000 146333  90  36 112 314 179  42
45711 1351 2031 40 88 174445150 100190 80 20201 123333 120  86 330 392 294  24
45811 1422 2011 35 76 128140140  90175 80 02100 124323 128  52 223 373 145  61
45911 1532 2031 64 87 154540136  78200 40 20210 122333 128  67 307 344 341   0
46011 1581 2000 50 97 152300160  88210 85 20211 138333 140  55 319 330 223  32
46111 1381 2031 30 84 134510110  80140 80 01210 999999  98  50 187 384 326  15
46211 1352 2031 14 90 140235128  90142 92 22200 119323 122  50 198 392 191  51
46311 1342 3011 40 84 152395120  90160 90 22000 124333 115  68 243 395 270  31
46411 1683 2001 20 72 100110130  60150 80 12300 143313 100  28 149 303 130  57
46511 1493 2011 40 72 164385120  70160 70 02000 117323 128  92 262 354 265  25
46611 1552 2031 40 70 120285150  95190 95 02301 126323 120  50 227 338 216  36
46711 1580 2000 27 70 120360118  82145 80 22310 242333 104  50 173 330 252  23
46811 1432 2011 12 63 112150140  82152 75 22000 123313 112  49 170 370 149  59
46911 1392 1011 25 66 118305115  85140 90 02300 129323 114  52 165 412 225  45
47011 1663 3031  0 61 122175200 100220100 02211 122323 116  61 268 309 162  47
47111 1502 2031 24 66 134222120  75150 80 22200 131323 116  68 200 352 185  47
47211 1453 3001 27 80 175310140  90140 90 01200 116323 170  95 244 365 228  37
47311 1530 0000  0 78 128540118  75170 70 02000 143313 120  50 217 344 341   0
47411 1564 3031 25 78 124280205 110230120 22201 1 8323 116  46 285 336 213  36
47511 1492 2031 15 75 154360165 122195152 22211 128311 114  79 300 354 252  28
47611 1490 0000 30 93 168575140  90170 90 20200 153333 117  75 285 354 358   0
47711 1562 2001  0 82 165290160 100200100 12000 168333 120  83 329 336 218  34
47811 1380 0000 26 77 120250134  86160 80 12010 999999 105  43 191 384 198  48
47911 1390 0000 10 88 120160130  90140100 22000 999999 120  32 167 381 154  59
48011 1622 2000 30 56 115 75190 100160 90 12100 230121 100  59 183 320 113  64
48111 1703 2001 20 95  98210120  86140 70 22211 252113  98   3 137 298 179  39
48211 1532 0000 30112 152220130  80160 90 12301 230199 142  40 243 344 184  46
48311 1682 2011 25 53 110285135  75160 80 22300 122323  90  57 175 303 216  28
48411 1502 0031 20 80 138208160  96180100 00211 126333 128  58 248 352 178  49
48511 1462 2031 45 70 136240115  80160 95 02200 126999 128  66 217 362 194  46
48611 1583 2031 28 82 130240100  52128 50 22010 136323 112  48 166 330 194  41
```

```
48711 1572 2031 50 62 154558120  76170 68 00200 110313  95  92 261 333 349  -4
48811 1553 3031 30 80 120120115  75145 65 20200 110333 118  40 173 338 135  60
48911 1520 0001 40 75 125150130  85170 90 12300 231299 110  50 212 346 149  56
49011 1612 2101 52 68 173610118  70170 70 00200 212311 125 105 294 322 375 -16
49111 1452 2031 50 86 142243150 100200100 02000 127323 126  56 283 365 195  46
49211 1512 0101 20 85 154355110  70130 90 22210 2 8111 140  69 200 349 250  28
49311 1553 3031 40 70 140540140  94180 90 21201 144213 103  70 251 338 341   0
49411 1511 0001 10 66 168295150  90170 90 02200 999999 120 102 285 349 221  36
49511 1461 0001 20 90 170360140  90160 80 22200 999999 118  80 271 362 252  30
49611 1691 2011 55 60 116240135  70190 70 02000 128999 114  56 220 301 194  35
49711 1513 3031 20 88 156240140 100160100 22211 124123 136  68 249 349 194  44
49811 1491 1001 70 77 173640140  90210 94 01201 125333 128  96 363 385 390  -1
49911 1583 3011 18 76 140240102  60124 70 12000 140913 115  64 173 330 194  41
50011 1383 3131 30 75 160570120  88150 95 22200 124323 116  85 239 384 355   7
50111 1501 1031 90 76 174580140 100230 90 20000 121333 124  98 400 383 360   5
50211 1381 3031 15 98 160 65135  75150 80 22200 129323 160  62 239 384 108  71
50311 1581 0001 30 87 160360140  88999999 20200 172333 125  73 999 330 252  23
50411 1690 0001 20 75 110 80170  90190110 02110 165333 102  35 208 301 115  61
50511 1660 0000 10 70 114240160  90150 90 02000 212999 110  44 170 309 194  37
50611 1492 2011 52 64 107180118  66170 60 02010 999999  98  43 181 354 164  53
50711 1620 0000  4100 150195116  90110 96 22000 218999 142  50 164 320 172  46
50811 1440 0001 20100 172215160  95180100 02001 159333 142  72 309 368 181  50
50911 1583 0101 40 78 142165140  80180100 22210 229111 124  64 255 330 157  52
51011 1452 2031 55 77 150390115  80170 80 12200 133313 999  73 254 365 267  26
51111 1583 3031  5 77 134100165  90170 90 22211 128323 126  57 227 330 125  62
51211 1542 2001 20 88 175 60130  90150100 12200 157323 110  87 262 341 105  68
51311 1552 2001 40 88 150260126  90166 90 21211 999999 116  62 248 338 203  39
51411 1682 2001 20 64 115120130  80150 90 12000 999999 110  51 172 303 135  55
51511 1682 2031 45 60 120315140  85185 70 22210 129999 100  60 221 303 230  24
51611 1471 2031 26 75 180745104  68130 70 22201 134333 110 105 233 360 441 -22
51711 1550 0000 40 84 140115130  70170 70 02000 145323 136  56 237 338 132  60
```

REFERENCES

Anderson, G. D. (1969). A comparison of methods for estimating a probability density function. Unpublished doctoral dissertation, University of Washington, Seattle.

Cencov, N.N. (1962). Evaluation of an unknown distribution density from observations. Soviet Math. 3, 1559-1562.

Kasser, I.S. and R.A. Bruce (1969). Comparative effects of aging and coronary heart disease on submaximal and maximal exercise. Circulation 39, 759-774.

Kronmal, R.A., L. Bender and J. Mortensen (1970). A conversational statistical system for medical records. J. Royal Statist. Soc. 19, 82-92.

Kronmal, R.A. and M. Tarter (1968). The estimation of probability densities and cumulatives by Fourier series methods. J. Amer. Statist. Assoc. 63, 925-952.

Tarter, M.E., R.L. Holcomb and R.A. Kronmal (1967). A description of new computer methods for estimating the population density. In Proceedings, Association for Computing Machinery 22, 511-519. Washington, D.C., Thompson Book Company.

Tarter, M.E. and R.A. Kronmal (1970). On multivariate density estimates based on orthogonal expansions. Ann. Math. Statist. 41, No. 2, 718-722.

Zygmund, A. (1959). Trigonometric Series, Vol. II. London, Cambridge Press.

DISCUSSION OF THE KRONMAL AND TARTER CHAPTER

D. M. Allen

I agree with the original premise that if data are generated by observing continuous random variables we ought to summarize the data using a continuous estimated density function and a continuous cumulative distribution function. However, I tend to think that an artist would do a good job of drawing a distribution function. If the artist draws his curve through the center of the "jumps" of the empirical distribution function, the curve will be a good continuous approximation of the empirical distribution function.

Density functions present a more difficult problem. Dr. Edwin Chen made some histograms using the data presented in the Kronmal-Tarter paper. Some of these histograms were multimodal while the corresponding polynomial approximations were unimodal. The length and location of the intervals has long been a problem when making histograms. It appears that the choice of the degree of the approximating polynomial is also a difficult problem. I have been listening to Dr. Tukey's comments at the end of each paper. My interpretation of each of his critiques has been "hang loose and try something else." The something else I suggest in this case is to present a succession of estimated densities. Start with something that is as flexible as possible, then apply successive degrees of smoothing. One would still have to choose one of the approximations and I hesitate to suggest a criterion for this purpose. However, I think one could be guided by the extent of the difference between successive densities. I feel that the technique of polynomial approximation of densities has potential and that additional effort in this area is worthwhile.

One of the formulas in the paper appeared to be similar to the characteristic function of the empirical distribution function. I would be interested to know if there is a relationship between them.

An empirical Bayesian might find these techniques useful since an analytic representation of the density is given with minimal assumptions.

J. Tukey

The issues here seem to me to be issues of how we assess believability, how much will the global override the local, and how we deal with the negative densities which will inevitably occur for some sets of data. The first and the third lead us to want to look at $\sqrt{f}$ rather than f. Both have tendencies to bring in comparison distributions. The second leads us to look at the actual smoothing kernel.

I feel that the simplest and most effective approach to the first issue is that associated with hanging or suspended rootograms. The basic idea is simple. Fit some sort of comparison distribution. Divide the line or plane up into cells. For each cell find the value of

$$2(\sqrt{\text{observed}} - \sqrt{\text{fitted}})$$

or, if we feel persnickety, the value of

$$\sqrt{2 + 4 \cdot \text{observed}} - \sqrt{1 + 4 \cdot \text{fitted}}$$

and display these values, which, except for a small allowance for the fitting of constants that we are usually wise to neglect, are moderately Gaussian (0, 1). The appearance of trends that do not correspond to misfitting can then be taken as speaking for the existence of corresponding deviations from fit.

The combination of such a plot, using a Gaussian reference, and figure 12 would to me offer a very much clearer picture of how seriously I was to take bimodality (and the hump near 198.91) than does figure 12 alone.

This hump (near 198.91) can focus our attention on the possibility of the global overriding the local. While it is presumably at least quite difficult for such things to happen, what do we know about the possibility that just where the other data values fall has produced this hump?

The controlling entity is the function

$$\phi(y, x) = \sum_k e^{-2\pi i k' x} e^{2\pi i k' y} = \phi(y-x)$$

which relates the appearance of an observation at x to n times the estimated density at y. With a limitation on the summation, such functions do have ripples near their cutoff, which can often be reduced by replacing the "zero-or-one" nature of the summation, that is, by inserting intermediate weights near the cutoff.

The fact that the fitting process is expressible as

$$f(y) = \frac{1}{C_n} \sum_x \phi(y - x)$$

may also be important in itself, since we may want to truncate ϕ at some of its zeroes, as well as wishing to make its minor lobes smaller.

Fourier expansion is, as we have just seen, an elegant way to do translation-invariant smoothing, but it has its peculiarities.

Granting, as we should, that the natural quantity to expand is $\sqrt{f}$, particularly since it is about equally well locally determined everywhere, we should notice that, if we expand $\sqrt{f}$ and then square the result, the final result will be nonnegative, thus avoiding the difficulties with the negative regions for "fitted f" that I suspect are actually present in figures 13 and/or 17. But how can one use observations to expand $\sqrt{f}$?

The best answer seems to be: iteratively. Suppose first that the initial smoothing f_o has no difficulties with negative or over-small values. (Either because the data are nicely arranged or because we have improved the smoothing function.) Then we can try to reach an approximation, $\sqrt{f_1}$, to $\sqrt{f}$ by taking each observation with weight $1/\sqrt{f_o}$, correlating weighted Fourier coefficients, squaring up, and then renormalizing the resulting f to integrate to 1. This can then be iterated.

The zero-order iterative argument is simple, asymptotically. If a small interval Δ has density f, then the number of observations will be, asymptotically, $n f \Delta$. The contribution to an expansion of $\sqrt{f}$ ought to be proportional to $n\sqrt{f}\Delta$. If f_o is close to f, then

$$\frac{n f \Delta}{\sqrt{f_o}} \sim \frac{n f \Delta}{\sqrt{f}} = n\sqrt{f}\,\Delta \,.$$

The first-order iterative argument is naturally more complicated. Let us not bother with the first iterative step, but rather look at the step from f_i to f_{i+1}, where f_i already comes from a smoothing for $\sqrt{f_i}$. Suppose $\sqrt{f_i}$ to be too big, over a range comparable with, or larger than, the smoothing distance, by a factor of $1 + \sigma$. Then the contributions to the smoothed value of $\sqrt{f_{i+1}}$, near the center of this range, will be too small in the ratio

$$\frac{1}{1+\sigma} \sim 1 - \sigma \,.$$

The best iterate, in the center of such a region, would be

$$\frac{1}{2}\sqrt{f_{i+1}} + \frac{1}{2}\sqrt{f_i}$$

rather than

$$\sqrt{f_{i+1}} \,.$$

Where the initial deviation is concentrated in a small region, of course, the opposite is true.

The simplest way to deal with such a situation is probably to iterate blithely until we have at least 3 iterates and then apply some simple acceleration process, either to the Fourier coefficients or to the smoothed values, as may please us.

There remains the question of what to do if the initial f_o is not well behaved, being either negative or too small at some observations. I would recommend, as probably effective, the following rules for avoiding the really bad consequences of such:

- take $\sqrt{|f_o|}$, not $\sqrt{f}$, thus avoiding the worst aspects of negative densities; and

- find, for a Gaussian distribution of the appropriate dimensionality and a sample of n, the values of

$$g_{j|n} = \frac{\text{lower 10\% point of } \sqrt{f} \text{ for the } j^{th} \text{ smallest of } n}{3 \cdot \text{maximum value of } \sqrt{f}} ,$$

order the values of $\sqrt{f_o}$, and replace them by

$$\text{larger } \{\sqrt{f_o}, (\sqrt{f_o})_{max} \cdot g_{j|n}\}$$

thus eliminating any excrutiatingly small values for $1/\sqrt{f_o}$.

Whether the latter process is helpful enough for i = 1 or beyond, helpful enough to counterbalance its overriding of very small $\sqrt{f_i}$ where these are deserved, is unclear.

CITATION INDEX

Alker, H.R. (1964). Dimensions of conflict in the General Assembly. Amer. Pol. Sci. Rev. 58, 642-657.

Allen, D.M. (1971). The prediction sum of squares as a criterion for selecting predictor variables. Technical Report No. 23, Department of Statistics, University of Kentucky.

Amorocho, J. and A. Brandstetter (1967). The representation of storm precipitation fields near ground level. J. Geophys. Res. 72, 1145.

Anderson, G.D. (1969). A comparison of methods for estimating a probability density function. Unpublished doctoral dissertation, University of Washington, Seattle.

Anderson, R.L., D.M. Allen and F.B. Cady (1972). Selection of predictor variables in linear multiple regression. In Statistical Papers in Honor of George W. Snedecor, T.A. Bancroft, Ed., Iowa State University Press.

Andrews, D.F., P.J. Bickel, F.R. Hampel, P.J. Huber, W.H. Rogers and J.W. Tukey (1972). Robust Estimates of Location: Survey and Advances, Princeton, New Jersey, Princeton University Press.

Armitage, P., C.K. McPherson and J.B. Copas (1969). Statistical studies of prognosis in advanced breast cancer. J. Chron. Dis. 22, 343-360.

Atkins, H., R.D. Bulbrook, M.A. Falconer, J.L. Hayward, K.S. Maclean and P.H. Schurr (1964). Urinary steroid estimations in the prediction of response to adrenalectomy or hypophysectomy. Lancet, Nov. 28, 1133-1136.

Atkins, H., R.D. Bulbrook, M.A. Falconer, J.L. Hayward, K.S. Maclean and P.H. Schurr (1968). Ten years' experience of steroid assays in the management of breast cancer, a review. Lancet, Dec. 14, 1255-1260.

Atkins, H., M.A. Falconer, J.L. Hayward, et al (1966). The timing of adrenalectomy and of hypophysectomy in the treatment of advanced breast cancer. Lancet, April 16, 827-830.

Atkins, H.J., M.A. Falconer, J.L. Hayward, K.S. Maclean, P.H. Schurr and P. Armitage (1960). Adrenalectomy and hypophysectomy for advanced cancer of the breast. Lancet 1, 1148-1153.

Bahadur, R.R. (1960). On the asymptotic efficiency of tests and estimates. Sankhya 22, 229-252.

Ball, M.M. (1951). Bloc voting in the General Assembly. Internat. Organ. 5, 3-31.

Bartlett, F. and E.R. John (1970). Reply to Schwartz's comments. Science 169, 304-305.

Bingham, C., M.D. Godfrey and J.W. Tukey (1967). Modern techniques of power spectrum estimation. I.E.E.E. Transactions on Audio and Electroacoustics, AU15, No. 2, 56-66.

Brier, G.W. and R.A. Allen (1951). Verification of weather forecasts. In Compendium of Meteorology, T.F. Malone, Ed., Boston, Amer. Meteorol. Soc., 841-848.

Bulbrook, R.D., F.C. Greenwood and J.L. Hayward (1960). Selection of breast cancer patients for adrenalectomy or hypophysectomy by determination of urinary 17-hydroxycorticosteroid and aetiocholanolone. Lancet 1, 1154.

Cacoullos, T. (1966). Estimation of a multivariate density. Ann. Inst. Statist. Math. Tokyo 18, 179-189.

Cady, F.B. and D.M. Allen (1972). Combining experiments to predict future yield data. Agron. J. 64, 211-214.

Cencov, N.N. (1962). Evaluation of an unknown distribution density from observations. Soviet Math. 3, 1559-1562.

Cover, T.M. (1968). Estimation by the nearest-neighbor rule. I.E.E.E. Trans. Infor. Theory IT-14, 50-55.

Cox, D.R. (1966). Some procedures connected with the logistic qualitative response curve. In Research Papers in Statistics: Essays in Honour of J. Neyman's 70th Birthday, F.N. David, Ed., London, Wiley.

Cutler, S.J., M.M. Black, G.H. Fridell, et al (1966). Prognostic factors in cancer of the female breast. II. Reproducibility of histopathologic classification. Cancer 19, 75-82.

Daniel, C. and F.S. Wood (1970). Fitting Equations to Data. New York, Wiley.

Dickey, J.M. (1968a). Smoothed estimates for multinomial cell probabilities. Ann. Math. Statist. 39, 561-566.

Dickey, J.M. (1968b). Estimation of disease probabilities conditioned on symptom variables. Math. Biosci. 3, 249-265.

Dickey, J.M. (1969). Smoothing by cheating. Ann. Math. Statist. 40, 1477-1482.

Dickey, J.M. and B.P. Lientz (1970). The weighted likelihood ratio, sharp hypotheses on chances, the order of a Markov chain. Ann. Math. Statist. 41, 214-226.

Dixon, W.J. (1969). BMD Biomedical Computer Programs, X-Series Supplement. Berkeley and Los Angeles, University of California Press.

Dixon, W. J. (1970). BMD Biomedical Computer Programs. 2nd ed., 3rd printing, revised. Berkeley and Los Angeles, University of California Press.

Dixon, W. J. (1971). BMD Biomedical Computer Programs. Berkeley, Los Angeles, London, University of California Press.

Donchin, E. (1969). Data analysis techniques in average evoked potential research. In Average Evoked Potentials, E. Donchin and D. B. Lindsley, Eds., Washington, D. C. Govt. Printing Offices, NASA SP-191, 199-217.

Draper, N. R. and H. Smith (1966). Applied Regression Analysis. New York, Wiley.

Fisher, R. A. (1936). The use of multiple measurements in taxonomic problems. Ann. Eugen. 7, 179-188.

Fix, E. and J. L. Hodges, Jr. (1951). Discriminatory analysis, nonparametric discrimination. USAF School of Aviation Med., Randolph Field, Texas, Project 21-49-004, Report 4, Contract AF41(128)-31, February, 1951.

Freeman, W. J. (1964). A linear distributed feedback model for prepyriform cortex. Exp. Neurol. 10, 525-547.

Freeny, A. E. (Mrs.)(1969). Statistical treatment of rain gauge calibration data. Bell Sys. Tech. J. 48, 1757.

Freeny, A. E. (Mrs.) and J. D. Gabbe (1969). A statistical description of intense rainfall. Bell Sys. Tech. J. 48, 1789.

Gilbert, E. S. (1968). On discrimination using qualitative variables. J. Amer. Statist. Assoc. 63, 1399-1412.

Goldman, R., J. Walrath, E. Jacobson and J. Dickey (1971). First Look at Graphs. Research Report 41, revised. Statistics Department, State Univ. of New York at Buffalo.

Good, I. J. (1950). *Probability and the Weighting of Evidence*. New York, Hafner.

Good, I. J. (1965). *The Estimation of Probabilities*. Cambridge, M.I.T. Press.

Gunel, E. and J. Dickey (1972). Bayes factors for independence in contingency tables. (In preparation)

Hartigan, J. A. (1972). Direct clustering of a data matrix. *J. Amer. Statist. Assoc.* 67, 123-129.

Hayward, J. L. (1966). Assessment of response to treatment at Guy's Hospital Breast Clinic. In *Clinical Evaluation in Breast Cancer*, New York, Academic Press.

Hills, M. (1967). Discrimination and allocation with discrete data. *Appl. Statist.* 16, 237.

Hovet, T. (1960). *Bloc Politics in the United Nations.* Cambridge, Harvard University Press.

Jacobson, E. (1971). SHOP: a computer program for printer plots and histograms of sub-classes. Research Report 52, Statistics Department, State University of New York at Buffalo.

Jacquez, J. A. (1964). The diagnostic process. In *Computer Diagnosis and Diagrammatic Methods*, J. A. Jacquez, Ed., Springfield, Illinois, Charles C Thomas.

Jeffreys, H. (1961). *Theory of Probability*, 3rd ed., Oxford, Clarendon Press.

Jenden, D. J., M. D. Fairchild, M. R. Mickey, R. W. Silverman and C. Yale (1972). A multivariate approach to the analysis of drug effects on the electroencephalogram. *Biometrics* 28, 73-80.

Joiner, B. L., J. R. Rosenblatt and J. W. Dean (1970). OMNITAB - and an example in data analysis. (Preliminary draft for discussion - not for publication), Revised 9/21.

Kasser, I. S. and R. A. Bruce (1969). Comparative effects of aging and coronary heart disease on submaximal and maximal exercise. Circulation 39, 759-774.

Kiang, N. Y. S. and T. T. Sandel (1961). Off-responses from the auditory cortex of unanesthetized cats. Arch. Ital. Biol. 99, 121-134.

Kronmal, R. A. and M. Tarter (1968). The estimation of probability densities and cumulatives by Fourier series methods. J. Amer. Statist. Assoc. 63, 925-952.

Kronmal, R. A., L. Bender and J. Mortensen (1970). A conversational statistical system for medical records. J. Royal Statist. Soc. 19, 82-92.

Laird, R. J. and F. B. Cady (1969). Combined analysis of yield data from fertilizer experiments. Agron. J. 61, 829-834.

Ledley, R. S. and L. B. Lusted (1959). Reasoning foundations of medical diagnosis. Science 130 (3366), 9-21.

Lijphart, A. (1963). The analysis of voting in the General Assembly. Amer. Pol. Sci. Rev. 57, 902-917.

Lusted, L. B. (1968). Introduction to Medical Decision Making. Springfield, Illinois, Charles C Thomas.

Marquardt, D. W. (1963). An algorithm for least squares estimation of nonlinear parameters. J. Soc. Indust. Appl. Math. 2, 431.

Medhurst, R. G. (1965). Rainfall attenuation of centimeter waves: comparison of theory and measurement. I. E. E. E. Trans. on Antennas and Propagation 550.

Morse, A.P. and F.E. Grubbs (1947). The estimation of dispersion from differences. _Ann. Math. Statist._ 18, 194-214.

Parzen, E. (1962). On estimation of a probability density function and mode. _Ann. Math. Statist._ 33, 1065-1076.

Radhakrishna, S. (1964). Discrimination analysis in medicine. _Statistician_ 14, 147-167.

Riggs, R.E. (1958). _Politics in the United Nations_. Champaign, University of Illinois Press.

Russett, B.M. (1966). Discovering voting groups in the United Nations. _Amer. Pol. Sci. Rev._ 66, 327-339.

Sandel, T.T. and N.Y.S. Kiang (1961). Off-responses from the auditory cortex of anesthetized cats. _Arch. Ital. Biol._ 99, 105-120.

Sanders, F. (1958). The evaluation of subjective probability forecasts. Scientific Report No. 5, Contract AFCRC-TN-58-465. Cambridge, Mass. Inst. of Tech.

Sarfaty, G. and M. Tallis (1970). Probability of a woman with advanced breast cancer responding to adrenalectomy or hypophysectomy. _Lancet_, Oct. 3, 685-687.

Schwartz, M. (1970). Means and variances of average-response wave forms. _Science_ 169, 303-304.

Semplak, R.A. (1966). Gauge for continuously measuring rate of rainfall. _Rev. Sci. Instr._ 37, 1554.

Semplak, R.A. and H.E. Keller (1969). A dense network for rapid measurement of rainfall rate. _Bell Sys. Tech. J._ 48, 1745.

Semplak, R.A. and R.H. Turrin (1969). Some measurements of attenuation by rainfall at 18.5 GHz. Bell Sys. Tech. J. 48, 1767.

Specht, D.F. (1971). Series estimation of a probability density function. Technometrics 13, 409-424.

Tarter, M.E., R.L. Holcomb and R.A. Kronmal (1967). A description of new computer methods for estimating the population density. In Proceedings, Association for Computing Machinery 22, 511-519, Washington, D.C., Thompson Book Company.

Tarter, M.E. and R.A. Kronmal (1970). On multivariate density estimates based on orthogonal expansions. Ann. Math. Statist. 41, No. 2, 718-722.

Truett, J., J. Cornfield and W. Kannel (1967). A multivariate analysis of the risk of coronary heart disease in Framingham. J. Chron. Dis. 20, 511-524.

Tukey, J.W. (1962). The future of data analysis. Ann. Math. Statist. 33, 1-67.

Tukey, J.W. (1949). One degree of freedom for non-additivity. Biometrics, 5, No. 3.

Tukey, J.W. (1970). Exploratory Data Analysis (Limited Preliminary Edition), Vols. I, II, III. Reading, Mass., Addison-Wesley.

Van Ryzin, J. (1965). Non-parametric Bayesian decision procedures for (pattern) classification with stochastic learning. Trans. Fourth Prague Conf. Information Theory, Statistical Decision Functions and Random Processes.

Van Ryzin, J. (1966). Bayes risk consistency of classification procedures using density estimation. Sankhya A 28, 261-270.

Walker, S.H. and D.B. Duncan (1967). Estimation of the probability of an event as a function of several independent variables. Biometrika 54, 167-179.

Walrath, J., R. Goldman and J. Dickey (1971). Nonparametric discriminators: computer programs for research in computer assisted medical diagnosis. Research Report 48, revised. Statistics Department, State Univ. of New York at Buffalo.

Warner, H.R., A.F. Toronto, L.G. Veasey and R. Stephenson (1961). A mathematical approach to medical diagnosis: application to congenital heart disease. J. Amer. Med. Assoc. 177, 177-183.

Whittle, P. (1958). On the smoothing of probability density functions. J. Roy. Statist. Soc., B 20, 334-343.

Zippin, C. and N. Petrakis (1971). Identification of high risk groups in breast cancer. Cancer 23, No. 6.

Zygmund, A. (1959). Trigonometric Series, Vol. II, London, Cambridge Press.